SOIL BIOCHEMISTRY

BOOKS IN SOILS, PLANTS, AND THE ENVIRONMENT

Series Editor
G. STOTZKY
Department of Biology
New York University
New York, New York

Soil Biochemistry, Volume 1, edited by A. D. McLaren and G. H. Peterson

Soil Biochemistry, Volume 2, edited by A. D. McLaren and J. Skujiņš

Soil Biochemistry, Volume 3, edited by E. A. Paul and A. D. McLaren

Soil Biochemistry, Volume 4, edited by E. A. Paul and A. D. McLaren

Soil Biochemistry, Volume 5, edited by E. A. Paul and J. N. Ladd

Soil Biochemistry, Volume 6, edited by Jean-Marc Bollag and G. Stotzky

Soil Biochemistry, Volume 7, edited by G. Stotzky and Jean-Marc Bollag

Organic Chemicals in the Soil Environment, Volume 1, edited by C. A. I. Goring and J. W. Hamaker

Organic Chemicals in the Soil Environment, Volume 2, edited by C. A. I. Goring and J. W. Hamaker

Humic Substances in the Environment, by M. Schnitzer and S. U. Khan

Microbial Life in the Soil: An Introduction, by T. Hattori

Principles of Soil Chemistry, by Kim H. Tan

Soil Analysis: Instrumental Techniques and Related Procedures, edited by Keith A. Smith

Soil Reclamation Processes: Microbiological Analyses and Applications, edited by Robert L. Tate III and Donald A. Klein

Symbiotic Nitrogen Fixation Technology, edited by Gerald H. Elkan

Soil-Water Interactions: Mechanisms and Applications, edited by Shingo Iwata, Toshio Tabuchi, and Benno P. Warkentin

Soil Analysis: Modern Instrumental Techniques, Second Edition, edited by Keith A. Smith

Soil Analysis: Physical Methods, edited by Keith A. Smith and Chris E. Mullins

Growth and Mineral Nutrition of Field Crops, by N. K. Fageria, V. C. Baligar, and Charles Allan Jones

Semiarid Lands and Deserts: Soil Resource and Reclamation, edited by J. Skujiņš

Plant Roots: The Hidden Half, edited by Yoav Waisel, Amram Eshel, and Uzi Kafkafi

Plant Biochemical Regulators, edited by Harold W. Gausman

Additional Volumes in Preparation

SOIL BIOCHEMISTRY

Volume 7

edited by

G. STOTZKY

Department of Biology
New York University
New York, New York

JEAN-MARC BOLLAG

Department of Agronomy
The Pennsylvania State University
University Park, Pennsylvania

MARCEL DEKKER, INC.   New York • Basel • Hong Kong

Library of Congress Cataloging-in-Publication Data
(Revised for vol. 7)

Soil biochemistry.

 (v.2: Books in soil science) (v. 3-5: Books
in soils and the environment) (v. 6- :Books
in soils, plants and the environment)
    Vol. 2-   edited by A. Douglas McLaren and
others.
    Vol. 7: edited by G. Stotzky, Jean-Marc
Bollag.
    Includes bibliographical references and
indexes.
    1. Soil biochemistry. I. McLaren, Arthur
Douglas. II. Peterson, George H.
III. Series. IV. Series: Books in soil science.
V. Series: Books in soils and the environment.
VI. Series: Books in soils, plants and the
environment.
S592.7.S64   631.4'17   66-27705
ISBN-0-8247-8575-4 (v. 7)

This book is printed on acid-free paper.

MARCEL DEKKER, INC.
270 Madison Avenue, New York, New York, 10016

Current printing (last digit):
10 9 8 7 6 5 4 3 2 1

PRINTED IN THE UNITED STATES OF AMERICA

Preface

As indicated in the preface to Volume 6, we are publishing in the series Soil Biochemistry on a more frequent basis after an unfortunate hiatus of almost ten years. The field of soil biochemistry in all its ramifications is continuously evolving, and research in this area is steadily providing new results, concepts, and advances. The series has exposed its readers—and will continue to do so in the future—to new techniques and approaches with which to answer both old and new questions, as well as to reviews of the latest developments on intensively studied topics.

The enhanced research activity in soil biochemistry is to a large extent the result of the increasing concern about environmental quality in addition to the constant striving for greater agricultural productivity. More and more chemicals, both organic and inorganic, are being introduced into soils through spills, disposal processes, and even agricultural practices that are considered by many to constitute proper usage. Although soils in many parts of the world now produce more food and fiber than ever before, soils are now also being expected to serve as repositories for all sorts of wastes, many of them toxic, and to mitigate the detrimental impacts of these wastes. Consequently, there is an increasing need to understand better the fundamental biological, chemical, and physical processes that occur in soil, not only to delineate the limits of what soils can tolerate and process, but also to enhance the efficacy of new techniques for maintaining and improving the quality and productivity of soils.

This series should be especially useful as both an introduction and a reference for scientists and students in related fields who want to extend their knowledge of and research into soil biochemistry. As has been emphasized in previous volumes, soil biochemistry is truly an interdisciplinary field of inquiry and therefore attracts interest from researchers in a spectrum of disciplines including soil science, agronomy, forestry, horticulture, plant pathology, microbiology, biochemistry, colloid chemistry, and environmental engineering and science.

In this volume, a particularly wide variety of topics is purposely discussed. The topics range from recent advances in understanding the biochemistry of sulfur cycling, the processes of humification, the extraction of soil enzymes, and the interactions between soil minerals and microorganisms to more specialized areas such as the formation of desert varnishes, the role of nematophagous fungi in soil, the movement of microorganisms in soil, new techniques for the biochemical analysis of biomass, community structure, and microbial activity in soil, and the application of molecular techniques to soil microbial ecology and biotechnology.

We extend our sincere thanks to the authors who contributed to this volume. We encourage other specialists in the field to contribute to future volumes.

G. Stotzky
Jean-Marc Bollag

Contents

v

Contributors

John B. Adams Department of Geological Sciences, University of Washington, Seattle, Washington

Claire Chenu Station de Science du Sol, Institut National de la Recherche Agronomique, Versailles, France

Malcolm S. Cresser Department of Plant and Soil Science, University of Aberdeen, Aberdeen, Scotland

Carin Dackman Department of Microbial Ecology, Lund University, Lund, Sweden

James T. Fleming The Center for Environmental Biotechnology, University of Tennessee, Knoxville, Tennessee

Minhong Fu* Department of Agronomy, Iowa State University, Ames, Iowa

Shimna M. Gammack Department of Plant and Soil Science, University of Aberdeen, Aberdeen, Scotland

James J. Germida Department of Soil Science, University of Saskatchewan, Saskatoon, Saskatchewan, Canada

Vadakattu V. S. R. Gupta[†] Department of Soil Science, University of Saskatchewan, Saskatoon, Saskatchewan, Canada

---

Present affiliations:
*Department of Soil, Crop, and Atmospheric Sciences, Cornell University, Ithaca, New York.
[†]Division of Plant Industry, Commonwealth Scientific and Industrial Research Organization (CSIRO), Canberra, Australia.

Konrad Haider Institut für Pflanzenernährung und Bodenkunde, Bundesforschungsanstalt für Landwirtschaft (FAL), Braunschweig, Germany

Hans-Börje Jansson Department of Microbial Ecology, Lund University, Lund, Sweden

Jane S. Kemp Department of Plant and Soil Science, University of Aberdeen, Aberdeen, Scotland

Kenneth Killham Department of Plant and Soil Science, University of Aberdeen, Aberdeen, Scotland

Kave Nikbakht The Center for Environmental Biotechnology, University of Tennessee, Knoxville, Tennessee

Birgit Nordbring-Hertz Department of Microbial Ecology, Lund University, Lund, Sweden

Janet Packard The Center for Environmental Biotechnology, University of Tennessee, Knoxville, Tennessee

Fred E. Palmer Departments of Microbiology and Geological Sciences, University of Washington, Seattle, Washington

Eric Paterson Department of Plant and Soil Science, University of Aberdeen, Aberdeen, Scotland

Michel Robert Station de Science du Sol, Institut National de la Recherche Agronomique, Versailles, France

Gary S. Sayler The Center for Environmental Biotechnology, University of Tennessee, Knoxville, Tennessee

James T. Staley Department of Microbiology, University of Washington, Seattle, Washington

M. Ali Tabatabai Department of Agronomy, Iowa State University, Ames, Iowa

Anders Tunlid Laboratory of Ecological Chemistry, Lund University, Lund, Sweden

Milton Wainwright Department of Molecular Biology and Biotechnology, University of Sheffield, Sheffield, England

David C. White Departments of Microbiology and Environmental Toxicology, University of Tennessee, and Environmental Science Division, Oak Ridge National Laboratory, Knoxville, Tennessee

SOIL
BIOCHEMISTRY

1

Biochemistry of
Sulfur Cycling in Soil

JAMES J. GERMIDA *University of Saskatchewan, Saskatoon, Saskatchewan, Canada*

MILTON WAINWRIGHT *University of Sheffield, Sheffield, England*

VANDAKATTU V. S. R. GUPTA* *University of Saskatchewan, Saskatoon, Saskatchewan, Canada*

I. INTRODUCTION

Sulfur (S) is an essential element for the growth and activity of organisms. It is abundant throughout the earth's crust (ca. 0.1%), and in soil it is derived from the atmosphere, weathered rock, fertilizers, pesticides, irrigation water, and such [1]. Since the industrial revolution, the increased burning of fossil fuels has resulted in a greater input of atmospheric S to the soil budget [2—4]. In addition, volatilization of S (as hydrogen sulfide, carbon disulfide, carbonyl sulfide, methyl mercaptan, dimethyl sulfide, dimethyl disulfide, sulfur dioxide) from marine algae, marsh lands, mud flats, plants, and soils contributes to the global circulation of S through the atmosphere.

Sulfur exists in a number of oxidation states (+6 to -2), and for biological systems, it is the most oxidized and most reduced states that are important. The most oxidized form is a component of the

---

Present affiliation: Commonwealth Scientific and Industrial Research Organization (CSIRO), Canberra, Australia.

nervous system (sulfatides) and connective tissue (sulfated poly-saccharides), whereas the most reduced form is required by all microorganisms for storage and transformations of energy, synthesis of amino acids and proteins, enzyme reactions, and as a constituent of coenzymes, ferridoxins, vitamins, and others [5].

Sulfur is considered to be a macronutrient in most ecosystems, but in some ecosystems, the availability of S to the biota is limiting. For example, sulfur-deficient soils (>100 million ha) are found in many parts of the world [6], and over 8 million ha in western Canada are either deficient or potentially deficient [7]. Plants need substantial amounts of S for growth and grain production, with their requirements varying according to species. As a result, to manage effectively the S needs of crops, it is important to understand the nature and quantities of different S pools in soil and the various transformation processes of the S cycle (Fig. 1). This review discusses the major pools and transformations of the soil S cycle, with particular emphasis given to biological and biochemical processes.

II. NATURE AND FORMS OF SULFUR IN SOIL

The nature and quantities of various S pools in soil are influenced by pedogenic factors, such as climate, regional vegetation, and local topography. The total S content of soils ranges from 0.002 to 10% [9], with the highest levels being found in tidal flats and in saline, acid sulfate, and organic soils. The total S concentrations in the surface layers of soils from various parts of the world range from 18 to 6400 μg S g^{-1} for African soils; 42 to 6450 μg S g^{-1} for Asian soils, 24 to 2000 μg S g^{-1} for Australian soils; 20 to 4210 μg S g^{-1} for European soils, 145 to 8000 μg S g^{-1} for soils in the USSR; 32 to 2300 μg S g^{-1} for North American soils; and 27 to 1104 μg S g^{-1} for Central and South American soils [6,7]. The total S content in western Canadian agricultural soils ranges from 0.01 to 0.1% [6,7,10]. Roberts and Bettany [11] reported that the total S in surface horizons increases from semiarid (Brown Chernozems) to more humid soil zones (Black Chernozems), and from the upper to lower slope portions of catenas in Saskatchewan, Canada.

Organic S constitutes more than 90% of the total S present in most surface soils [9,11,12]. Thus, a close relation exists between the organic C, total N, and organic S contents of soil. Freney and Williams [9] summarized information on the C:N:S ratios for a wide range of soils from all over the world and reported ratios ranging from 50:4:1 (cultivated brown Chernozems of western Canada) to 271:13:1 (virgin Luvisols of Canada), depending on various pedogenic factors. The mean worldwide C:N:S ratio is

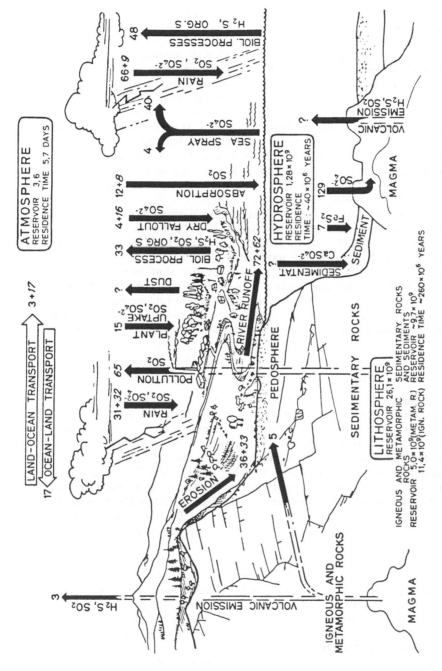

Figure 1 The global sulfur cycle. The fluxes are given in millions of tonnes sulfur per year (tg S yr⁻¹). Numbers in italics indicate the amounts that anthropogenic activities have added; nonitalic numbers denote the transfers estimated to have prevailed before anthropogenic activities had a significant influence on the sulfur cycle. Adapted from Ref. 8.

approximately 130:10:1 for agricultural soils and 200:10:1 for native grassland and woodland soils [6,7,10]. The differences in the organic C:N:S ratios of various regions and landscapes have only recently been stressed [11,12]. Organic C:N:S increased from 68:7:1 in Brown Chernozemic soils to 145:11:1 in Gray soils, with a similar trend being noted downslope within a catena [11]. Native grassland soils generally exhibit wider C:N:S ratios than their cultivated counterparts. For example, the C:N:S ratios of the native and cultivated gray wooded soils of western Canada are 271:13:1 and 129:11:1, respectively [7]. In contrast to C:S ratios, the variation in N:S ratios as the result of cultivation and between soil groups is less, and most agricultural soils have N:S ratios in the range of 6:1 to 10:1. In agricultural soils, the N:S ratio usually tends to decrease with increasing soil depth.

Cultivation of native pasture soils causes a more rapid loss of N than of S, but the loss of carbon-bonded S (C-S) follows more closely that of N [13]. This results in higher HI-S:C-S ratios in cultivated soils than in native pastures. The hydriodic acid-reducible sulfur (HI-S) fraction is organic sulfur that is directly reducible to H_2S by hydriodic acid; it is believed to comprise mainly ester sulfates [14]. Similar observations have been reported by other workers [12,15—17] and might be explained by the difference between biochemical and biological mineralization of organic compounds in soil.

A. Inorganic Sulfur

Inorganic forms of S account for less than 25% of the total S in most agricultural soils [18]. Sulfide (S^{2-}), elemental S (S^0), sulfite (SO_3^{2-}), thiosulfate ($S_2O_3^{2-}$), tetrathionate ($S_4O_6^{2-}$) and sulfate (SO_4^{2-}) are the main forms of inorganic S in agricultural soils. In well-drained soils sulfides account for less than 1% of the total S [19], and measurable quantities of $S_2O_3^{2-}$ and $S_4O_6^{2-}$ are detected only in soils treated with S^0 fertilizer or exposed to pollutants [20, 21, J. R. Lawrence, Microbial oxidation of elemental sulfur in agricultural soils, PhD dissertation, University of Saskatchewan, Saskatoon, Saskatchewan, Canada, 1987]. There are several forms of SO_4^{2-}, including easily soluble SO_4^{2-}, adsorbed SO_4^{2-}, insoluble SO_4^{2-}, and SO_4^{2-} coprecipitated/cocrystallized with $CaCO_3$. Water-soluble salts of Mg, Ca, and $NaSO_4$ account for less than 5% of the total S in surface horizons of most well-drained soils, although higher levels may accumulate under arid conditions [10]. Sulfate adsorption is influenced by soil pH, nature of colloidal surfaces, presence of amorphous Fe and Al oxides, organic ligands, and high concentrations of soluble SO_4. Subsurface soils usually contain more adsorption sites [22,23]. Coprecipitated/cocrystallized S is usually abundant in calcareous soils [19].

B. Organic Sulfur

Inasmuch as the precise nature of organic S compounds in soil cannot be identified [9], organic S is grouped into two broad groups i.e., organic sulfates and C-S. Organic sulfates include sulfate esters (C-O-S-), sulfamates (C-N-S), and sulfated thioglycosides (N-O-S). This fraction is measured (i.e., HI-S) after conversion to H_2S following treatment with hydriodic acid [24]. Organic sulfates constitute 30 to 75% of the total organic S in soil [9,25]. Carbon-bonded S includes the S present in amino acids, proteins, polypeptides, heterocyclic compounds (e.g., biotin and thiamine), sulfinates, sulfones, sulfonates, and sulfoxides [18,26]. A large portion of C-S present in soil has yet to be identified. Amino acid-S, however, may constitute up to 30% of the organic S in soil [9], and trace amounts of sulfated polysaccharides, sulfolipids, and vitamins also have been detected [27−30]. These compounds are susceptible to decomposition and do not accumulate in the uncombined form. For example, most of the amino acid-S (cysteine, cystine, methionine) is bound to mineral and humus fractions or in protein polymers. Various methods and extractants that are being evaluated for the isolation, separation, and identification of soil organic S have been reviewed elsewhere [7,31].

III. TRANSFORMATIONS OF SULFUR COMPOUNDS IN SOIL

Many of the transformations of S compounds (i.e., oxidation and reduction, volatilization, decomposition and mineralization of plant and microbial residues, immobilization, and such) are mediated by microbial activity. A conceptual model that integrates the main forms of S (inorganic and organic) and the ways by which they are transformed in soil was proposed by Maynard et al. [32,33]. In this model (Fig. 2), it was suggested that microbial biomass has a crucial role in mediating transformations between organic S and sulfate in solution. A conceptual model that integrates the various components and pathways involved in the microbial component of the sulfur cycle is shown in Figure 3. This model includes known transformation processes (i.e., oxidation, reduction, mineralization, immobilization, and microbial interactions) involved in the soil S cycle. From the studies used to develop this model, it was evident that microbial biomass acted as a source or sink for inorganic sulfate, whereas microbial activity influenced both the movement of S between different pools (inorganic sulfate, labile organic S, resistant organic S pools) and losses of S from these pools (e.g., conversion of complex organic S compounds into mobile forms, which may be lost by leaching). Interactions among different

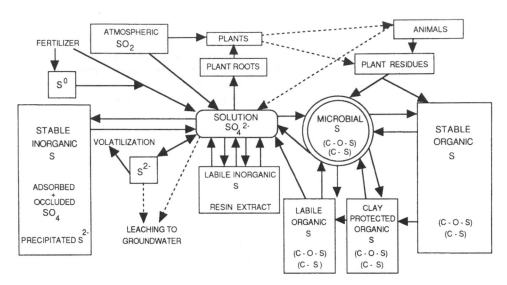

Figure 2   A conceptual diagram of the main forms and transformations of sulfur in the soil—plant system.   Both soil inorganic and
organic S are divided into labile and stable forms.   C–O–S refers
to ester sulfates, and C–S refers to carbon-bonded S.   Adapted
from Ref. 32.

microbial groups also increased the turnover rate of microbial biomass S and even influenced the rate of movement of S between different pools.

A. Sulfate Reduction

The reduction of SO_4^{2-} to H_2S is a process that is mediated mainly
by anaerobic bacteria and, therefore, it occurs to any notable extent only in anaerobic soils.   The process is not important in well-
aerated agricultural soils, except in anaerobic microsites.   Sulfate
reduction, however, is a major component of the S cycle in soils exposed to waterlogging or periodic flooding, expecially when readily
decomposable plant residues are present.

Microorganisms reduce oxidized S molecules by either an assimilatory or dissimilatory process.   During assimilatory sulfate reduction, the ion is reduced to the thiol (SH) group of organic compounds, thereby enabling the organism to meet its sulfur requirements.   In dissimilatory sulfate reduction, large quantities of H_2S
are released by bacteria that use electrons from organic compounds
to reduce SO_4^{2-} to H_2S.   For a long time, the organisms involved

were thought to represent a narrow physiological and ecological group that belongs to one of two bacterial genera: namely, *Desulfovibrio* and *Desulfotomaculum* [34]. The former is a gram-negative, heterotrophic, anaerobic vibrio, whereas *Desulfotomaculum* is a gram-negative, heterotropic, obligately anaerobic spore-forming rod. The early isolates of sulfate-reducing bacteria used only a limited range of compounds for both C and energy sources, including ethanol and other alcohols, as well as organic acids, such as lactate; that is, compounds that were oxidized to acetate, rather to CO_2. These bacteria were thought to generate energy primarily by substrate level phosphorylation and, therefore, required a fermentable substrate for growth. Later studies showed, however, that sulfate-reducing bacteria are capable of a far more diverse metabolism than was indicated by these initial studies.

The overall equation that describes the reactions involved in the process is

$$2CH_3CHOHCOONa + MgSO_4 \rightarrow H_2S + 2CH_3COONa +$$
$$CO_2 + MgCO_3 + H_2O$$

The range and diversity of bacteria that have been shown to be capable of dissimilatory sulfate reduction have recently been increased considerably, and two groups of sulfate reducers are now recognized. The first group contains bacteria in which the carbon and energy source is not completely oxidized to CO_2. This group includes a range of species of *Desulfovibrio* and *Desulfotomaculum*, the principal metabolic products of which are acetate and H_2S, although some strains can also grow fermentatively on pyruvate in the absence of SO_4^{2-}. A number of species of *Desulfovibrio* are also capable of using H_2S and SO_4^{2-} as the sole energy source.

The second group of SO_4^{2-} reducers includes those species that can completely oxidize organic carbon to CO_2. All members of this group can oxidize acetate to CO_2 under anaerobic conditions by using SO_4^{2-} as the terminal electron acceptor and, thus, bring about the complete oxidation of organic matter under anaerobic conditions. This group of SO_4^{2-} reducers is far more diverse than the former and, in addition to *Desulfotomaculum acetoxidans*, it includes species of *Desulfobacter*, *Desulfococcus*, *Desulfosarcina*, and *Desulfonema*. Most of this diverse collection of sulfate-reducing bacteria have been isolated from aquatic environments, such as marine sediments, and little is known of their role in SO_4^{2-} reduction in soils [35].

Heterotrophic microorganisms can also liberate H_2S from organic matter under anaerobic conditions, and heterotrophs, such as fungi, produce H_2S when growing in media supplemented with S^0 or $S_2O_3^{2-}$. Here, the S appears to serve mainly as a hydrogen acceptor in

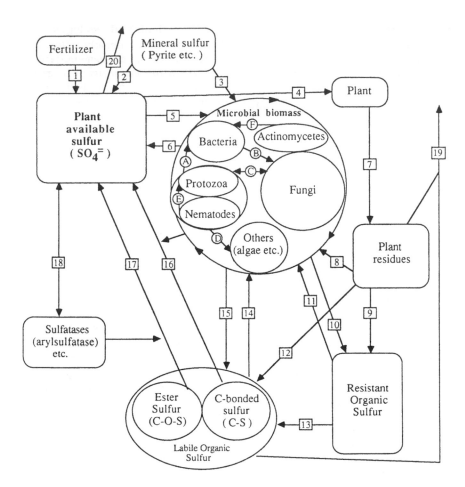

Figure 3   Conceptual model for various microbial components and processes of the sulfur cycle in soil. *Microbial components*:   (A) protozoa and nematodes feeding on bacteria; (B) bacteria feeding on damaged fungal mycelium and on compounds released from dead hyphae; (C) protozoa and nematodes feeding on fungi and the action of nematophagous fungi; (D) protozoa grazing on algae; (E) nematodes grazing on protozoa; and (F) actinomycetes feeding on bacteria. *Microbial processes*:   (1) fertilizer dissolution (biological and abiological) and oxidation of elemental S fertilizers; (2) oxidation of reduced S in minerals and the release of SO_4^{2-} sulfur; (3) microbial assimilation of nutrients from S-containing minerals; (4) plant uptake of SO_4^{2-} sulfur; (5) microbial assimilation and immobilization of SO_4^{2-} sulfur; (6) mineralization of microbial

dehydrogenations and alcoholic fermentations of hexoses and pentoses, leading to the production of SO_4^{2-} and H_2S [36]:

$$S^0 + 2H^+ \rightarrow H_2S$$

$$2H_2S + O_2 \rightarrow S^0 + 2H_2O$$

$$2H_2O + H_2S \rightarrow SO_4^{2-} + 2H_3O^+$$

Factors Influencing Sulfate Reduction

When a soil is flooded, oxygen disappears, and this is followed by the reduction of NO_3^-, NO_2^-, manganic compounds, ferric compounds and SO_4^{2-}. However, the reduction of one of these components does not need to be complete before reduction of the next commences, although sulfate is not reduced in the presence of oxygen and nitrate. Because of this sequence of reduction reactions, sufficient ferrous ions are generally available to react with any H_2S produced and as a result, free H_2S is rarely liberated from soils. Sulfate reduction increases with the period of soil submergence and, after the addition of organic matter, sufficient organic substrates to stimulate the process are also liberated from seeds and from the rhizosphere, with the result that in paddy soils, blackening caused

---

Figure 3 (continued)

biomass sulfur; (7) dead and residual straw, and such; (8) microbial decomposition of plant material; (9) humification and (or) stabilization of resistant plant residues; (10) humification and (or) stabilization of resistant microbial residues; (11) microbial utilization of resistant organic S; (12) release of labile S compounds during decomposition of plant residues; (13) microbial decomposition and organic matter turnover resulting from other environmental factors, such as drying and wetting cycles, temperature fluctuations, and such; (14) microbial assimilation of S from labile organic S compounds; (15) release of labile microbial S-containing components from dead microorganisms or resulting from grazing by predators; (16) biological mineralization; (17) biochemical or enzymatic mineralization; (18) end—product-controlled activity of sulfatase enzymes; (19) microbial decomposition of organic sulfur compounds and production of volatile sulfur compounds, such as H_2S, dimethyl sulfide, dimethyl disulfide, SO_2, carbon disulfide, and carbonyl sulfide; (20) microbial reduction of inorganic sulfate. Adapted from Gupta, V.V.S.R. Microbial biomass sulfur and biochemical mineralization of sulfur in soils. (PhD dissertation, University of Saskatchewan, Saskatoon, Saskatchewan, Canada, 1989.)

by ferrous sulfide (FeS) deposits often occurs in the root region or in the spermosphere [37]. The distribution of sulfate-reducing bacteria in these soils is generally correlated with the distribution of organic matter [37,38]. The factors that influence SO_4^{2-} reduction in the environment have been discussed at length elsewhere [4,35, 37,39]. Sulfate reducers appear to tolerate high concentrations of both salt and H_2S [39] and function best above a pH of 5. The process can, however, occur in acid mine drainage, even at a pH less than 3.5, presumably because of the existence of acid-tolerant sulfate reducers. Sulfate reduction occurs over the temperature range of 10 to 37°C, with Q_{10} values of 2.9 to 3.5. Redox potentials (E_h) below 0 millivolts (mV) at pH 7 are necessary for the activity of sulfate-reducing bacteria. In general, the rate of SO_4^{2-} reduction increases with decreasing E_h, with the optimum being a function of soil pH (e.g., about -300 mV at pH 7). Connell and Patrick [40] found that little or no sulfide accumulated in soils with an E_h above -150 mV or with a pH outside the range of 6.5 to 8.5.

Although H_2S is generally formed anaerobically by the microbial reduction of SO_4^{2-}, it can also be produced aerobically in cysteine-amended soils by a combination of cystathionine γ-lyase activity and the nonenzymatic degradation of thiocysteine [41]. In this series of reactions, thiocysteine is produced by the action of the enzyme on cysteine. Hydrogen sulfide is then produced when thiocysteine reacts nonenzymatically with cysteine.

B. Production of Volatile Sulfur Compounds

Relatively small amounts of S-containing gases, including H_2S, are released from aerobic, agricultural soils, even when waterlogged [9,42,43], although substantial amounts of H_2S are liberated from salt marsh soils [44]. Other S gases produced in soils include carbon disulfide (CS_2), which results from the decomposition of cysteine and cystine; carbonyl sulfide (COS), released during the decomposition of thiocyanates and isothiocyanates; and methyl mercaptan (CH_3SH), dimethyl sulfide (CH_3SCH_3), and dimethyl disulfide (CH_3SSCH_3), which result from the breakdown of methionine and related compounds [45]. Bremner and Steele [43] listed the S-containing gases released from soils, as well as the organisms involved. Carbon disulfide is liberated not only from cystine and cysteine, but also from sulfur oxyanions such as $S_2O_3^{2-}$ and $S_4O_6^{2-}$ [46], which are products of microbial S oxidation in soils.

C. Sulfur Oxidation

Although abiotic oxidation of reduced sulfur compounds may occur to a limited extent in soil, microbial reactions clearly dominate the process. The major reduced forms of inorganic sulfur found in soils include S^0, S^{2-}, and the oxyanions, $S_2O_3^{2-}$, $S_4O_6^{2-}$, together with traces of SO_3^{2-}. These ions can be oxidized, ultimately to SO_4^{2-}. Elemental sulfur appears to be oxidized in soils by the following sequence of reactions [20]:

$$S^0 \rightarrow S_2O_3^{2-} \rightarrow S_4O_6^{2-} \rightarrow SO_4^{2-}$$

However, it is unclear whether this reaction occurs directly from microbial reactions in soils or whether some of the products result from abiotic side reactions.

The microorganisms involved in S oxidation in soils can be divided into the following groups: (1) chemolithotrophs, such as members of the genus *Thiobacillus*; (2) photoautotrophs, including species of purple and green sulfur bacteria; and (3) heterotrophs, including a wide range of bacteria and fungi. In most aerobic soils, members of groups 1 and 2 are largely responsible for oxidizing reduced forms of sulfur. Phototrophic bacteria are the predominant organisms oxidizing sulfides at the soil—water interface in flooded rice fields and in the rice rhizosphere.

The majority of the members of the genus *Thiobacillus* are obligate aerobes, although some, such as *T. denitrificans*, can grow anaerobically by using NO_3^- as a terminal electron acceptor. Other species of thiobacilli use electron donors such as ferrous iron (*T. ferrooxidans*) and thiocyanate (SCN^-) (*T. thioparus*), rather than sulfur. Members of the thiobacilli differ in their characteristics and, as a result, can be divided into the following groups: (1) *T. thioparus*, which aerobically oxidizes S_2^-, S^0, $S_2O_3^{2-}$, and SCN^-. This bacterium is a strict autotroph and grows in the pH range of 4.5 to 7.8. (2) *Thiobacillus thiooxidans* essentially resembles *T. thioparus*, but it grows optimally at acid pH, even as low as pH 1. This species represents the classic sulfur oxidizer, which was first reported by Waksman and Joffe [47], the biochemistry of which has been extensively studied [48]. (3) *Thiobacillus denitrificans* is typified by its ability to oxidize reduced forms of sulfur, as well as reducing NO_3^- to N_2; (4) *T. ferrooxidans* can utilize Fe^{2+} iron as an energy source by oxidizing it to Fe^{3+}. In other characteristics, this bacterium is similar to *T. thiooxidans*, although it is unable to oxidize S_2^-. (5) This group contains thiobacilli that are obligate chemolithotrophs, but are facultative heterotrophs (e.g., *T. neopolitanus*). (6) The last group includes thiobacilli that are facultatively chemolithotrophic, but obligately heterotrophic (e.g., *T. perometabolis*).

The biochemistry of sulfur oxidation by thiobacilli growing in vitro has been extensively reviewed [48—50] and, therefore, need not be discussed in detail here. The central pathway in the oxidation of inorganic S used by these organisms appears to be a system operating with single S units, with larger molecules of S being peripheral to the main energy-yielding oxidative process. The most probable pathway is shown in Figure 4. Free SO_3^{2-} is rarely observed, indicating either that SO_4^{2-} is produced directly or that SO_3^{2-} is not released from the enzyme complex after cleavage. Polythionates are thought not to have a central role in the in vitro oxidation of inorganic S by thiobacilli, as some strains are unable to either produce or to metabolize them. On the other hand, a thiosulfate-oxidizing enzyme has been isolated from various thiobacilli by Trudinger [51]. Electrons from S oxidation are transferred into electron transport chains, thereby producing ATP. The reducing power that is needed for CO_2 fixation is provided as NADH and is produced by a reversal of this electron transport chain, with electrons going from cytochromes to NAD + ATP.

Other Sulfur Bacteria

The gliding group of sulfur oxidizers includes bacteria the cells of which are arranged in trichomes and that show a gliding motion on the substrate. The most important members of this group in relation to S oxidation in soils are species of *Beggiatoa*, bacteria that participate in sulfide oxidation in the root zone of rice. All strains of *Beggiatoa* deposit sulfur in the presence of H_2S, and although they appear to be heterotrophs, there continues to be some debate about their physiology [52]. Phototrophic bacteria, such as *Chromatium* and *Chlorobium*, also have an important role in sulfide oxidation in rice paddies, but not in aerobic agricultural soils. A number of nonfilamentous chemolithotrophic sulfur-oxidizing bacteria have been isolated. *Sulfolobus*, for example, occurs in hot acid soils, whereas species of *Thiospira*, *Thiomicrospira*, and *Macromonas* have been isolated from waterlogged woodland soils. The importance of these bacteria in S oxidation has yet to be determined, however.

Heterotrophic Sulfur Oxidizers

A wide range of heterotrophic microorganisms can oxidize elemental and reduced forms of sulfur in vitro: bacteria (including actinomycetes), fungi, and even protozoa [53]. Sulfur-oxidizing fungi [54] and bacteria [55,56] have been isolated from a wide variety of soils. However, most of the information on these organisms originates from in vitro studies with nutrient-rich laboratory media, and it could be argued that heterotrophs have a major role in S oxidation in soils only when carbon substrates, such as plant residues, are present. On the other hand, recent evidence (reviewed later) suggests that mixed populations of heterotrophs probably have the

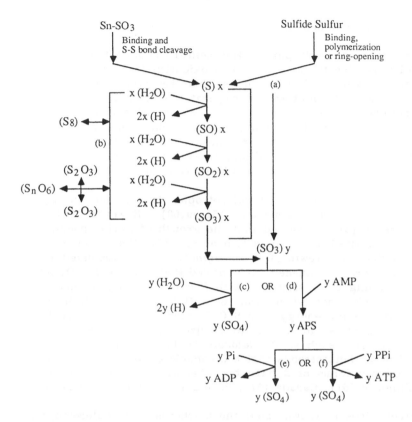

Figure 4 A schematic diagram that describes the essential features of the oxidation of inorganic sulfur by most thiobacilli. The scheme accounts for the oxidation of all sulfur substrates and explains the accumulation of polythionates as intermediates. For simplicity, charges are not indicated on ionic sulfur species or intermediates; (S), (SO), and (SO_2) indicate enzyme-bound sulfur and partially oxidized intermediates. The bracketed sequence is regarded as occurring on one enzyme or enzyme complex from which free intermediates do not normally arise. APS, adenylate sulfate; Pi, orthophosphate; PPI, pyrophosphate; (SO_3), sulfite; (SO_4), sulfate; (S_8), elemental sulfur; (SnO_6), polythionates. (H) represents H^+; e^- produced by dehydrogenations. In most cases, the electrons are accepted by c-type cytochromes for transfer to terminal acceptors (e.g., O_2, nitrate). Letters indicate enzymes or reaction types as follows: (a) thiosulfate cleavage (rhodanese, thiosulfate reductase); (b) polysulfur dehydrogenase acting in the manner of a reversed sulfite reductase; (c) sulfite dehydrogenase; (d) APS reductase; (e) ADP sulfurylase; and (f) ATP sulfurylase. Adapted from Ref. 49.

dominant role in S oxidation in many aerobic, agricultural soils [57, 58].

Reports from the early part of this century showed that bacteria and fungi were able to participate in S oxidation [54], although following the discovery of *T. thiooxidans*, their importance in the process was generally thought to be comparatively limited. More recently, however, several reports on the ability of heterotrophs to oxidize S have appeared, including one by Schook and Berk [59] that showed that *Pseudomonas aeruginosa* oxidizes a variety of S compounds, including sodium sulfide, thiosulfate, tetrathionate, dithionite, metabisulfite, and sulfite when grown in succinate-based medium.

It is generally assumed that heterotrophs do not gain energy from oxidizing reduced S compounds [51,59,60]. However, Tuttle and co-workers [61,62] showed that the growth of marine pseudomonads is stimulated during oxidation of $S_2O_3^{2-}$. Mason and Kelly [63] found that the growth of *P. aeruginosa* is also stimulated by $S_2O_3^{2-}$. Gommers and Kuenen [64] showed that *Thiobacillus* Q, an organism showing close similarity with *Pseudomonas alcaligenes*, can grow chemolithoheterotrophically on $S_2O_3^{2-}$ under acetate limitation, thereby gaining energy from $S_2O_3^{2-}$ oxidation. Actinomycetes have also been shown to be able to oxidize S in vitro [56,65], as have some yeasts, such as *Rhodotorula* [66]. A wide range of fungi and other heterotrophs that are capable of oxidizing elemental and reduced forms of S have been isolated from soils in the United Kingdom [67], Canada [57], Poland [68], and Czechoslovakia [58].

Relatively little is known about the biochemistry of heterotrophic S oxidation, although the process has been suggested to be enzymatic in filamentous fungi [69], yeast [70,71], and bacteria [62]. Kurek [70,71] showed that the yeast *Rhodotorula* possessed an enzyme complex consisting of a thiosulfate and a sulfite oxidase. Lettl [72] also demonstrated these enzymes, together with rhodanese, in heterotrophic sulfur-oxidizing bacteria isolated from soil. Information [60,70,71] in the literature also suggests the operation of a polythionate pathway in the heterotrophic oxidation of reduced S compounds. However, experimental evidence to determine whether a polythionate pathway is active or to determine the importance of SO_3^{2-} as an intermediate is lacking. Furthermore, both chemical and biological reactions could be involved in the oxidation of reduced S compounds in soil. A schematic diagram integrating the known and possible biological and chemical reactions of the various inorganic S compounds in soil is shown in Figure 5.

Ecology of Thiobacilli in Soil

Despite the fact that members of the genus *Thiobacillus* have generally been thought to be the organisms largely responsible for S

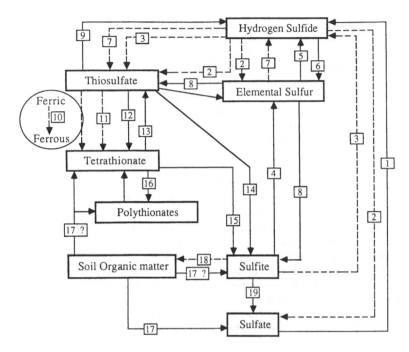

Figure 5 A schematic diagram integrating the possible biological and chemical reactions involving inorganic sulfur compounds in soil. The numbers indicate the reactions that may occur in soil: (1) dissimilatory sulfate reduction; (2) chemical oxidation of hydrogen sulfide at low concentration, resulting in the production of elemental sulfur, sulfite, and thiosulfite; (3) chemical reaction of sulfite with sulfide, producing thiosulfate; (4) biological sulfite reduction; (5) biological reduction of elemental sulfur; (6) biological oxidation of hydrogen sulfide to elemental sulfur; (7) chemical reaction of sulfur with hydrogen sulfide, producing thiosulfate and polythionates; (8) biological oxidation of elemental sulfur to thiosulfate or sulfite; (9) thiosulfate reductase activity; (10) chemical oxidation of thiosulfate to tetrathionate; (11) chemical condensation of thiosulfate to form tetrathionate; (12) biological oxidation of thiosulfate to tetrathionate; (13) biological reduction of tetrathionate; (14) rhodanese cleavage of thiosulfate; (15) biological oxidation of tetrathionate to sulfite; (16) chemical condensation reactions forming polythionates; (17) mineralization reactions (? indicates hypothetical intermediates); (18) biological or chemical reactions of free sulfite with soil organic matter; and (19) biological oxidation of sulfite via APS or SCOR pathways. Adapted from J. R. Lawrence. Microbial oxidation of elemental sulfur in agricultural soils. (PhD dissertation, University of Saskatchewan, Saskatoon, Saskatchewan, Canada, 1987.)

oxidation in soils, surprisingly few studies of their ecology have been reported, a situation that contrasts markedly with the extensive literature devoted to the physiology and biochemistry of these organisms. A number of studies have, in fact, concluded that *T. thiooxidans* is a relatively rare bacterium in soils and that it is outnumbered by populations of *T. thioparus* and *T. denitrificans* [52]. In contrast, Lettl et al. [73] found that *T. thiooxidans* and *T. thioparus* were readily isolated from the uppermost horizons of spruce forest soils. Vitolins and Swaby [74] isolated *T. intermedius, T. neopolitanus, T. thiooxidans, T. denitrificans*, and *T. ferrooxidans* from Australian soils, whereas Hayman (J. M. Hayman, MSc Thesis. Lincoln College, Christchurch, New Zealand, 1964) and McCaskill and Blair [75] were unable to isolate *T. thiooxidans* from New Zealand and Australian soils, respectively. Lee et al. [76] found that the numbers of thiobacilli in soil were affected by the availability of S^0, the surface area of the element, and possibly soil and moisture conditions. They found that addition of S^0 stimulated growth of thiobacilli, resulting in a significant increase in cell numbers. However, there appears to be no correlation between rates of S oxidation and the incidence of thiobacilli, except that rates of S oxidation are generally low in soils that lack these organisms [53,77] and are accelerated in soil inoculated with thiobacilli [78,79].

Evidence for the Involvement of Heterotrophs
in Sulfur Oxidation in Soil

The apparent dominant role that thiobacilli have in oxidizing S in soils has meant that there have been few studies on the potentially important role that heterotrophs may have in the process. The fact that most, if not all, soil heterotrophs, to some extent, are capable of oxidizing S suggests that they probably participate in the oxidation of elemental and reduced forms of S in soil. However, substantial rates of heterotrophic S oxidation in soil would appear to be limited by low levels of available carbon present to support heterotrophic growth and activity [80]. There is, however, evidence that shows that fungi can use plant residues, such as cereal straw, as a carbon source to support S oxidation [67]. In contrast, species of wood-decomposing fungi that have been tested appear to be incapable of oxidizing S^0 when growing from inoculated wooden blocks into soil [81]. Some evidence, however, suggests that ectomycorrhizal fungi are able to oxidize S in culture and when growing from the roots of a higher plant symbiont into soil [81].
    Evidence that supports the view that heterotrophs might have an important role in S oxidation, at least in certain soils, has been provided by Lawrence and Germida [82]. By using a newly developed

most probable number (MPN) method, they showed that between 3 and 37% of the total heterotrophic population of soils of western Canada were capable of oxidizing S^0 to $S_2O_3^{2-}$. Populations of heterotrophs capable of participating in this reaction were ten thousand times greater than populations of heterotrophs that oxidized S^0 to SO_4^{2-}. Bacteria tended to oxidize S^0 only as far as $S_2O_3^{2-}$, whereas fungi could achieve the complete oxidation to SO_4^{2-}. In general, little $S_4O_6^{2-}$ was produced. Heterotrophic S oxidizers were also found to predominate in two Gray Luvisolic soils [83], which confirms earlier reports on the ubiquity of S-oxidizing heterotrophs in a wide variety of soils [53]. Lawrence and Germida [84] provided more conclusive evidence to implicate heterotrophs in S oxidation when they showed that the rate of S oxidation increased linearly with the measured soil microbial biomass (largely heterotrophs) and with increases in biomass following glucose amendment. By using selective antibiotics to inhibit fungal or bacterial populations, they further demonstrated the important and possibly predominant role of heterotrophs in S oxidation in soil. These studies appear to confirm earlier predictions concerning the importance of heterotrophs in S oxidation in soils [53]. McCready and Krouse [85], for example, concluded from isotope discrimination techniques that the S oxidation that occurred in their soils was mediated by heterotrophs, rather than by autotrophs. Nevertheless, it remains difficult to partition accurately rates of S oxidation between heterotrophic and chemolithotrophic oxidation. It seems likely, however, that mixed populations of microorganisms, including heterotrophs are responsible for bringing about S oxidation in most soils. In this way, for example, S^0 may be oxidized to $S_2O_3^{2-}$ by one group of organisms, and others may oxidize this intermediate to the end product, SO_4^{2-}. Such synergistic interactions between S-oxidizing microorganisms have been demonstrated in vitro [58].

Factors Influencing the Rate of Sulfur Oxidation in Soils

A large amount of information is available concerning the factors that influence S oxidation in soils. Details of these studies have been presented elsewhere [53], and they will only be summarized here. Unfortunately, many of the data concerning the factors affecting S oxidation in soils are conflicting, largely because suitable standard techniques for measuring S oxidation rates in soil are lacking. For example, Vitolins and Swaby [74] and Kittams and Atoe [86] concluded that soil properties did not markedly influence the rate of S oxidation, whereas Nor and Tabatabai [20] found that the process occurred faster in alkaline, than in acid soils.

The application of lime to soils has stimulated, retarded, or had no effect on S oxidation [87]. More consistent results have been reported on the effects of soil temperature, for which the expected mesophilic optimum in oxidation usually occurs, with oxidation rates being low at lower and higher temperatures [20,53]. The rate of S oxidation is generally optimal at soil moisture contents close to field capacity [53,88]. The particle size of S^0 has a major impact on the rate at which it is oxidized, with oxidation rate increasing as particle size is decreased [53]. The data on the effect of organic matter on S oxidation in soils are also conflicting [53]. If heterotrophic S oxidation is a dominant process in soil, then it would be expected that rates of S oxidation would be stimulated by the addition of organic matter. Although this has often been true, some studies have demonstrated the reverse or no effect of organic matter on the process [53].

It is hoped that much of this confusion concerning the effects of environmental variables on S oxidation rates should be avoided in the future, as the result of the introduction by Janzen and Bettany [88] of a method that enables absolute S oxidation rates to be determined. By using this approach, rates of S oxidation can be calculated that are independent of S^0 particle size and temperature. Janzen and Bettany [88] used this approach to show that S oxidation rates were parabolically related to water potential at all temperatures. Maximum rates of S oxidation were found at water potentials near field capacity. The rates of S oxidation in soil were also exponentially related to temperature, and mean Q_{10} values ranged from 3.2 to 4.3, indicating a marked sensitivity of the process to temperature. Lawrence et al. [83] reported that the response to S^0 application to soils can be biphasic, so that, although the response of initial rates of oxidation to S^0 application may be high, subsequent additions can cause a decline in oxidation rates.

McCaskill and Blair [75] pointed out that most of the literature concerning the effects of environmental variables on S^0 oxidation is based on measurements of extractable soil SO_4^{2-} and, as a result, has tended to exaggerate estimates of oxidation rates. As an alternative approach, they measured (1) S uptake by plants; (2) $^{35}SO_4$ dilution; and (3) acetone extraction and measurement of residual S^0, following an oxidation period. They confirmed, by using these methods, the importance of particle size as a determinant of S^0 oxidation, but they found no relation between soil texture and S^0 oxidation rate over a range of clay contents from 9 to 52%.

Unfortunately, the use of these novel approaches to measuring S oxidation rates does not mean that the influence of environmental factors on the process in soil will necessarily be clarified. For example, in contrast with the findings of McCaskill and Blair [75], Janzen and Bettany [88] found that the oxidative capacity of soils was negatively correlated with clay content. The slow or incomplete oxidation of S^0 fertilizers to SO_4^{2-}, together with a marked variability in rates of oxidation in various soils, has been reported in soils of the Canadian Prairie Provinces [89]. These findings further emphasize the need to determine accurately the factors that influence rates of S oxidation in soils.

Oxidation of Sulfide

The oxidation of iron sulfides in soil involves both chemical and microbial processes and, as a result, is a more complex process than is the oxidation of S^0. Both ferrous sulfide (FeS) and pyrite (FeS_2) follow an essentially similar series of oxidation steps [90]. First, ferrous sulfate is formed as the result of an abiotic oxidation step:

$$2FeS_2 + 2H_2O + 7O_2 \rightarrow 2FeSO_4 + 2H_2SO_4$$

This reaction is followed by the bacterial oxidation of ferrous sulfate, generally by *T. ferrooxidans*:

$$4FeSO_4 + O_2 + 2H_2SO_4 \rightarrow 2FE_2(SO_4)_3 + 2H_2O$$

Subsequently, ferric sulfate is reduced and pyrite oxidized by a strictly chemical reaction:

$$Fe_2(SO_4)_3 + FeS_2 \rightarrow 3FeSO_4 + 2S$$
$$2S + 6Fe_2(SO_4)_3 + 8H_2O \rightarrow 12FeSO_4 + 8H_2SO_4$$

The elemental sulfur produced is finally oxidized by *T. thiooxidans*, and the acidity produced helps the whole process to continue:

$$2S^0 + 3O_2 + 2H_2O \rightarrow 2H_2SO_4$$

Several S-oxidizing thiobacilli as well as heterotrophs can be isolated from acid sulfate soils in which pyrite is being oxidized, but with the exception of *T. ferrooxidans*, they appear not to have an important role in the process [61]. Attempts to implicate other iron oxidizers in pyrite oxidation (e.g., *Metallogenium* spp. or *Leptospirillum ferrooxidans*), have also been unsuccessful [90].

D. Immobilization and Mineralization Processes

Immobilization

Immobilization and mineralization are biological processes that occur concurrently and exhibit a strong relationship to the soluble SO_4^{2-}

pool in soil. To estimate the plant-available S status of soils, it is
necessary to understand the factors that influence these processes.

Microbial assimilation of inorganic SO_4^{2-} into organic S through
the assimilatory sulfate reduction pathway is known as immobilization. This process involves ATP-sulfurylase and two energy-rich
sulfate nucleotides: adenosine 5'-phosphosulfate (APS), and 3'-
phosphoadenosine-5'-phosphosulfate (PAPS). The overall reaction
of SO_4^{2-}—S incorporation into amino acids is as follows:

$$SO_4^{2-}\text{-S} \rightarrow APS \rightarrow PAPS \rightarrow \text{(active sulfite)} \rightarrow$$

$$\text{sulfide + serine} \rightarrow \text{cysteine}$$

Most of the intracellular S accumulated by microorganisms is in
the form of amino acids in proteins; however, microorganisms also
accumulate ester S, sulfonates, S-containing vitamins, and cofactors [91]. Fungi accumulate large amounts of ester S [92,
V. V. S. R. Gupta, Microbial biomass sulfur and biochemical mineralization of sulfur in soils. PhD dissertation, University of Saskatchewan, Saskatoon, Saskatchewan, 1989]. In laboratory and
field experiments, substantial amounts of added inorganic SO_4^{2-}-S
are quickly incorporated into organic S fractions. The rate and
magnitude of immobilization in soil are further increased in the presence of an energy source (e.g., presence of metabolizable organic
matter or addition of easily degradable C sources, such as glucose)
[93—99]. Laboratory incubation studies using $^{35}SO_4^{2-}$-S indicate
that a major portion of the S is incorporated into organic esters,
although C-S fractions also rapidly accumulate [93,94,96,100].
Much of this microbially synthesized S is eventually found in the
fulvic acid fractions, especially as organic sulfates. However, resistant humic fractions can also be detected in soil [93,94].

Biological and Biochemical Mineralization

Mineralization of organic S is largely mediated by microbial activity.
The various known pathways of mineralization are summarized in
Figure 6. The C-S is mineralized either through oxidative decomposition or desulfuration, where the various sulfatases are involved in the mineralization of ester S compounds. McGill and Cole
[101] proposed a dichotomous system to accommodate the various
patterns of C, N, S, and P cycling within soil humus. This concept suggests two pathways of mineralization: biological and biochemical. In the former process, elements (e.g., N. S) in direct
association with C are mineralized as microorganisms oxidize C to
obtain energy. Mineralization of S by this process, therefore, may
not be reflected by increases in the SO_4^{2-}-S pool in soil. Heterotrophic soil microorganisms decompose organic S compounds to grow,

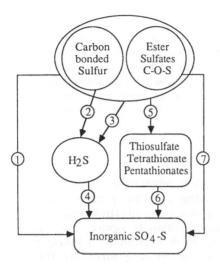

Figure 6 A schematic diagram of organic sulfur mineralization that indicates known pathways: (1) biological mineralization during the oxidation of carbon as an energy source; (2) hydrolysis of cysteine by cysteine desulfhydrolase; (3) anaerobic mineralization (desulfurization) of organic matter; (4) biological oxidation of hydrogen sulfide to sulfate through elemental sulfur and sulfite; (5) incomplete oxidation of organic S into inorganic S compounds; (6) biological oxidation of tetrathionate to sulfate through sulfite; and (7) biochemical mineralization when sulfate esters are hydrolyzed by sulfatases. Adapted from V. V. S. R. Gupta, Microbial biomass sulfur and biochemical mineralization of sulfur in soils. (PhD dissertation, University of Saskatchewan, Saskatoon, Saskatchewan, Canada, 1989.)

and as the C−S bond is broken, the S is released. As this process involves actively growing microorganisms, their requirement for S may be met by the S content of the substrate, and net S mineralization may be detected. If the S demand of the growing microorganisms is not met by the S content of the substrate, mineralization will not be detected and growth will be limited. In biochemical mineralization, those elements that exist as esters (e.g., P_0, ester S) are mobilized by periplasmic or extracellular hydrolases. This process, also known as enzymatic mineralization, occurs mainly outside the cell and is regulated by end-product inhibition [102].

Because soil organic S consists of C-S and R-O-S forms, its mineralization may be controlled by either biological or biochemical pathways. The differences in the nature of the substrates and the mechanisms involved in these two pathways suggest separate controls and mechanisms. For example, biological mineralization is controlled by the microbial need for C and energy sources, whereas biochemical mineralization is controlled by factors that influence enzyme synthesis, activity, and kinetics.

Factors Affecting the Mineralization of Sulfur in Soil

Mineralization is generally measured as net mineralization; that is, the amount of SO_4^{2-}-S accumulated during the time period under study, or as the difference between gross mineralization and immobilization. Thus, for high net mineralization to occur, the mineralization—immobilization process must be dominated by mineralization. Inasmuch as microbial activity is the driving force for these two processes, all factors, such as energy and nutrient supply, soil temperature, moisture, pH, or others that affect microbial activity, have a critical influence on net mineralization. The rates of net S mineralization observed in open (i.e., leached) incubation systems generally range from 0.5 to 2.0 mg kg^{-1} $week^{-1}$ in most agricultural soils. However, the rates of net S mineralization obtained in closed (i.e., nonleached) incubation systems are lower than those in open incubation systems. For example, Maynard et al. [33] observed net negative S mineralization in closed incubation systems (i.e., net immobilization), whereas in open incubation systems, they observed a net S mineralization of 0.5 to 1.5 mg kg^{-1} $week^{-1}$.

Net S mineralization in aerobic soils is reported to increase as temperature increases from 5 to 30°C, but declines at 50°C [98, 103]. Soil moisture has a substantial influence on S mineralization in soil. Chaudhry and Cornfield [104] reported that 60% moisture-holding capacity was optimum for S accumulation, whereas Williams [105] found that S mineralization decreased at moisture levels >40%. Air drying, oven drying, and heating soils causes the release of SO_4^{2-}-S from organic S [105—109]. Alternate drying-wetting cycles also stimulate the release of S from soil.

As C-S is mineralized during C oxidation, the C:S ratios of added organic substrates (e.g., crop residues) significantly influence S mineralization. Alfalfa residues stimulated the process, whereas corn and sawdust decreased SO_4^{2-}-S production over a 26-week incubation [105]. Methionine mineralization in forest soils was influenced by temperature and antibiotics [110].

Nutrient ratios of soil organic matter also influence S mineralization in soil, with wider C:S ratios often coinciding with lower rates of S mineralization [12]. The amount of S mineralized has been shown to be high in Brown soils with narrow C:S ratios when compared with Gray soils with wider C:S ratios [11]. However, it was not possible to predict mineralization rates using C:S, N:S, C:HI-S, or N:C-S ratios [109]. Gupta and Germida [111] reported wider C:S ratios for soil macroaggregates (> 250 μm diameter) than for microaggregates (< 250 μm diameter) in both a native prairie grassland and it's cultivated counterpart. In laboratory studies, this difference was not directly reflected in the amount of S mineralized in these aggregate groups. Macroaggregates from both native and cultivated soils exhibited more S mineralization than microaggregates, which was attributed to the higher microbial biomass (especially fungal) observed in macroaggregates. Thus, C:S ratios of soils alone do not necessarily indicate the S mineralization potential of soils.

Nitrogen and sulfur mineralization do not follow similar trends, even though N and S are closely related with organic matter [10, 94,101,112]. The differences in N and S mineralization patterns may help to explain the differences in C:N:S ratios that occur during soil pedogenesis [113]. Roberts and Bettany [11] have reported the influence of pedogenic processes, such as climate, vegetation, and topography, on the C:N:P:S ratios of organic matter and on their mineralization patterns.

Plants significantly increase S mineralization in soils [26,33,106, 114]. The abundant supply of energy sources in the rhizosphere increases microbial activities, thus increasing S mineralization. However, the reassimilation of released inorganic sulfates by the growing microorganisms may result in no increase in the plant-extractable S pool and may even result in a reduction in the plant-extractable S pool when the microbial demand exceeds the rate of mineralization. Speir et al. [115] observed higher activities of sulfatases in rhizosphere soil, compared with nonrhizosphere soil, and these enzyme activities would influence S mineralization. Studies on the seasonal variations in the sulfate S status in relation to organic S levels have indicated that ester S may serve as a readily available source of S that can be mineralized in response to biological demand [93,116].

IV. ENZYMES INVOLVED IN SULFUR CYCLING IN SOIL

Although the literature on soil enzyme activity is extensive [117], relatively few reports have been devoted to enzymes involved in sulfur transformations. Arylsulfatase is by far the most frequently studied soil sulfur enzyme, followed by rhodanese. In contrast, the activity of other enzymes involved in S cycling, such as cysteine disulfohydrase, have only occasionally been studied.

A. Sulfatases

Types, Detection, and Occurrence

Sulfatases (sulfohydrolases, EC 3.1.5) are enzymes that hydrolyze sulfuric acid esters, for which the linkage with sulfate is in the form of $R-C-O-SO_3^-$, and R represents diverse groups of parent moieties [102]:

$$R-C-O-SO_3^- + H_2O \rightarrow R-C-OH + SO_4^{2-} + H^+$$

The different ester S compounds that serve as substrates for microbial sulfatases are given in Table 1. Sulfatases are classified according to the type of organic S esters that they hydrolyze, with the following main groups being recognized: arylsulfatases, alkylsulfatases, steroid sulfatases, glucosulfatases, chondrosulfatases, and mycosulfatases [118]. Of these, the activity of only arylsulfatases have been measured in soils. These enzymes are sometimes referred to as "phenolsulfatases" or, more correctly, arylsulfohydrolases. Arylsulfatases are the most widely distributed sulfatases in nature, being found in a wide variety of agricultural [109,119–130] and forest soils [131–133], marine and freshwater sediments [134,135], intertidal sands [136], salt marshes [137], and elsewhere [138,139]. Many soil microorganisms, including fungi and bacteria, produce arylsulfatases [118]. Bacteria and fungi are the chief sources of sulfatases in soil, although plants and animals also produce these enzymes [102,140,141]. Microbial sulfatases can be grouped into three categories: constitutive, inducible, and derepressible. Most sulfatases, however, are not constitutive. Detailed information about the physiological and genetic control of sulfatases has been reviewed by Dodgson et al. [102].

Arylsulfatases were first detected in soils by Tabatabai and Bremner [142], who suggested that, together with other sulfatases, they have an important role in the mineralization of soil organic S. The method devised by these workers is based on the use of *p*-nitrophenyl sulfate which, after hydrolysis to *p*-nitrophenol, is quantitatively extracted from soil with dilute alkali. However, as the alkali can lead to excessively colored extracts, this method is of limited use for measuring activity of the enzymes in organic soils. This problem can be overcome by using a modification of the technique in which ether is used in place of NaOH [126].

Despite the wide use of *p*-nitrophenyl sulfate as the substrate with which to assay arylsulfatases, there is no evidence that such

Table 1   Some Sulfate Esters That Are Substrates for Various Microbial Sulfatases[a]

Type A	Type B
Adenosine 5'-phosphosulfate	α-Carrageenan
Arylsulfates	Chondroitin-4-sulfate
Algal arylsulfates	
Arylsulfates based on indole nucleus	Chondroitin-6-sulfate
Catecholamine-*O*-sulfates	Dermatan sulfate
Naphthoquinone sulfates	
Naphthopyrone sulfates	
Plant flavonoid sulfates	Heparan sulfate
Simple phenyl sulfates	
Thyroid hormone sulfates	
Tyrosine-*O*-sulfate	
Vitamin B$_6$ derivatives	
Choline sulfate	
Primary and secondary alkylsulfates	
D-Glucose-6-sulfate	
D-Gluconate-6-sulfate	
D-Glycerate-3-sulfate	
3'-Phosphoadenosine-5'-phosphosulfate	
Steroid sulfates	Keratan sulfate
Sulfamates	

[a]Type A substrates are of small molecular weight and may be desulfated by one sulfatase without any prior enzymatic action. Type B substrates are polymers and must undergo depolymerization before desulfation; more than one sulfatase may be involved.
Source: Adapted from Ref. 102.

a chromogenic ester occurs in nature. A more recent technique, developed by Fitzgerald et al. [132], employs the release of [^{35}S]-SO$_4{}^{2-}$ from tyrosine [^{35}S] sulfate as a measure of the enzyme's activity and, thereby, uses a substrate that is considered environmentally more relevant than *p*-nitrophenyl sulfate. In this method, the ^{35}SO$_4{}^{2-}$-S released from the ^{35}S-labeled tyrosine sulfate was extracted sequentially with 1 M LiCl and 1 M NaH_2PO_4, and pooled fractions hydrolyzed in 6 M HCl to recover residual $^{35}SO_4^{2-}$. The $^{35}SO_4^{2-}$ was separated by electrophoresis in barium acetate. With this method, Fitzgerald et al. [132] observed the arylsulfatase activity in all horizons of a forest soil without a lag period for the release of $^{35}SO_4^{2-}$-S. This method, however, is cumbersome and, in contrast with the p-nitrophenyl sulfate method, has not been tested on a wide range of soils.

Bremner and Zantua [143] showed that p-nitrophenyl sulfate is not hydrolyzed in autoclaved soils, indicating that the effect is biological. The pH optimum of this enzyme has been reported as 5.5 and 6.2 in soils [142,144] and 5.4 in marine sands [137], although Chandramohan et al. [135] reported that the enzyme in marine sediments possessed two pH optima, thereby implying the existence of two different enzymes. Speir and Ross [144] reported using an assay temperature of 30°C instead of 37°C, as recommended by Tabatabai and Bremner [142]. They also attributed the lower levels of enzyme activities in New Zealand pasture soils, compared with the general range of activities reported for United States soils, to this change in incubation temperature [144]. The temperature optimum of the enzyme is 10 to 20°C higher in the presence of soil [144]. Arylsulfatase activity in various soils was not affected by the addition of SO_4^{2-} [142], although Chandramohan et al. [135] found that, in marine sediments, 0.02 M sulfate reduced the activity of the enzyme by ca. 47%. Tabatabai and Bremner [142] found that arylsulfatase activity correlated with soil organic matter content, but not with various other soil properties, such as percentage N or clay and sand content. Cooper [119] reported that the activity of the enzyme in Nigerian soils correlated not only with total C, but also with total N and total organic S. Waterlogging of soils reduced arylsulfatase activity by 5 to 31% [145], and the addition of crop residues stimulated activity [119]. The total activity of the enzymes observed in soil is the sum of the activities of intracellular and extracellular enzymes (i.e., enzymes associated with proliferating and latent cells, cell debris, clay and humic colloids, and the soil aqueous phase) [146]. Sulfatases undoubtedly have an important role in S mineralization, particularly in forest soils, although their presence need not necessarily be associated with microbial activity [133]. These enzymes may also have an important role in S cycling in arctic soils [127,147].

Role in Sulfur Mineralization

As indicated previously, a considerable portion of the soil S belongs to the HI reducible fraction, which consists of ester S compounds [7,24,142]. In addition, inorganic SO_4^{2-}-S added to the soil is incorporated into organic matter as ester S, sulfonate S, and amino acids [93,148–150]. The incorporation of SO_4^{2-}-S into the ester sulfate pool (as phenolic SO_4-ester linkages) occurs in all horizons of soil S [132,150] and continues for 20 to 30 days after the addition of SO_4^{2-}-S. As the result of their universal existence and labile nature, sulfate esters constitute a major source of S for the inorganic SO_4^{2-}-S pool or plant-available S pool in soil.

Dodgson et al. [102] pointed out the need for more research on sulfatase activity in soils to better understand the role of sulfatases in S mineralization. Information is needed on (1) the capacity of soil microorganisms to produce sulfatases and the location of these enzymes in soil, along with the substrate transport mechanisms; (2) different sulfatases produced by micro- and macroorganisms in soil and their effectiveness; (3) the chemical nature of ester sulfates in soil; and (4) the capacity of soils to hydrolize different kinds of ester sufates.

Kinetics of Enzyme Activities in Soil

The kinetic and thermodynamic parameters of an enzyme in soil reflect the nature and stability of the enzyme and enzyme–substrate complexes [151]. Kinetic parameters of an enzyme are derived from its velocities at different substrate concentrations. From the Hanes–Woolf linear transformation of the Michaelis–Menton equation, Tabatabai and Bremner [152] calculated the K_m and V_{max} for arylsulfatase activity in 11 Iowa soils. The relatively small differences in these kinetic parameters in the different soils were attributed to the similarity in the origin and type of enzymes, or that the K_m values of different arylsulfatases were similar because of their association with soil constituents. However, Perucci and Scarponi [153] observed significant differences in the kinetic parameters in the two soils that they studied. Pettit et al. [121] calculated the kinetic parameters for arylsulfatases using three methods of data analysis (i.e., direct plot, Lineweaver–Burk, and Eadie–Hofstee plots) and found that the enzymes followed Michaelis–Menton kinetics in all three cases. The data in Table 2 indicate the general range of kinetic parameters observed for arylsulfatases in different soils.

The specific enzyme activity (K_m) decreases when an enzyme becomes attached or entrapped within an insoluble matrix or when there is significant adsorption of the substrate to the soil matrix. Perucci and Scarponi [153] observed that the addition of crop residues significantly changed the K_m and V_{max} values for soil arylsulfatases, and the changes depended both on soil properties and on the type of crop residues incorporated. They also reported that any increase in the temperature of the system increased V_{max} values in soils, suggesting differential microbial response to the temperature and crop residues. Dick et al. [154] found that manure application

Table 2 Kinetic Parameters of Arylsulfatases in Different Soils, from the United States, Canada, United Kingdom, and Italy[a]

Location/soil	K_m $(10^{-3}$ M)	V_{max}[b]	Ref.
United States			
Iowa soils			164
Thurman, sand	1.37	14	
Lindley, silty clay loam	5.69	271	
Shelby, clay loam	3.41	114	
Clarion, sandy loam	2.86	143	
Webster, clay loam	3.53	196	
Hayden, sandy loam	5.10	680	
Primghar, silty clay loam	3.00	300	
Harpster, clay loam	4.60	460	
Glencoe, silty clay	3.61	328	
Canada			
Gray Luvisol (Waitville)			
Bromegrass—alfalfa pasture			Gupta (1989)
Control	1.85	101.0	
AS-22[c]	1.98	86.7	
AS-44[c]	3.84	83.7	
Gray-Luvisol (Loon River)			Gupta (1989)
Native	7.60	1591.1	
Cleared-Unseeded	9.40	771.3	
Cultivated (5 y)	2.60	237.6	
Cultivated (40 y)	1.70	110.9	
Gray Luvisol (Waitville)			Gupta (1989)
Native: O horizon	7.03	1739.1	
Native: Ae horizon	4.13	320.2	
Alfalfa	3.31	408.8	
Cultivated (70 y)	3.90	259.7	
United Kingdom			
Kent silt loam soil	4.98	0.88[d]	121
Italy			
Perugia soils[e]			
Sand clay: loam soil			153
No amendment	0.488	32.09	
Tobacco residue added	0.735	55.68	

(continued)

Table 2 (Continued)

Location/soil	K_m (10^{-3} M)	V_{max}[b]	Ref.
[Sand clay: loam soil]			
Sunflower residue added	0.663	50.50	
Wheat straw residue added	0.467	29.78	
Maize residue added	0.422	30.14	
Clay: loam soil			153
No amendment	0.950	116.98	
Tobacco	0.705	189.06	
Sunflower	0.800	179.86	
Wheat straw	0.788	211.87	
Maize	0.880	206.18	

[a]Soils from Canada collected from 0—15 cm of field plots subjected to various management practices; other soils from 0—15 cm of unspecified fields. All assays conducted at 37° unless otherwise specified.
[b]μg g^{-1} h^{-1} of *p*-nitrophenol unless otherwise indicated.
[c]Elemental sulfur (kg ha^{-1} y^{-1}) was applied as Agri-Sul for 5 years.
[d]μmol g^{-1} h^{-1} of *p*-nitrophenol; assay conditions at 20°C.
[e]V_{max} for this soil series reported as nmol g^{-1} h^{-1} of *p*-nitrophenol.
Source: V. V. S. R. Gupta. Microbial biomass sulfur and biochemical mineralization of sulfur in soils, PhD dissertation, University of Saskatchewan, Saskatoon, Saskatchewan, Canada, 1989.

(22.4 t ha^{-1} y^{-1}) significantly increased (twofold) V_{max} values, but showed no significant changes in K_m values. Tabatabai and Bremner [152] did not find any significant relation between K_m and soil properties, such as pH, cation-exchange capacity, percentage organic C, and percentage clay and sand.

The rate of enzyme-catalyzed reactions increase with increasing temperature, until a temperature is reached at which the enzyme is inactivated [123]. Perucci and Scarponi [153] found that the influence of crop residues on calculated thermodynamic parameters, such activation energy (E_a) and changes in enthalpy (ΔH_a) and entropy (ΔS_a), and free energy of activation (ΔG_a), was variable between the two soils studied. The ΔG_a was correlated with the catalytic efficiency of the enzyme system, but was not affected by substrate addition.

Factors Affecting Enzyme Activities in Soil

The relative importance of soil arylsulfatases in S mineralization is
dependent on the factors governing their synthesis, location, and
stability in soil. As indicated before, most arylsulfatases are not
constitutive, and their synthesis in microorganisms is controlled by
the C and S contents of the system [102]. Fitzgerald and Ash [155]
reported that irrespective of the S source, C sources, such as acetate, oxaloacetate, malate, and α-ketoglutarate, significantly reduced the enzyme activity of *Pseudomonas* $C_{12}B$, when compared
with such substrates as citrate. Similarly, with few exceptions,
synthesis was stimulated when growth was limited by low levels of
SO_4^{2-} or by growth on nonlimiting levels of L-methionine; growth
on excess SO_4^{2-} or with L-cystine as the S source repressed arylsulfatase synthesis [156].

Similar to microbial biomass and other biochemical properties,
maximum arylsulfatase activity was observed in the surface soils
(top 15 cm) of cultivated, grassland, and forest soils and exhibited a sharp decline with increasing depth [129,142,157]. The O1-
horizon of both hardwood and conifer soils showed the maximum
arylsulfatase activity [129]. The pH optimum of this enzyme was
reported as 5.5 to 6.2 [158]. Gupta et al. [130] observed a strong
negative correlation between pH and arylsulfatase activity in S^0-
treated soils (brome grass—alfalfa pasture and canola—summer fallow fields). Significant correlations between arylsulfatase activity
and various chemical, microbiological, and biochemical properties
are reported elsewhere [151]. Arylsulfatase activity in soils is
significantly correlated with content of organic C, percentage N,
percentage clay, microbial biomass, microbial activity (CO_2 evolution and O_2 uptake), activity of amidase, invertase, α-galactosidase,
and urease, and the soil S status [111,119,129,142,154,157,159,160—
163]. The various correlations reported with the different soil properties may be attributed to the indirect influence of these properties
on the microbial biomass and its activity. Furthermore, significant
positive correlations were observed with levels of total S, organic
S, C-S, and ester S in soils [119,125,129,159]. As the organic S
pool includes various ester S compounds that are substrates for
arylsulfatases, these positive correlations indicate a substrate-in-
duced response of this enzyme. However, no significant relationships were observed between arylsulfatase activity and S mineralization in laboratory studies in soil without plants [109,164]. In
contrast, Speir [122] and Lee and Speir [125] reported positive
correlations between plant uptake of S and arylsulfatase activity,
particularly at low soil S levels. Thus, arylsulfatase activity might
provide an indicator for mineralized S that can be taken up by
plants. As arylsulfatases are only one of the many types of

sulfatase enzymes involved in the mineralization of ester S compounds, it is unlikely that arylsulfatase activity alone completely explains the variations in S mineralization in soils under different cultivation practices, crops, and seasons.

As with other microbiological properties, temporal fluctuations in soil arylsulfatase levels have been reported [119,163]. Cooper [119] observed significant seasonal changes in arylsulfatase activities in a fallow tropical soil. Arylsulfatase activity significantly increased during the rainy season when the soils were continually moist, but when the soils dried at the end of the rainy season, a reduction in enzyme activity occurred. Repeated wetting and drying cycles additionally reduced enzyme activity. Ross et al. [163] attributed their observation of no temporal changes in arylsulfatase activity to the more uniform climate of New Zealand and suggested that seasonal effects may be relatively unimportant in New Zealand. The soil moisture status has a substantial influence on arylsulfatase activity. Neal [127], for example, attributed the higher arylsulfatase activities he observed near or under snow patches to the soil moisture status. He also suggested that within an environmental gradient, soil moisture and water movement influence enzyme activities, with waterlogging of soils significantly reducing arylsulfatase activity. Pulford and Tabatabai [146] also reported a significant relation between arylsulfatase activity with Eh_7 (i.e., activity increased as Eh_7 increased).

Arylsulfatase activity in soil is notably influenced by the presence of plants [125,127,138,150,164–166]. Speir et al. [150] reported that arylsulfatase activity in planted soils maintained a constant level, because any loss of enzyme activity (e.g., through denaturation) was replaced by new enzyme(s) from plants or microorganisms. Temperature-dependent denaturation caused a reduction in enzyme activity in fallow soil, however. Skiba and Wainwright [138] also observed higher arylsulfatase activities in the rhizospheres of climax vegetation in coastal sands, whereas lack of vegetation cover resulted in lower activities. In addition to the effects of the presence of plants, the magnitude of the rhizosphere influence was also influenced by the type of vegetation [127,166]. For example, arylsulfatase activity was significantly higher in soils from a permanent pasture and an alfalfa field than in cultivated wheat fields [V. V. S. R. Gupta, Microbial biomass sulfur and biochemical mineralization of sulfur in soils. PhD dissertation, University of Saskatchewan, Saskatoon, Saskatchewan, Canada, 1989]. Clearing and cultivation of native grassland and forest soils significantly decreased the arylsulfatase activity in soil [165].

The influence of temperature (18 to 25°C) on arylsulfatase activity in the presence of plants does not correspond to mineralization of organic S [125]. Activity of the enzymes is also influenced by tillage treatments and crop rotation. Dick [167] observed significantly higher arylsulfatase activity in no-till plots than in conventional-till plots, with the highest activity occurring in a corn—oat—alfalfa rotation and the lowest in a corn—soybean rotation. In addition, these effects were most pronounced in the top (0 to 7.5 cm) soil and were largely related to organic C levels. Neal and Herbein [147] observed a reduction in arylsulfatase activity in arctic tundra soils subjected to vehicle disturbance. Similarly, compacted skid trails in forest soil exhibited lower enzyme activity than the control soils [154]. The presence of earthworms, however, increased arylsulfatase activity in rye grass pastures in New Zealand [164].

Tabatabai and Bremner [142] reported that mercuric (4 mM), phosphate (5 mM), cyanide, and sulfite ions inhibited enzyme activity, whereas chloride ions and the common extractants involved in sulfatase analysis had little effect.

The synthesis and activity of microbial arylsulfatases in vitro are influenced by the amount of sulfur in the medium and the presence of specific S ions (e.g., SO_4^{2-}). For example, inorganic SO_4^{2-}, cystine, and various intermediates of cystine metabolism repress the synthesis of the enzymes, whereas the presence of methionine or an abundant supply of energy (ATP) derepressed synthesis of the enzymes [102]. Bettany and Stewart [6] suggested that the accumulation of the end product, SO_4^{2-}, and its possible inhibitory influence was the reason for reduced S mineralization in closed incubation experiments. Correlation studies, however, did not indicate any significant relation between the SO_4^{2-} status in soil and arylsulfatase activity [109,123,158]. In contrast, high concentrations of S ions (i.e., HSO_3^-, SO_3^{2-}, SO_4^{2-}) resulting from deposition of acid precipitation can significantly reduce arylsulfatase activity in agricultural, forest, and peat soils [103, 139,168,169]. Similarly, Gupta et al. [130] found that repeated application of S^0 fertilizers, even at agricultural rates, significantly decreased arylsulfatase activity in field soils. This was attributed to a substantial decline in the major source of the enzymes (i.e., microbial biomass) and to end-product inhibition caused by large quantities (~ 60 µg S g^{-1}) of SO_4^{2-} in these S^0-treated soils.

B. Rhodanese

Detection and Occurrence

Rhodanese (thiosulfate-cyanide sulfotransferase, EC 2.8.1.1) is involved in the transformation (cleavage) of $S_2O_3^{2-}$. It catalyzes the formation of thiocyanate and sulfite from thiosulfate and cyanide [170]:

$$S_2O_3^{2-} + CN^- \rightarrow SO_3^{2-} + SCN^-$$

This enzyme is detected in animal [171] and plant tissue [172], bacteria [173], and soils [83,174]. The method used to detect rhodanese involves colorimetric determination of the SCN^- produced when soil (pretreated with toluene) is incubated with buffered $S_2O_3^{2-}$ and CN^- solutions at 37°C for 1 h [174]. Rhodanese activities range from 120 to 875 and from 38 to 130 nmol SCN^- produced per gram per hour in Iowa and Saskatchewan soils, respectively [83,174]. Rhodanese activity is correlated with the organic C content of soils and is affected by a variety of soil pretreatments and inorganic ions [93].

Role in Sulfur Oxidation

Thiosulfate is produced during the oxidation of S^0, and the activity of rhodanese has been widely associated with this process. Activity of the enzyme in soil increases during oxidation of S^0 [176] and in soils subjected to atmospheric sulfur deposition [139]. Long-term application of S^0 fertilizer significantly increased rhodanese activity in soil, and this increase was significantly correlated with an increase in the numbers of autotrophic and heterotrophic $S_2O_3^{2-}$ oxidizers [83]. In contrast, there was no relationship between rhodanese activity and S^0 oxidation in flooded rice soils, and enzyme activity varied with soil type [177]. Rhizosphere soils, both flooded and nonflooded, exhibited higher enzyme activities than their non-rhizosphere counterparts.

Tabatabai and Singh [175] calculated the kinetic and thermodynamic parameters of rhodanese in Iowa soils. The K_m values for rhodanese with $S_2O_3^{2-}$ and CN^- as substrates ranged from 1.20 to 10.3 mM (AV 5.46) and from 2.48 to 10.2 mM (AV 5.81), respectively. The V_{max} values ranged from 0.51 to 1.43 mM SCN^- produced per gram per hour. These kinetic parameters for the soil enzyme were similar to those for rhodanese isolated from other biological systems [174]. The variation in substrate affinities observed in eight different soils suggested several sources of the enzyme in soils, or that soil constituents had a significant effect on the catalyzed reaction. The activation energy values for soil rhodanese ranged from 21.6 to 34.0 kJ mole^{-1} from 10 to 60°C, and the average Q_{10} under these conditions was 1.37 [174,175]. The variations in K_m found by Tabatabai and Singh [174,175] suggest that soil rhodanese originates from a variety of sources, such as plants, animals, and microorganisms. The similarity between the nature and catalytic activity of rhodanese from different sources is unknown.

C. Cysteine Desulfhydrase

Cysteine desulfhydrase (L-cysteine hydrogen sulfide-lyase [de-aminating] EC 4.4.1.1) is a multifunctional pyridoxal-containing enzyme that catalyzes the following elimination reactions:

Homoserine $\rightarrow$ H_2O + NH_3 + 2 oxybutyrate

L-Cysteine $\rightarrow$ micocysteine + pyruvate + NH_3

L-Cysteine $\rightarrow$ pyruvate + NH_3 + H_2S

The desulfhydration of cysteine is, thus, an indirect reaction catalyzed by the system. The enzyme is catalytically inactive in the absence of pyridoxal phosphate. The enzyme has been found in animals and microorganisms, but apparently not in plants [91,178, 179]. Cysteine desulfhydration may be catalyzed by this enzyme, leading to the liberation of equimolar quantities of pyruvate, H_2S, and NH_3. The activity of this enzyme was measured in various soils by determining the formation of pyruvate from cysteine [138]. This enzyme requires pyridoxal phosphate to function; hence, although its activity can be detected in soils in laboratory assays in which pyridoxal phosphate is added, it is debatable whether it can function in soil in the absence of this cofactor.

D. Miscellaneous Enzymes Involved in Sulfur Cycling in Soil

With the exception of the foregoing enzymes, relatively little is known about other soil enzymes that may be involved in S cycling. Abramyan and Galshyan [180] measured the activity of arylsulfatase cysteine dehydrogenase, sulfide oxidases and reductases, and sulfate oxidases and reductases in soil. They concluded that arylsulfatases predominate in organic soils, whereas oxidoreductases typically occur in inorganic soils. Most of the studies pertaining to sulfatases in soils has been directed to arylsulfatases, and few studies have been conducted on alkylsulfatases and other enzymes involved in the hydrolysis of ester sulfates such as choline sulfate, dodecyl sulfate, glucose-6-sulfate, or others [120]. Inasmuch as ester sulfates are composed of different groups of compounds in soil, information on the enzymes involved in their hydrolysis would not only help to better understand the extent of biochemical mineralization of these compounds, but also to predict the contribution of biochemical mineralization to the overall mineralization (i.e., biochemical plus biological) of S in different soils.

V. LAND USE PROBLEMS AND THE SULFUR CYCLE

A. Acid Sulfate Soils

Acid sulfate soils contain sulfides, mainly in the form of pyrites, which on drying are oxidized to yield free and adsorbed sulfates. They are characterized by yellow mottling which is the result of jarasite, and have a pH that is typically below 4 [181]. Although these soils cover large areas of the tropics, they are usually of only local importance in temperate regions. The acidification of these soils results from the abiotic and microbial oxidation of pyrite [53]. Problems with crop production on these soils occur as the result of Al and Mn toxicity, rather than of the direct effects of acidity. Acid sulfate soils can be reclaimed by (1) control of the water table; (2) the addition of lime; (3) the selection of crops tolerant of Al, Mn, and Fe; and (4) by generally improving soil fertility.

B. Sodic Soils

Many soils in arid or semiarid regions are sodic and, as a result, are largely unproductive. In these soils, the exchangeable Na percentage is more than 15%, and they are typically alkaline, with a pH higher than 8.5 [182]. The relevance of the S cycle in sodic soils lies in the potential use of S^0, the oxidation of which yields sufficient sulfuric acid to reduce the pH and, thereby, improve their fertility. Sulfur-containing compounds that can be used in place of S^0 include ammonium polysulfide, sulfur dioxide, ammonium thiosulfate, ammonium bisulfite, and pyrites. Pyrite, for example, when oxidized produces sulfuric acid that reacts with $CaCO_3$ or $CaMg(CO_3)_2$, if they are present, to form Ca, Mg, and $CaSO_4$. Then H, Ca, Mg in the free system are exchanged with Na in exchange sites in the clay lattice [183]. Exchangeable sodium falls as a result, and a reduction in pH occurs. If the free Na and excess SO_4^{2-} in the soil are not removed, the concentration of Na_2SO_4 may result in a saline soil. Some sodic soils contain insufficient S-oxidizing microorganisms to enable them to achieve rapid rates of oxidation of added S. In such cases, inoculation of the added S source, generally with species of thiobacilli, has been successful [184,185].

C. Sulfur Transformations in Soils Polluted with Atmospheric Sulfur

Soils subjected to atmospheric pollution receive S from the atmosphere largely in the form of dilute H_2SO_4. Sulfate, therefore, is the major S ion entering soils from the atmosphere, although small amounts of SO_3^{2-} and HSO_3^{2-} may also enter, either in solution or as the result of dry deposition. Particulate atmospheric pollutants that consist largely of soot may also be deposited, particularly close to sources of heavy atmospheric pollution such as steel works and coking plants [186,187].

As SO_4^{2-} is the most oxidized form of the element, it cannot be further oxidized by microorganisms. However, microbes play a role in S cycling in atmospheric-polluted soils by (1) immobilizing and reducing SO_4^{2-}; (2) oxidizing sulfite and bisulfite, and (3) oxidizing the products formed during the reduction of sulfate that enters the soil as acid rain. Reduced forms of S also enter polluted soils in soot [188], and these are rapidly oxidized to SO_4^{2-}.

Vegetation and soils at atmospherically polluted sites invariably contain higher total S and SO_4^{2-} contents than samples obtained from similar sites that have not been exposed to high levels of pollution. Large amounts of S may be present in organic forms, such that, occasionally, S mineralization may be a more important source of SO_4^{2-} than exogenous inputs. The amounts of inorganic S, such as SO_4^{2-}, SO_3^{2-}, and $S_2O_3^{2-}$, are often higher in atmospherically polluted soils and decrease with distance from point sources of pollution [188]. Associated with these increases in inorganic forms of S are increases in the numbers and species composition of S-oxidizing microorganisms. Lettl [189], for example, showed that thiobacilli generally occur at low frequency or are absent in nonpolluted soils, whereas the numbers increase substantially in soils that receive sulfur emissions. Increases in the number of obligate chemolithotrophs, such as *T. thiooxidans*, suggest that the latter soils receive reduced S that is being oxidized and contributes to the soil SO_4^{2-} content [190]. This production of SO_4^{2-}, which then results in sulfuric acid, contributes to the acidification of atmospherically polluted soils. Soils that receive S emissions also oxidize added elemental S^0 more rapidly than do corresponding nonpolluted soils [191], again emphasizing that microbial S oxidation is important in the acidification of soils subject to air pollution. Further evidence for the involvement of such oxidative processes is provided by the fact that rhodanese activity may increase in such polluted soils [192]. Arylsulfatase activity in peats has also been shown to correlate directly with increasing levels of deposition of atmospheric S, suggesting that soil enzyme activity measurements may be useful indicators of air pollution [193,194].

Despite that SO_4^{2-} is the principal form of S that enters soils exposed to atmospheric pollution, few reports have appeared discussing the possibility that it may be cycled by microorganisms. However, sulfate can be reduced to sulfides, but not free H_2S in waterlogged soils [195]. When these soils are then dried, any S^{2-} that is formed is then available for oxidation back to SO_4^{2-}. Thus a S reduction—oxidation cycle may operate in soils, depending upon their water status. Although such a cycle could operate in nonpolluted soils, it will be more obvious in atmospherically polluted soils because of the elevated levels of SO_4^{2-} present. In this way, S may be immobilized as S^{2-} and then remobilized on oxidation, to form SO_4^{2-}. Even though the total amount of S present would not change, the form in which it exists would alter and, as a result, potentially influence soil pH and the rate of cation leaching. The ability of microorganisms to both mineralize organic S and immobilize SO_4^{2-} in atmospherically polluted soils was emphasized by Swank et al. [149]. Therefore, it seems that microbial processes may be more important than is generally assumed in influencing both the form and the mobility of sulfur in soils that are subjected to atmospheric pollution. Finally, in this section mention should be made of the fact that the microbial oxidation of blown S^0 from stockpiles (e.g., where produced as the result of sour gas treatment) can cause localized, but marked, acidification of soils [196,197].

D. Potential Negative Effects of Added Sulfur on Soils and Soil Processes

Despite the beneficial effects of sulfur fertilization on crop growth, the oxidation of S^0 invariably results in soil acidification that will eventually cause a range of undesirable effects on soils. The effects of atmospheric S deposition on soils has been extensively studied [198—204] and, although there is no consensus on the degree of negative impact of acid rain, it is clear that some detrimental effects on soil processes can result from this form of pollution. Surprisingly, there have been few studies on the impact of S^0 on soil properties, although S^0 is widely used as a fertilizer. Gupta et al. [130] showed that, although the addition of S^0 to soil significantly decreased the pH of the two soils examined, it also caused a 29 to 45% and a 2 to 51% decline in microbial biomass carbon. Other soil changes included a decrease in organic C, a narrowing of the C:N ratio and increases in levels of total S and SO_4^{2-}. Repeated S applications reduced the activity of a variety of soil enzymes and populations of protozoa, algae, and nitrifiers. Further evidence of the negative impact of sulfur fertilization on soils was provided by Gupta and Germida [205], who showed that repeated applications of S^0 over a 5-year period reduced the microbial biomass and the numbers of predatory protozoa. They concluded that such adverse effects on soil microbial biomass and protozoal numbers was long-lasting. Clearly, the potential negative effects on soils of repeated application of S^0 need to be considered before a recommendation is made that this form of fertilizer be applied to S-deficient soils. Similar considerations should be given to the long-term impact of atmospheric S deposition on such soils.

VI. CONCLUSIONS

Our understanding of the S cycle has increased dramatically over the past 25 years. The recognition that S plays an important function in crop nutrition has stimulated research on the forms and amounts of S in soil and the processes that control the supply of S to plants. Concern over the environmental impact of S, whether in acid rain or particulate deposits, has provided impetus for studies on the fate of S, most notably the SO_4^{2-} ion, in soils. This has led to a noticeable increase in our understanding of how S is immobilized (chemically and biologically) and mineralized in soils. Only recently has attention been given to soil—atmosphere S fluxes and the processes involved. Much work remains to be done to accurately predict S fluxes to, from, and within ecosystems.

  Microbial population dynamics and activity play a key role in the S cycle of ecosystems. Identification of specific microbial communities responsible for sulfur oxidation, reduction, and volatilization will help delineate factors that ultimately affect S fluxes in agroecosystems. The use of new techniques, such as stable isotopes, offers the opportunity to trace the flow of S through these microbial communities and the various S pools in soil. The integration and use of new analytical techniques to assess microbial ecology and activity in soil will provide a better understanding of microbial S transformations.

ACKNOWLEDGMENTS

Contribution No. R620, Saskatchewan Institute of Pedology, University of Saskatchewan, Saskatoon, Saskatchewan, Canada. The support of the Natural Sciences and Engineering Research Council of Canada is appreciated.

REFERENCES

1. Brasted, R. C. 1961. Sulfur, selenium, tellurium, polonium, and oxygen, p. 231—235. *In* M. C. Sneed and R. C. Brasted (ed.), Comprehensive inorganic chemistry, Vol. 8.D. Van Nostrand Co., New York.

2. Kellog, W. W., R. D. Cadle, E. R. Allen, L. L. Lazrus, and
    E. A. Martell. 1972. Sulfur cycle. Science 175:587–596.
3. Hitchcock, D. R. 1976. Atmospheric sulphates from biologi-
    cal sources. J. Air Pollut. Control Assoc. 26:210–216.
4. Brown, K. A. 1982. Sulphur in the environment: a review.
    Environ. Pollut. (Ser. B). 3:47–80.
5. Trudinger, P. A., and R. E. Loughlin. 1981. Metabolism of
    simple sulfur compounds, p. 165–256. *In* A. Neuberger (ed.),
    Comprehensive biogeochemistry and sulfur metabolism, 19A.
    Elsevier Scientific, Amsterdam.
6. Bettany, J. R., and J. W. B. Stewart. 1983. Sulphur cycling
    in soils, p. 767–785. *In* A. I. More (ed.), Proceedings of the
    1982 international sulphur conference, Vol. 2. British Sulphur
    Corp., London.
7. Tisdale, S. L., R. B. Reneau, Jr., and J. S. Platou. 1982.
    Atlas of sulfur deficiencies, p. 295–322. *In* M. A. Tabatabai
    (ed.), Sulfur in agriculture. American Society of Agronomy,
    Madison, Wisconsin.
8. Zehnder, A. J. B., and S. H. Zinder. 1980. The sulfur cycle,
    p. 105–145. *In* O. Hutzinger (ed.), The handbook of environ-
    mental chemistry, Vol. 1, Part A. The natural environment and
    the biogeochemical cycles. Springer-Verlag, New York.
9. Freney, J. R., and C. H. Williams. 1983. The sulfur cycle in
    soil, p. 129–201. *In* M. V. Ivanov and J. R. Freney (eds.),
    The global biogeochemical sulfur cycle, SCOPE 19. John Wiley
    & Sons, New York.
10. Biederbeck, V. O. 1978. Soil organic sulfur and fertility, p.
    273–310. *In* M. Schnitzer and S. U. Khan (eds.), Soil or-
    ganic matter. Elsevier Scientific Publishing Co., New York.
11. Roberts, T. L., and J. R. Bettany. 1985. The influence of
    topography on the nature and distribution of soil sulfur across
    a narrow environmental gradient. Can. J. Soil Sci. 65:419–434.
12. Bettany, J. R., J. W. B. Stewart, and E. H. Halstead. 1973.
    Sulfur fractions and carbon, nitrogen and sulfur relationships
    in grassland, forest and associated transitional soils. Soil Sci.
    Soc. Am. Proc. 37:915–918.
13. Bettany, J. R., S. Saggar, and J. W. B. Stewart. 1980. Com-
    parison of the amounts of sulfur in soil organic matter fractions
    after 65 years of cultivation. Soil Sci. Soc. Am. J. 44:70–75.
14. Williams, C. H. 1975. The chemical nature of sulphur com-
    pounds in soils, p. 21–30. *In* K. D. McLachlan (ed.), Sul-
    phur in Australasian agriculture. Sydney University Press.
    Sydney.
15. Scott, N. M., and G. Anderson. 1976. Organic sulfur frac-
    tions in Scottish soils. J. Sci. Food Agric. 27:358–366.

16. Tabatabai, M. A., and J. M. Bremner. 1972. Forms of sul-
     fur, and carbon, nitrogen and sulfur relationships, in Iowa
     soils. Soil Sci. 114:380–386.
17. Neptune, A. M. L., M. A. Tabatabai, and J. J. Hanway.
     1975. Sulfur fractions and carbon, nitrogen, phosphorus and
     sulfur relationships in some Brazilian and Iowa soils. Soil Sci.
     Soc. Am. Proc. 39:51–55.
18. Halstead, R. L., and P. J. Rennie. 1977. The effects of
     sulfur on soils in Canada, p. 181–220. *In* N. E. Cooke (ed.),
     Sulfur and its inorganic derivatives in the Canadian environ-
     ment. Natl. Res. Counc. Can., Publ. No. NRCC 15015.
     Ottawa, Canada.
19. Williams, C. H. 1972. Sulfur deficiency in Australia. Sulfur
     Inst. J. 8:5–8.
20. Nor, Y. M., and M. A. Tabatabai. 1977. Oxidation of ele-
     mental sulfur in soils. Soil Sci. Soc. Am. J. 41:736–741.
21. Wainwright, M., and S. J. Grayston. 1989. Accumulation
     and oxidation of metal sulphides by fungi, p. 119–130. *In*
     R. K. Poole and G. M. Gadd (eds.), Metal–microbe interac-
     tions. Special Publication of the Society of General Micro-
     biology, Vol. 26. IRL Press, New York.
22. Barrow, N. J. 1975. Reactions of fertilizer sulfate in soils,
     p. 50–57. *In* K. D. McLachlan (ed.), Sulfur in Australian
     agriculture. Sydney University Press, Sydney.
23. Metson, A. J. 1979. Sulfur in New Zealand soils. I. A re-
     view of sulfur in soils with particular reference to adsorbed
     sulfate sulfur. N.Z. J. Agric. Res. 22:95–114.
24. Freney, J. R. 1961. Some observations on the nature of or-
     ganic sulfur compounds in soil. Aust. J. Agric. Res. 12:424–
     432.
25. Lowe, L. E. 1969. S fractions of selected Alberta profiles of
     the Gleysolic order. Can. J. Soil Sci. 49:375–381.
26. Freney, J. R., G. E. Melville, and C. H. Williams. 1975.
     Soil organic matter fractions as sources of plant available sul-
     fur. Soil Biol. Biochem. 7:217–221.
27. Lowe, L. E. 1968. Soluble polysaccharide fractions in se-
     lected Alberta soils. Can. J. Soil Sci. 48:215–217.
28. Kowalenko, C. G. 1978. Organic nitrogen, phosphorus and
     sulfur, p. 95–136. *In* M. Schnitzer and S. U. Khan (eds.),
     Soil organic matter. Elsevier Scientific Publishing Co., New
     York.
29. Hardwood, J. L., and R. G. Nicholls. 1979. The plant sul-
     pholipid—a major component of the sulphur cycle. Biochem.
     Soc. Trans. 7:440–447.

30. Chae, Y. M., and L. E. Lowe. 1980. Distribution of lipid sulfur and total lipids in soils of British Columbia. Can. J. Soil Sci. 60:633–640.

31. Scott, N. M. 1985. Sulfur in soils and plants, p. 379–401. *In* D. Vaugham and R. E. Malcohm (eds.), Soil organic matter and bioavailability. Martinus–Nijhoff, Dordrecht.

32. Maynard, D. G., J. W. B. Stewart, and J. R. Bettany. 1984. Sulfur cycling in grassland and parkland soils. Biogeochemistry 1:97–111.

33. Maynard, D. G., J. W. B. Stewart, and J. R. Bettany. 1985. The effects of plants on soil sulfur transformations. Soil Biol. Biochem. 17:127–134.

34. Konopka, A. E., R. H. Miller, and L. E. Sommers. 1986. Microbiology of the sulfur cycle, p. 23–56. *In* M. A. Tabatabai (ed.), Sulfur in agriculture. American Society of Agronomy, Madison, Wisconsin.

35. Postgate, J. R. 1979. The sulfate reducing bacteria. Cambridge University Press, Cambridge.

36. Sciarini, L. J., and F. J. Nord. 1943. On the mechanisms of enzyme action. Part 22. Elementary sulphur as a hydrogen acceptor in dehydrogenations by living fusaria. Arch. Biochem. Biophys. 3:261–267.

37. Freney, J. R., V. A. Jacq, and J. F. Baldensperger. 1982. The significance of the biological sulfur cycle in rice production, p. 272–317. *In* Y. R. Dommergues and H. G. Diem (eds.), Microbiology of tropical soils and plant productivity. Martin Nijhoff, The Hague.

38. Wakao, N., and C. Furusaka. 1976. Influence of organic matter on the distribution of sulphate-reducing bacteria in a paddy field soils. Soil. Sci. Plant Nutr. 22:203–205.

39. Freney, J. R., and J. Boonjawat. 1983. Sulfur transformations in wetland soils, p. 28–38. *In* G. J. Blair and A. R. Till (eds.), Sulfur in S.E. Asia and S. Pacific, agriculture. UNE, Indonesia.

40. Connell, W. E., and W. H. Patrick. 1967. Sulfate reduction in soil: Effects of redox potential and pH. Science 159:86–87.

41. Morra, M. J., and W. A. Dick. 1985. Production of thiocysteine (sulfide) in cysteine amended soils. Soil Sci. Soc. Am. J. 49:882–886.

42. Banwart, W. L., and J. M. Bremner. 1976. Evolution of volatile sulfur compounds from soils treated with sulfur containing organic materials. Soil Biol. Biochem. 8:439–443.

43. Bremner, J. M., and C. C. Steele. 1978. Role of micro-
    organisms in the atmospheric sulfur cycle. Adv. Microb.
    Ecol. 2:155–201.

44. Jorgensen, B. B., M. H. Hansen, and K. Ingvarson. 1978.
    Sulfate reduction in coastal sediments and release of hydrogen
    sulfide to the atmosphere, p. 245–253. *In* W. E. Krumbein
    (ed.), Environmental biogeochemistry and geomicrobiology,
    Vol. 1. Ann Arbor Science, Ann Arbor, Michigan.

45. Francis, A. J., J. M. Duxbury, and M. Alexander. 1975.
    Formation of volatile organic products in soils under anerobi-
    osis. II. Metabolism of amino acids. Soil Biol. Biochem.
    7:51–56.

46. Minami, K., and S. Fukushi. 1981. Detection of carbon di-
    sulfide among gases produced by thiosulfate and tetrathionate
    addition to soils. Soil Sci. Plant Nutr. 27:541–543.

47. Waksman, S. A., and J. S. Joffe. 1922. Microorganisms
    concerned in the oxidation of sulfur in soils: II. *Thiobacil-
    lus thiooxidans*, a new sulfur oxidizing organism isolated from
    soil. J. Bacteriol. 7:239–256.

48. Aleem, M. I. H. 1975. Biochemical reaction mechanisms in
    sulfur oxidation by chemosynthetic bacteria. Plant Soil 43:
    587–607.

49. Kelly, D. P. 1985. Physiology of the thiobacilli: elucidating
    the sulphur oxidation pathway. Microbiol. Sci. 2:105–109.

50. Kuenen, J. G., and R. F. Beudeker. 1982. Microbiology of
    thiobacilli and other sulphur-oxidizing autotrophs, mixotrophs
    and heterotrophs. Phil. Trans. R. Soc. Lond. Ser. B. 298:
    473–497.

51. Trudinger, P. A. 1967. Metabolism of thiosulphate and tetra-
    thionate by heterotrophic bacteria from soil. J. Bacteriol. 93:
    550–559.

52. Gude, H., W. R. Strohl, and J. M. Larkin. 1981. Mixotro-
    phic and heterotrophic growth of *Beggiatoa alba* in continuous
    culture. Arch. Microbiol. 129:357–360.

53. Wainwright, M. 1984. Sulphur oxidation in soils. Adv. Agron.
    37:349–396.

54. Wainwright, M. 1988. Inorganic sulphur oxidation by fungi,
    p. 71–89. *In* L. Body, R. Marchant, and D. J. Read (eds.),
    Nitrogen, phosphorus, and sulphur utilization by fungi. Cam-
    bridge University Press, Cambridge.

55. Lettl, A. 1984. Soil heterotrophic bacteria in transformations
    of inorganic sulphur. Folia Microbiol. 29:131–137.

56. Wainwright, M., U. Skiba, and R. P. Betts. 1984. Sulphur
    oxidation by a *Streptomyces* sp. growing in a carbon-deficient
    medium and autoclaved soil. Arch. Microbiol. 139:272–276.

57. Germida, J. J., J. R. Lawrence, and V. V. S. R. Gupta. 1984. Microbial oxidation of sulphur in Saskatchewan soils, p. 703–710. *In* J. W. Terry (ed.), Proceedings of the sulphur 84 conference. Sulphur Development Institute of Canada, Calgary.

58. Germida, J. J. 1985. Modified sulfur-containing media for studying sulfur-oxidizing microorganisms, p. 333–344. *In* D. E. Caldwell, J. A. Brierley, and C. L. Brierley (eds.), Planetary ecology. Van Nostrand Reinhold, New York.

59. Schook, L. B., and R. S. Berk. 1978. Nutritional studies with *Pseudomonas aeroginosa* grown on inorganic sulfur sources. J. Bacteriol. 133:1377–1382.

60. Van Niel, C. B. 1953. Introductory remarks on the comparative biochemistry of microorganisms. J. Cell Comp. Physiol. 41(Suppl. 1):11–38.

61. Tuttle, J. H. 1980. Organic carbon utilization by resting cells of thiosulfate-utilizing marine heterotrphs. Appl. Environ. Microbiol. 40:516–521.

62. Tuttle, J. H., J. H. Schwartz, and G. M. Whited. 1983. Some properties of thiosulfate-oxidizing enzyme from marine heterotroph 16B. Appl. Environ. Microbiol. 46:438–445.

63. Mason, J., and D. P. Kelly. 1988. Thiosulfate oxidation by obligately heterotrophic bacteria. Microbial. Ecol. 15:123–124.

64. Gommers, P. J. F., and J. G. Kuenen. 1988. *Thiobacillus* strain Q, a chemolithoheterotrophic sulphur bacterium. Arch. Microbiol. 150:117–125.

65. Yagi, S., S. Kitai, and T. Kimura. 1971. Oxidation of elemental sulfur to thiosulfate by *Streptomyces*. Appl. Microbiol. 22:157–159.

66. Kurek, E. 1983. Oxidation of inorganic sulphur compounds by yeast. Acta Microbiol. Pol. 28:169–172.

67. Grayston, S. J., W. Nevell, and M. Wainwright. 1986. Sulphur oxidation by fungi. Trans. Br. Mycol. Soc. 87:193–198.

68. Krol, M. 1983. Occurrence in soils and activity of sulphur oxidizing microorganisms. Pameit. Pulawski Prace Iung Zeszyt. 79:45–62.

69. Wainwright, M., and K. Killham. 1980. Sulphur oxidation by *Fusarium solani*. Soil Biol. Biochem. 12:555–558.

70. Kurek, E. 1983. An enzymatic complex active in sulphite and thiosulphate oxidation by *Rhodotorula* sp. Arch. Microbiol. 134:143–147.

71. Kurek, E. 1985. Properties of an enzymatic complex active in sulphite and thiosulphate oxidation by *Rhodotorula* sp. Arch. Microbiol. 143:277–282.

72. Lettl, A. 1983. Occurrence of thiosulphate sulphurtransfer-
    ase producers in populations of mesophilic heterotrophic bac-
    teria and microfungi of spruce humus. Folia Microbiol. 28:
    106–111.
73. Lettl, A., O. Langkramer, V. Lochman, and M. Jaks. 1981.
    Thiobacilli and sulphate production from inorganic sulphur
    compounds in upper horizons of some spruce forest soils.
    Folia Microbiol. 26:29–36.
74. Vitolins, M. I., and R. J. Swaby. 1969. Activity of sul-
    phur-oxidizing microorganisms in some Australian soils. Aust.
    J. Soil Res. 7:171–183.
75. McCaskill, M. R., and G. J. Blair. 1987. Particle size and
    soil texture effects in elemental sulfur oxidation. Agron. J.
    79:1079–1083.
76. Lee, A., C. C. Boswell, and J. H. Watkinson. 1988. Effect
    of particle size on the oxidation of elemental sulphur, thio-
    bacilli numbers, soil sulphate and its availability to pasture.
    N.Z. J. Agric. Res. 31:179–186.
77. Swaby, R. J., and R. Fedel. 1973. Microbial production of
    sulphate and sulphide in some Australian soils. Soil Biol. Bio-
    chem. 5:773–781.
78. Li, P., and A. C. Caldwell. 1966. The oxidation of elemental
    sulfur in soil. Soil Sci. Soc. Am. Proc. 30:370–372.
79. Pepper, I. L., and R. H. Miller. 1978. Comparison of the
    oxidation of thiosulfate and elemental sulfur by two hetero-
    trophic bacteria and *Thiobacillus thiooxidans*. Soil Sci. 126:
    9–14.
80. Lynch, J. M. 1982. Limits to microbial growth in soils. J.
    Gen. Microbiol. 128:405–410.
81. Grayston, S. J., and M. Wainwright. 1988. Sulphur oxida-
    tion by soil fungi, including species of mycorrhizae and wood-
    rotting Basidiomycetes. FEMS Microbiol. Ecol. 53:1–8.
82. Lawrence, J. R., and J. J. Germida. 1988. Most probable
    number procedure to enumerate S^0-oxidizing, thiosulfate-pro-
    ducing heterotrophs in soil. Soil Biol. Biochem. 20:577–578.
83. Lawrence, J. R., V. V. S. R. Gupta, and J. J. Germida.
    1988. Impact of elemental sulfur fertilization on agricultural
    soils. II. Effects on sulfur-oxidizing populations and oxida-
    tion rates. Can. J. Soil Sci. 68:475–483.
84. Lawrence, J. R., and J. J. Germida. 1988. Relationship be-
    tween microbial biomass and elemental sulfur oxidation in agri-
    cultural soils. Soil Sci. Soc. Am. J. 52:672–677.
85. McCready, R. G. L., and H. R. Krouse. 1982. Sulfur iso-
    tope fractionation during the oxidation of elemental sulfur by
    thiobacilli in a solonetzic soil. Can. J. Soil Sci. 62:105–110.

86. Kittams, H. A., and O. J. Attoe. 1965. Availability of phosphorus in rock phosphate—sulfur fusions. Agron. J. 57:331–334.

87. Weir, R. G., B. Barkus, and W. T. Atkinson. 1963. The effect of particle size on the availability of brimstone sulfur to white clover. Aust. J. Exp. Agric. Anim. Husb. 3:314–318.

88. Janzen, H. H., and J. R. Bettany. 1987. The effect of temperature and water potential on sulfur oxidation in soil. Soil Sci. 144:81–89.

89. Solberg, E. D., M. Nyborg, D. H. Laverty, and S. S. Malhi. 1982. Oxidation of elemental sulphur used as a fertilizer, p. 241–252. *In* Proceeding 19th annual soil science workshop. Edmonton, Alberta.

90. Arkesteyn, G. J. M. W. 1980. Pyrite oxidation in acid sulphate soils. Plant Soil 54:119–134.

91. Zinder, S. H., and Brock, T. D. 1978. Microbial transformations of sulfur in the environment, p. 445–466. *In* J. O. Nriagu (ed.), Sulfur in the environment. Wiley Interscience Publishers.

92. Saggar, S. K., J. R. Bettany, and J. W. B. Stewart. 1980. Measurement of microbial biomass sulfur in soil. Soil Biol. Biochem. 13:493–498.

93. Freney, J. R., G. E. Melville, and C. H. Williams. 1971. Organic sulfur fractions labelled by addition of $^{35}SO_4$ to soil. Soil Biol. Biochem. 3:133–141.

94. Saggar, S. K., J. R. Bettany, and J. W. B. Stewart. 1981. Sulfur transformations in relation to carbon and nitrogen in incubated soils. Soil Biol. Biochem. 13:499–511.

95. Goh, K. M., and P. E. H. Gregg. 1982. Field studies on the fate of radioactive sulfur fertilizer applied to pasture. Fert. Res. 3:337–351.

96. McLaren, R. G., J. I. Keer, and R. S. Swift. 1985. Sulfur transformations in soils using S^{35} labelling. Soil Biol. Biochem. 17:73–79.

97. Strickland, T. C., and J. W. Fitzgerald. 1985. Incorporation of sulfate sulfur into organic matter extracts of litter and soil: involvement of ATP sulphurylase. Soil Biol. Biochem. 17:779–784.

98. Swift, R. S. 1985. Mineralization and immobilization of sulfur in soil, 9:200–204. *In* R. J. Morris (ed.), Sulfur in agriculture. Sulphur Institute, Washington, D.C.

99. Watwood, M. E., J. W. Fitzgerald, and J. R. Gosz. 1986. Sulfur processing in forest soil and litter along an elevational and vegetative gradient. Can. J. For. Res. 16:689–695.

100. Strickland, T. C., J. W. Fitzgerald, and W. T. Swank. 1986
In situ measurements of sulfate incorporation into forest floor
and soil organic matter. Can. J. For. Res. 16:549–553.

101. McGill, W. B., and C. V. Cole. 1981. Comparative aspects
of cycling of organic carbon, nitrogen, sulfur and phosphorus
through soil organic matter. Geoderma 26:267–286.

102. Dodgson, K. S., G. F. White, and J. W. Fitzgerald. 1982.
Sulfatases of microbial origin, Vols. I and II. CRC Press,
Boca Raton, Florida.

103. Chaudhry, I. A., and A. H. Cornfield. 1967. Effect of
temperature of incubation on sulphate levels in aerobic and
sulphide levels in anaerobic soils. J. Sci. Food Agric. 18:
82–84.

104. Chaudhry, I. A., and A. H. Cornfield. 1967. Effect of
moisture content during incubation of soil treated with organic materials on changes in sulphate and sulphide levels.
J. Sci. Food Agric. 18:38–40.

105. Williams, C. H. 1967. Some factors affecting the mineralization of organic sulfur in soils. Plant Soil 26:205–223.

106. Barrow, N. J. 1967. Studies on the adsorption of sulfate
by soils. Soil Sci. 104:342–349.

107. Freney, J. R. 1967. Sulfur containing organics, p. 220–259.
In A. D. McLaren and G. H. Peterson (eds.), Soil biochemistry, Vol. 1. Marcel Dekker, New York.

108. Tabatabai, M. A., and J. M. Bremner. 1972. Distribution of
total and available sulfur in selected soils and soil profiles.
Agron. J. 64:40–44.

109. Kowalenko, C. G., and L. E. Lowe. 1975. Mineralization of
sulfur from four soils and its relationship to soil C, N and P.
Can. J. Soil Sci. 55:9–14.

110. Fitzgerald, J. W., and T. L. Andrew. 1984. Mineralization
of methionine sulfur in soils and forest floor layers. Soil
Biol. Biochem. 16:565–570.

111. Gupta, V. V. S. R., and J. J. Germida. 1988. Distribution
of microbial biomass and its activity in different soil aggregate
size classes as affected by cultivation. Soil Biol. Biochem. 20:
777–786.

112. Maynard, D. G., J. W. B. Stewart, and J. R. Bettany. 1983.
Sulfur and nitrogen mineralization in soils compared using two
incubation techniques. Soil Biol. Biochem. 15:251–256.

113. Stewart, J. W. B., and A. N. Sharpley. 1987. Controls on
dynamics of soil and fertilizer phosphorus and sulfur, p. 101–
121. *In* R. F. Follett (ed.), Soil fertility and organic matter
as critical components of production systems. SSSA Spec.
Publ. 19. American Society of Agronomy, Madison, Wisconsin.

114. Tsuji, T., and K. M. Goh. 1979. Evaluation of soil sulfur fractions as sources of plant available sulfur using radioactive sulfur. N.Z. J. Agric. Res. 22:281—291.

115. Speir, T. W., R. Lee, E. A. Pansier, and A. Cairns. 1980. A comparison of sulphatase, urease and protease activities in planted and in fallow soils. Soil Biol. Biochem. 12:281—291.

116. Strickland, T. C., J. W. Fitzgerald, J. T. Ash, and W. T. Swank. 1987. Organic sulfur transformation and sulfur pool sizes in soil and litter from a southern Appalachian hardwood forest. Soil Sci. 143:453—458.

117. Burns, R. G. (ed.). 1978. Soil enzymes. Academic Press, London.

118. Roy, A. B., and P. A. Trudinger. 1970. The biochemistry of inorganic compounds of sulphur. Cambridge University Press, Cambridge.

119. Cooper, P. J. M. 1972. Arylsulfatase activity in northern Nigerian soils. Soil Biol. Biochem. 4:333—337.

120. Houghton, C., and F. A. Rose. 1976. Liberation of sulfate from sulfate esters by soils. Appl. Environ. Microbiol. 31: 969—976.

121. Pettit, N. M., L. J. Gregory, R. B. Freedman, and R. G. Burns. 1977. Differential stabilities of soil enzymes. Assay and properties of phosphatase and arylsulphatase. Biochim. Biophys. Acta 485:357—366.

122. Speir, T. W. 1977. Studies on a climosequence of soils in tussock grasslands. II. Urease, phosphatase, and sulphatase activities of top soils and their relationships with other properties including plant available sulphur. N.Z. J. Soil Sci. 20:159—166.

123. Speir, T. W., and D. J. Ross. 1978. Soil phosphatase and sulphatase, p. 198—250. *In* R. G. Burns (ed.), Soil enzymes. Academic Press, London.

124. Al-Khafaji, A. A., and M. A. Tabatabai. 1979. Effects of trace elements on arylsulfatase activity in soils. Soil Sci. 127:129—133.

125. Lee, R., and T. W. Speir. 1979. Sulphur uptake by ryegrass and its relationship to inorganic and organic sulphur levels and sulphatase activities in soil. Plant Soil 53:407—425.

126. Sarathchandra, S. U., and K. W. Perroto. 1981. Determination of phosphatase and arylsulfatase in soils. Soil Biol. Biochem. 13:543—545.

127. Neal, J. L. 1982. Abiotic enzymes in arctic soils: influence of predominant vegetation upon phosphomonoesterase and sulfatase activities. Commun. Soil Sci. Plant Anal. 13: 863—878.

128. Perucci, P., and L. Scarponi. 1983. Effect of crop residue addition on arylsulfatase activity in soils. Plant Soil 73:323–326.

129. David, M. B., S. C. Schindler, M. J. Mitchell, and J. E. Strick. 1983. Importance of organic and inorganic sulfur to mineralization processes in a forest soil. Soil Biol. Biochem. 15:671–677.

130. Gupta, V. V. S. R., J. R. Lawrence, and J. J. Germida. 1988. Impact of elemental sulfur fertilization on agricultural soils. I. Effects on microbial biomass and enzyme activities. Can. J. Soil Sci. 68:463–473.

131. Strickland, T. C., J. W. Fitzgerald, and W. T. Swank. 1984. Mobilization of recently formed forest soil organic sulfur. Can. J. For. Res. 14:63–67.

132. Fitzgerald, J. W., M. E. Watwood, and F. A. Rose. 1985. Forest floor and soil arylsulfatase hydrolysis of tyrosine sulphate, an environmentally relevant substrate for the enzyme. Soil Biol. Biochem. 17:885–887.

133. Fitzgerald, J. W., and T. C. Strickland. 1987. Mineralization of organic sulfur in the O_2 horizon of a hardwood forest: involvement of sulphatase enzymes. Soil Biol. Biochem. 19:779–781.

134. King, G. M., and M. J. Klug. 1980. Sulfhydrolase activity in sediments of Wintergreen Lake, Kalamazoo County, Michigan. Appl. Environ. Microbiol. 39:950–956.

135. Chandramohan, D., K. Devendran, and R. Natarajan. 1974. Arylsulfatase activity in marine sediments. Marine Biol. 27:89–92.

136. Wainwright, M. 1981. Enzyme activity in intertidal sands and salt marsh soils. Plant Soil. 59:357–363.

137. Oshrain, R. L., and W. J. Wiebe. 1979. Arylsulfatase activity in salt marsh soils. Appl. Environ. Microbiol. 38:337–340.

138. Skiba, K., and M. Wainwright. 1983. Assay and properties of some sulphur enzymes in coastal sands. Plant Soil 70:125–132.

139. Press, M. C., J. Henderson, and J. A. Lee. 1985. Arylsulfatase activity in peat in relation to acidic deposition. Soil Biol. Biochem. 17:99–103.

140. Nissen, P. 1968. Choline sulfate permease: transfer of information from bacteria to higher plants? Biochem. Biophys. Res. Commun. 32:696–703.

141. Fitzgerald, J. W. 1978. Naturally occurring organosulfur compounds in soil, p. 391–443. *In* J. O. Nriagu (ed.), Sulfur in the environment. Ecological impacts, Part II. John Wiley & Sons, New York.

142. Tabatabai, M. A., and J. M. Bremner. 1970. Factors affecting soil arylsulfatase activity. Soil Sci. Soc. Am. Proc. 34:427–429.

143. Bremner, J. M., and M. I. Zantua. 1975. Enzyme activity in soils at subzero temperatures. Soil Biol. Biochem. 7:383–387.

144. Speir, T. W., and D. J. Ross. 1981. A comparison of the effects of air-drying and acetone dehydration on soil enzyme activities. Soil Biol. Biochem. 13:225–229.

145. Pulford, I. D., and M. A. Tabatabai. 1988. Effect of waterlogging on enzyme activities in soils. Soil Biol. Biochem. 20:215–219.

146. Skujins, J. J. 1967. Enzymes in soil, p. 371–414. *In* A. D. McLaren and S. H. Peterson (eds.), Soil biochemistry, Vol. I. Marcel Dekker, New York.

147. Neal, J. L., and S. A. Herbein. 1983. Abiotic enzymes in arctic soils: changes in sulphatase activity following vehicle disturbance. Plant Soil 70:423–427.

148. Fitzgerald, J. W., J. T. Ash, T. C. Strickland, and W. T. Swank. 1983. Formation of organic sulfur in forest soils: a biologically mediated process. Can. J. For. Res. 13:1077–1082.

149. Swank, W. T., J. W. Fitzgerald, and J. T. Ash. 1984. Microbial transformations of sulfate in forest soils. Science 223:182–184.

150. Schindler, S. C., M. J. Mitchell, T. J. Scott, R. D. Fuller, and C. T. Driscoll. 1986. Incorporation of ^{35}S-sulfate into organic and organic constituents of two forest soils. Soil Sci. Soc. Am. J. 50:457–462.

151. Burns, R. G. 1978, Enzymes in soil: some theoretical and practical considerations, p. 295–339. *In* R. G. Burns (ed.), Soil enzymes. Academic Press, London.

152. Tabatabai, M. A., and J. M. Bremner. 1971. Michaelis constants of soil enzymes. Soil Biol. Biochem. 3:317–323.

153. Perucci, P., and L. Scarponi. 1984. Arylsulfatase activity in soils amended with crop residues: kinetic and thermodynamic parameters. Soil Biol. Biochem. 16:605–608.

154. Dick, R. P., D. D. Myrold, and E. A. Kerle. 1988. Microbial biomass and soil enzyme activities in compacted and rehabilitated skid trail soils. Soil Sci. Soc. Am. J. 52:512–516.

155. Fitzgerald, J. W., and J. T. Ash. 1982. Influence of carbon source on arylsulfatase derepression in *Pseudomonas* $C_{12}B$. Can. J. Microbiol. 28:383–388.

156. Fitzgerald, J. W., and M. E. Cline. 1977. The occurrence of an inducible arylsulfatase in *Comamonas terrigena*. FEMS Microbiol. Lett. 2:221–224.

157. Speir, T. W., D. J. Ross, and V. A. Orchard. 1984. Spa-
      tial variability of biochemical properties in a taxonomically uni-
      form soil under grazed pasture. Soil Biol. Biochem. 16:153–
      160.
158. Tabatabai, M. A., and J. M. Bremner. 1970. Arylsulfatase
      activity of soils. Soil Sci. Soc. Am. Proc. 34:225–229.
159. Stott, D. E., and C. Hagedorn. 1980. Interrelations be-
      tween selected soil characteristics and arylsulfatase and
      urease activities. Commun. Soil Sci. Plant Anal. 11:935–955.
160. Ross, D. J., T. W. Speir, D. J. Giltrap, B. A. McNeilly, and
      L. F. Molly. 1975. A principal components analysis of some
      biochemical activities in a climosequence of soils. Soil Biol.
      Biochem. 7:349–355.
161. Frankenberger, W. T., Jr., and W. A. Dick. 1983. Rela-
      tionship between enzyme activities and microbial growth and
      activity indices in soil. Soil Sci. Soc. Am. J. 47:945–951.
162. Sarathchandra, S. U., K. W. Perott, and M. P. Upsdell.
      1984. Microbiological and biochemical characteristics of a
      range of New Zealand soils under established pasture. Soil
      Biol. Biochem. 16:177–183.
163. Ross, D. J., T. W. Speir, J. C. Cowling, and K. N. Whale.
      1984. Temporal fluctuations in biochemical properties of soil
      under pasture: II. N mineralization and enzyme activities.
      Aust. J. Soil Res. 22:319–330.
164. Ross, D. J., and A. Cairns. 1982. Effects of earthworms
      and ryegrass on respiratory and enzyme activities of soil.
      Soil Biol. Biochem. 14:583–587.
165. Gupta, V. V. S. R., and J. J. Germida. 1986. Effect of
      cultivation on the activity of soil enzymes, p. 332–343. *In*
      Proceedings of the 1986 soils and crops workshop, "research
      in agriculture." University of Saskatchewan, Saskatoon.
166. Speir, T. W. 1976. Studies on a climosequence of soils in
      tussock grasslands: 8. Urease, phosphatase and sulphatase
      activities of tussock plant materials and of soil. N.Z. J. Sci.
      19:383–387.
167. Dick, R. A. 1984. Influence of long-term tillage and crop
      rotation combinations on soil enzyme activities. Soil Sci. Soc.
      Am. J. 48:569–574.
168. Brown, K. A. 1981. Biochemical activities in peat sterilized
      by gamma-irradiation. Soil Biol. Biochem. 13:469–474.
169. Killham, K., M. K. Firestone, and J. G. McColl. 1983. Acid
      rain and soil microbial activity: effects and their mechanisms.
      J. Environ. Qual. 12:13–137.
170. Lang, K. 1933. Die Rhodanbildung im Tierkörper. Biochem.
      Z. 259:243–256.

171. Westly, J. 1973. Rhodanese, p. 327–368. *In* A. Meister (ed.), Advances in enzymology, Vol. 1. John Wiley & Sons, New York.

172. Chew, M. Y. 1973. Rhodanese in higher plants. Phytochemistry 12:2365–2367.

173. Brown, T. J., P. J. Butler, and F. C. Happold. 1965. Some properties of the rhodanese system of *Thiobacillus denitrificans*. Biochem. J. 97:651–657.

174. Tabatabai, M. A., and B. B. Singh. 1976. Rhodanese activity of soils. Soil Sci. Soc. Am. J. 40:381–385.

175. Tabatabai, M. A., and B. B. Singh. 1979. Kinetic parameters of the rhodanese reaction in soils. Soil Biol. Biochem. 11:9–12.

176. Wainwright, M. 1978. Microbial sulphur oxidation in soil. Sci. Prog. 65:459–475.

177. Ray, R. C., N. Behera, and N. Sethunathan. 1985. Rhodanese activity of flooded and nonflooded soils. Soil Biol. Biochem. 17:159–162.

178. Kumagai, H., S. Sejima, Y. J. Choi, H. Tanaka, and H. Yamada. 1975. Crystallization and properties of cysteine disulfhydrase from *Aerobacter aerogenes*. FEBS Lett. 52:304–307.

179. Ohkishi, H., D. Nishikawa, H. Kumagai, and H. Yamada. 1981. Distribution of cysteine desulfhydrase in microorganisms. Agric. Biol. Chem. 45:253–257.

180. Abramyan, S. A., and A. S. H. Galshyan. 1986. Regulation of the activity of sulphur metabolism enzymes in soil. Soviet Soil Sci. 18:28–37.

181. Bloomfield, C., and J. K. Coulter. 1973. Genesis and management of acid sulfate soils. Adv. Agron. 25:265–326.

182. Mermut, A. R., D. Curtin, and H. P. W. Rostad. 1985. Micromorphological and submicroscopical features related to pyrite oxidation in an inland marine shale from east central Saskatchewan. 49:256–261.

183. Mermut, A. R., and M. A. Arshad. 1987. Significance of sulfide oxidation in soil salinization in southeastern Saskatchewan, Canada. Soil Sci. Soc. Am. J. 51:247–251.

184. Rupela, O. P., and P. Tauro. 1973. Isolation and characterization of *Thiobacillus* from alkali soils. Soil Biol. Biochem. 5:891–897.

185. Rupela, O. P., and P. Tauro. 1973. Utilization of *Thiobacillus* to reclaim alkali soils. Soil Biol. Biochem. 5:899–901.

186. Killham, K., and M. Wainwright. 1981. Microbial release of sulphur ions from atmospheric pollution deposits. J. Appl. Ecol. 18:889–896.

187. Wainwright, M., and K. Killham. 1982. Microbial transformation of some particulate pollution deposits in soil—a source of plant-available nitrogen and sulphur. Plant Soil 65:297–301.

188. Wainwright, M., and W. Nevell. 1984. Microbial transformations of sulphur in atmospheric-polluted soils. Rev. Environ. Health 4:339–356.

189. Lettl, A. 1981. The effect of emissions on the microbiology of the sulphur cycle. Commun. Inst. For. Czech. 12:27–50.

190. Wainwright, M. 1978. Sulphur oxidising micro-organisms on vegetation and in soils exposed to atmospheric polution. Environ. Pollut. 17:167–174.

191. Nevell, W., and M. Wainwright. 1987. Influence of soil moisture on sulphur oxidation in a brown earth soil exposed to atmospheric pollution. Biol. Fert. Soils 5:209–214.

192. Wainwright, M. 1979. Microbial S-oxidation in soils exposed to heavy atmospheric pollution. Soil Biol. Biochem. 11:95–98.

193. Jarvis, B. W., G. E. Land, and R. K. Wieder. 1987. Arylsulfatase activity in peat exposed to acid precipitation. Soil Biol. Biochem. 19:107–109.

194. Nevell, W., and M. Wainwright. 1987. Changes in sulphate concentration in an atmospheric-polluted brown earth following a waterlogging-drying cycle. Z. Pflanzenernähr. Bodenk. 150:147–150.

195. Killham, K., and M. Wainwright. 1984. Chemical and microbiological changes in soil following exposure to heavy atmospheric pollution. Environ. Pollut. (Ser. B) 33:121–131.

196. Maynard, D. G., P. A. Addison, and K. A. Kennedy. 1983. Elemental sulphur dust deposition on soils and vegetation of lodgepole pine stands in west-central Alberta, p. 458–464. *In* R. W. Weir, R. R. Riewe, and I. R. Methven (eds.), Resources and dynamics of the boreal zone. Association of Canadian Universities for Northern Studies, Ontario.

197. Maynard, D. G., J. J. Germida, and P. A. Addison. 1986. The effect of elemental sulfur on certain chemical and biological properties of surface organic horizons of a forest soil. Can. J. For. Res. 16:1050–1054.

198. Tabatabai, M. A. 1985. Effect of acid rain on soils. CRC Crit. Rev. Environ. Contr. 15:65–110.

199. Alexander, M. 1980. Effects of acidity on microorganisms and microbial processes in soil, p. 341–362. *In* T. C. Hutchinson and M. Havas (eds.), Effects of acid precipitation on terrestrial ecosystems. Plenum Press, New York.

200. Babich, H., and G. Stotzky. 1978. Atmospheric sulfur compounds and microbes. Environ. Res. 15:513–531.

201. Bryant, R. D., E. A. Gordy, and E. J. Laishley. 1979.
      Effect of soil acidification on the soil microflora. Water Air
      Soil Pollut. 11:437–445.
202. Bewley, R. J. F., and G. Stotzky. 1983. Simulated acid
      rain (H_2SO_4) and microbial activity in soil. Soil Biol. Bio-
      chem. 15:425–429.
203. Bewley, R. J. F., and D. Parkinson. 1984. Effects of sul-
      phur dioxide pollution on forest soil microorganisms. Can. J.
      Microbiol. 30:179–185.
204. McColl, J. G., and M. K. Firestone. 1987. Cumulative ef-
      fects of simulated acid rain on soil chemical and microbial
      characteristics and conifer seedling growth. Soil Sci. Soc.
      Am. J. 51:794–800.
205. Gupta, V. V. S. R., and J. J. Germida. 1988. Populations
      of predatory protozoa in field soils after 5 years of elemental
      S fertilizer application. Soil Biol. Biochem. 20:787–791.

2

Problems Related
to the Humification Processes
in Soils of Temperate Climates

KONRAD HAIDER   *Bundesforschungsanstalt für Landwirtschaft
(FAL), Braunschweig, Germany*

I. INTRODUCTION

Soil organic matter (SOM) has been extensively examined because of
its importance for soil fertility and productivity, and as a critical
component in agricultural production systems [1]. Furthermore, its
importance for establishing favorable physical conditions and archi-
tecture in soil has been emphasized. Because biological, chemical,
and physical processes are involved in the balance between concur-
rent mineralization and immobilization of organic residues and nutrients, soils must always be considered in connection with their microflora and microfauna. The biochemical and enzymatic activity of the
microbiota, as well as its buffering capacity, in regulating humification and the flow of nutrients, influence the processes related to
the formation and transformation of SOM. It is, therefore, important to understand that soil humus is a dynamic system of both active and more passive components. These differ in biological availability, stability, and time of residence. Parton et al. [2] and Doran
and Smith [3] suggested that short- and long-term changes in nutrient cycling and their response to organic management practices
are associated with changes in the relative quantities of both the
more labile and the more stable organic matter fractions. These relative quantities depend upon climate, soil type, and management.
Furthermore, Janssen [4] indicated that differences in carbon and

nitrogen mineralization after 25 years of various management practices were more closely related to the amount of young soil organic
matter than to the total C and N contents of a soil, indicating that
qualitative changes are more important than gross quantitative
changes. The latter changes, however, are also important. Scerbakow and Kislych [5] found a high correlation between the humus
contents and the yields of agricultural plants in Chernozem soils.
As the result of intensive management, these soils sometimes have
considerable decreases in their C and N contents. Furthermore, a
positive correlation between these contents and the activities of
saccharase, protease, urease, and dehydrogenase were indicated.
Beck [6,7] also observed a positive correlation among biomass, its
enzymatic activities, and the humus content in soil samples from
field plots under different forms of continuous management practices. By combining biomass and several enzymatic activities into
an index, he predicted a long-term decrease in humus contents by
specific management practices, if the index was smaller than that
calculated from the actual organic carbon content. If this index
was greater than that calculated from the actual humus content, a
long-term increase in humus was predicted.

Investigations on the chemical structure of SOM fractions have
also made considerable progress in the recent years. For instance,
^{13}C-nuclear magnetic resonance (NMR) spectroscopy has detected
large differences in humic substances isolated from marine, aquatic,
or terrestrial environments, including soils [8,9]. These differences usually involve the relative amounts of aliphatic and aromatic
carbon contents. Such differences (e.g., in the structure of fulvic
and humic acids from various soils) were not as obvious in previous
degradative studies [10].

In soils, primary particles combine into aggregates of varying
size. Tisdall and Oades [11] suggested that these aggregates are
important factors in retarding soil organic matter decomposition. It
was also shown that long-chain aliphatic or lipid compounds, together
with polysaccharides, influence soil structure and the dynamics of
nutrients [12,13]. Aggregates, therefore, are important factors
for soil stability. Generally, the former view that humic compounds
are stabilized mainly as the result of their complex and recalcitrant
structure is only partly true. More important for stabilization are
probably associations with metal ions and clays and aggregation
[14,15]. Once this association is disturbed by climatic alterations
or changes in soil management practices, the decay rates of SOM
sometimes increase dramatically until a new equilibrium is reached.
The last section of this chapter discusses these phenomena.

II. PLANT RESIDUES AS A SOURCE OF SOIL HUMUS

A. Whole-Plant Residue Materials and Their Transformation Rates

Plant residues introduced into soil are the main sources of humus. They consist of a wide range of different components, as shown in Table 1. On an average, they contain 15 to 60% cellulose, 10 to 30% hemicellulose, 5 to 30% lignin, and 2 to 15% protein. Additionally, they also contain minor amounts of phenols, sugars, and amino acids. During soil incubation of plant residues in a temperate climate, about 70% of the total residue carbon is released as CO_2 during the first year. The remainder decomposes more and more slowly with time and becomes steadily incorporated into soil humus. During the period of rapid decomposition, the morphological and chemical structures of the plant residues are drastically altered and the C:N ratio is narrowed by evolution of CO_2. Although complex processes are involved during the decomposition of the various ingredients, the overall degradation rate follows, reasonably well, first-order kinetics [16]. Residue decomposition in the field [17, 18] or in the laboratory [19] can be adequately described by

Table 1 Average Contents of Major Components in Plant Materials

Material	Percentage of Dry Weight			
	Cellulose	Hemi-cellulose	Lignin	Protein[a]
Ryegrass (mature)	19–26	16–23	4–6	12–20
Lucerne (stem)	13–33	8–11	6–16	15–18
Wheat straw	27–33	21–26	18–21	3
Pinus sylvestris (sawdust)	42–49	24–30	25–30	0.5–1
Beech wood	42–51	27–40	18–21	0.6–1

[a]Nitrogen × 6.25.
Source: Authors data and Ref. 36 and 36a.

assuming that plant materials consist of readily decomposable fractions that turn over in less than 1 year. Another, more stable, fraction has a turnover time of a few years. The decay of both the labile and the more stable fraction can be expressed as logarithmic functions, which can be combined to describe the overall process.

The composition of the organic residues incorporated into soil influences decomposition, as does temperature, soil moisture and texture, and other climatic and soil-derived characteristics. Kolenbrander [20] compared the decomposition of different organic residues under field conditions in the temperate climate. After the first year, 20% of green manure, 38% of straw, 60% of farmyard manure and 80 to 90% of peat remained as carbon residues in the soil. After 8 years, 3 to 10% of the plant residue and 20 and 50% of manure or peat carbon, respectively, were still present. Similar decomposition data were obtained in laboratory soil incubation experiments with different organic compounds and plant residues of increasing complexity, ranging from glucose, starch, and cellulose, to wheat straw, pine sawdust, almond shells, and cow manure [19]. As shown in Table 2, simple sugars or polysaccharides were readily utilized, and after 28 weeks, 90% of the glucose carbon and 85% that of the polysaccharides had been released as CO_2. With increasing complexity and lignin contents of the residues, degradation rates slowed. After 38 weeks of incubation, 32% of wheat straw carbon, about 50% of almond shell, and more than 65% of the ponderosa pine needle carbon remained in the soil.

The observation that the lignin contents of specific plant residues control degradation was presented by Herman et al. [21], who described the decomposition of root residue materials. This concept was further developed and used as a model to describe the decomposition rates of plant residues in soil by van Veen et al. [22] and Parton et al. [23]. The model developed by Parton et al. [23] is shown in Figure 1.

Plant residues consist of structural and metabolic materials that have, according to their lignin/N ratio, turnover times of 1 to 5 and 0.1 to 1 years. The metabolic C pool, defined by a low lignin/N ratio, is rapidly converted into a microbial C pool with a 0.1- to 1-year turnover time. Microbial and labile C together form the "active soil fraction," which consists of microbes and microbial products, with a short turnover time of 1.5 years. A larger pool of C and N forms a slowly transformable fraction that is physically protected and is in a chemical form that has more biological resistance to decomposition. The remainder of the active and the slowly transformable soil C is, with time, transformed into a fraction that is chemically recalcitrant and, additionally, is physically protected, and has the longest turnover time of 200 to 1500 years ("passive

Table 2 Decomposition of Various Organic Compounds and Plant Residues in Greenfield Sandy Loam Topsoil[a]

Substrate	Decomposition After Weeks[b]				
	1	4	12	20	28
Glucose	73	82	89	90	90
Starch	48	69	81	84	86
Cellulose	27	52	77	79	84
Green matter (corn 28 days)	27	45	69	73	82
Lima bean straw	36	57	75	78	79
Wheat straw	20	33	59	61	64
Corn straw	18	31	60	63	65
Cow farmyard manure	18	33	43	48	50
Prune wood (sawdust)	12	25	33	40	45
Almond shells	12	24	37	39	41
Douglas fir (sawdust)	2	5	15	29	34
Peat moss	<1	3	8	14	17
Soil humic acid	<1	<1	1	1	2

[a]Dry, powdered (1-mm) material was mixed (1000 ppm) with soil and incubated at the -33 kPa water potential at 22°C in the laboratory under continuous aeration.
[b]Percentage of added carbon evolved as CO_2.
Source: Data from author and Ref. 19.

soil fraction"). It can also be assumed that the lignin/N ratio controls the division into structural and metabolic parts of the plant residues and that most of the lignin in residues flows into the slowly transformable soil pool. This direct flow of lignin is based on data from laboratory incubations of labeled lignin-type material by Stott et al. [24]. They showed that lignin is slowly catabolized to CO_2, but very little lignin C is found in microbial biomass, whereas 70% or more was being stabilized in the soil. Further variables in this model are annual precipitation, soil temperature and texture, the amount of annual plant residue input, and its lignin/N ratio. The model was used to predict future organic

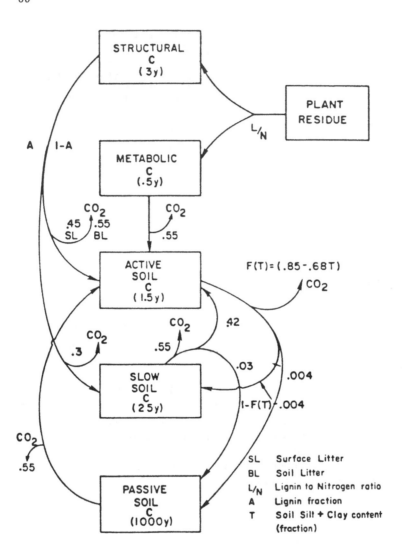

Figure 1   Diagram of the carbon flows through a readily available fraction (active soil C), a physically or chemically protected fraction (slow soil C), and a chemically recalcitrant and physically protected fraction (passive soil C).   Ciphers indicate carbon mineralization rates [23].

matter levels and the carbon and nitrogen flows in the Great Plains area of the United States.

Several authors [2,25,26] have emphasized the importance of young and easily decomposable humus in determining the flow of carbon and nutrients. Although there is an exchange of organic C from the passive into the active phase and vice versa, it is necessary that the active phase be continuously supplied with fresh organic material, which also sustains a vigorous microbial population [4].

B. Degradation and Transformation of the Main Plant Components: Cellulose and Lignins—Metabolism and Cometabolism

Cellulose and Hemicelluloses

Earlier studies [27,28] indicated that cellulose and hemicellulose components of plant residues are more rapidly decomposed than the lignin component. The ^{14}C-labeling of these components has considerably aided investigations of their decomposition rates and of the fate of the remaining carbon [24,29—31].

Cellulose loses about 75% of its carbon during the initial 4 months of soil incubation; after 1 and 2 years, only 18 and 16% of the initial carbon, respectively, was left in the soil (Table 3). About 20% of the residual carbon after 1 year of incubation was found in the microbial biomass. A greater portion (60 to 80%) was present in 6 N HCl hydrolyzable portion of the soil and, here, mostly in microbial metabolites, such as amino acids and sugars [31,32]. Sörensen [33] reported, based on field incubation experiments in which ^{14}C-labeled wheat straw was incubated for 20 years, that during the 8- to 20-year period, the proportion of residual ^{14}C in biomass or in amino acids from the hydrolyzable portion was nearly constant and averaged 3 to 22% of the remaining ^{14}C in soil.

Cellulose and hemicelluloses can be completely metabolized by numerous soil microorganisms and used as a sole carbon source for growth. These organisms include bacteria, actinomycetes, and fungi. The pathways and enzymes involved in the degradation of these polysaccharides are well characterized for several bacteria [34—36] and seem to be similar for most of the cellulolytic organisms. The degradation of crystal cellulose involves a random attack on the polysaccharide chain by *endo*-1,4-β-glucanases and the production of smaller units, which are afterwards cleaved from the free ends by *exo*-1,4-β-glucanases and glucosidases into cellobiose and glucose. The metabolic products can then be completely used by microbes for the production of energy and biomass.

Hemicelluloses are generally considered to be degraded faster than cellulose [37,38], probably as the result of a greater number of microorganisms that can use these compounds as a substrate.

Table 3 Biodegradation and Incorporation into Biomass and 6 N HCl Hydrolyzable Portions After Incubation of Soils (Typic Hapludalf and Mollic Haploxeralf) for Different Periods with Various ^{14}C-labeled Organic Compounds

Compound	% ^{14}C-CO_2 evolved				% [14]C in biomass		% [14]C hydrolyzed[a]	
	Months							
	1	6	12	24	12	24	12	24
Glucose (UL ^{14}C)[b]	70	85	89		19		72	
Wheat straw polysaccharide (UL ^{14}C)[b]	55	78	81	84	10	8	60	58
Wheat straw (UL ^{14}C)[b]	31	63	69	71	7	5	56	48
Lignin[c] ^{14}C-ring	5	26	33	45	0.5	0.4	19	18
Lignin[c] ^{14}C-side chain	7	28	34	42	0.5	0.4	21	19

[a]Percentage of residual ^{14}C in soil.
[b]Uniformly ^{14}C-labeled.
[c]Cornstalk material with ^{14}C label in the lignin portion at aromatic rings or at C-β of the side chains, respectively [30].
Source: Refs. 19,24,31.

Cheshire [39], however, observed that ^{14}C-labeled hemicelluloses are decomposed similarly to cellulose, but that parts of their sugars are recycled into other sugars of microbial polysaccharides. Furthermore, soil polysaccharides also can partially be directly derived from plants [39].

Lignin

The biodegradation of lignin is less well understood than that of cellulose. This is partly the result of the complicated structure of lignin, in which phenylpropanoid units—the lignin alcohols—are connected irregularly by C-O-C and C-C linkages. Lignin consists of large spheric molecules with a mean relative molecular mass (M_r) between 10,000 and 20,000 daltons (d). In the plant cell wall, lignin is additionally linked to hemicellulosic and, possibly, to cellulosic components [36]. The complex aromatic structure is only slowly attacked by microbes and, as it turns out [40], this attack is not directed toward a distinct type of linkage as in cellulose. Fewer species of microorganisms can degrade lignins than can degrade cellulose. The most active biodegraders of lignin belong to the white-rot fungi. These, however, are not common in arable soils, but can be found in forest soils. Lignin degradation and transformation in arable soils seem to be mostly a domain of the Fungi Imperfecti, the actinomycetes, and other bacteria [15,41]. The bacteria (including actinomycetes) exhibit only a limited activity in degrading high-molecular-mass lignin or the lignin portion of lignocelluloses. In arable soils, lignocelluloses are probably degraded by synergistic consortia of microbes that cannot completely use them individually.

Several observations highlight the peculiarities of the biodegradation of lignin, compared with that of cellulose or other polysaccharides: None of the organisms yet isolated can use lignin as a sole carbon source. An additional easily available carbon and energy source is always needed for degradation and, therefore, it is a *cometabolic* degradation. For this kind of degradation, Kirk and Farrell [40] coined the expression "enzymatic combustion," because no energy or metabolites can be gained from lignin for growth purposes of the ligninolytic organisms. Furthermore, degradation of lignin is much more vigorous under aerobic than under anaerobic conditions [42,43]. Colberg and Young [44] have reported that only lignin precursors with an M_r between 600 and 1200 d are degraded by mixed or pure cultures of anaerobic bacteria. This occurs by anaerobic cleavage of arylether linkages and conversion of the resulting phenols into CO_2 and CH_4. Higher molecular mass lignin was not converted at all.

Incubation of soil samples with ^{14}C-labeled lignin or lignocelluloses [24] indicated that during a 1-year period, about 15 to 30% of the lignin carbons were released as CO_2. Very little (< 1%) of the residual carbon was incorporated into the biomass, and on 6 N HCl hydrolysis of the soil, most of it remained in the nonhydrolyzable fraction (see Table 3). This is explained by the fact that microorganisms cannot use lignin as a carbon and energy source during degradation. Furthermore, lignin is vigorously catabolized only in well-aerated soils, whereas in poorly aerated soils or anoxic sediments and swamps, lignin degradation is very slow or not measurable [42]. This recalcitrance of lignin under anaerobic conditions is considered a major factor in the accumulation of peats, and most probably, of coal.

Information is now available about the biochemical pathways of lignin biodegradation and the specific enzymes involved. Characterization of a microbially attacked lignin by ^{13}C-NMR spectroscopy [45–47] or by mass spectrometry [48] showed a cleavage of bonds in both side chains and rings. This results in partly alphatic–aromatic degradation products, which, however, are still linked into a macromolecular matrix. Only at later stages of degradation are numerous monomer or dimer degradation products released, which originate from lignin by an extensive cleavage of side-chain and ring linkages [40]. Several of these products are shown in Figure 2. Furthermore, there is a considerable increase in polar groups, including keto, hydroxyl, and carboxyl groups.

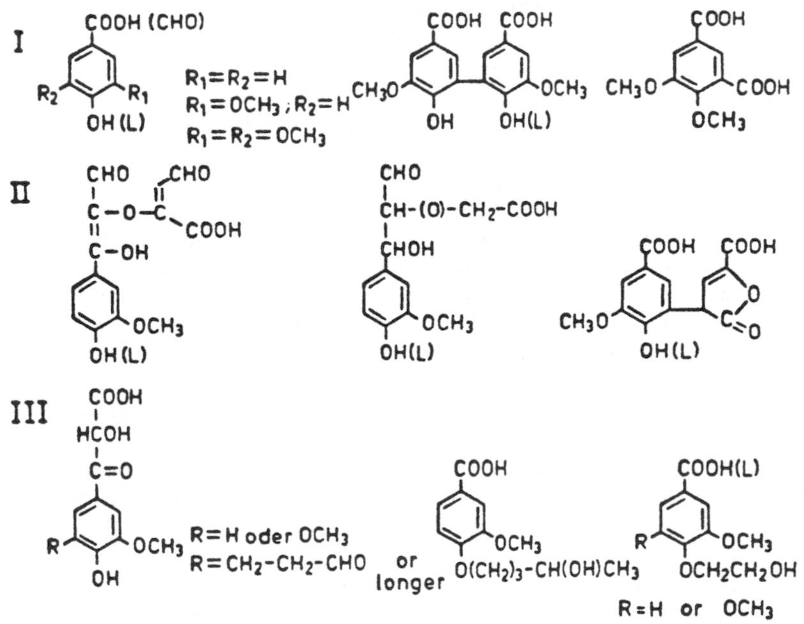

Figure 2 Degradation products detected in white rot-infected spruce or beech wood lignin, sometimes in connection with larger lignin residues (L). I. Derivatives of benzoic acid, benzaldehyde, or phthalic acid originating through cleavage reactions in lignin side chains; II. Derivatives originating through cleavage reactions of neighboring rings; III. Derivatives substituted by parts of the side chains from neighboring subunits. Adopted from [40,45,46,48].

Extracellular enzymes have been isolated from the culture fluids of ligninolytic fungi, mainly from those of *Phanerochaete chryso-sporium* that vigorously catalyze the degradation of dimeric and trimeric lignin models [49]. The enzymes function similar to H_2O_2-dependent peroxidases, but they have the specific capability to remove electrons from aromatic rings with completely etherified hydroxyl groups. This yields cationic radicals that stabilize by cleavage of C_α-C_β bonds and the formation of further radicals (Fig. 3; according to Schoemaker et al. [50]). The intermediate radicals can react with O_2 or water and become stabilized by the formation of hydroxy or keto derivatives. The enzymes also catalyze additional cleavage reactions, including bonds connecting C_α from the side chain to the ring or of bonds in the aromatic rings [51,52]. These reactions can occur with lignin models before they are degraded into smaller units.

The enzymes isolated from ligninolytic fungi catalyze only the degradation of low-molecular-mass lignin models, but have very little effect on the lignin itself [51,53]. Kirk [51] theorized that ligninolytic fungi that completely degrade lignin have additional enzymes or cofactors residing on cell surfaces. More probably, however, seems to be a transient binding of lignin on cell surfaces combined with the release of more water-soluble lignin products [54]. This binding is probably a prerequisite for lignin degradation, as it enables the enzymes involved to function cooperatively.

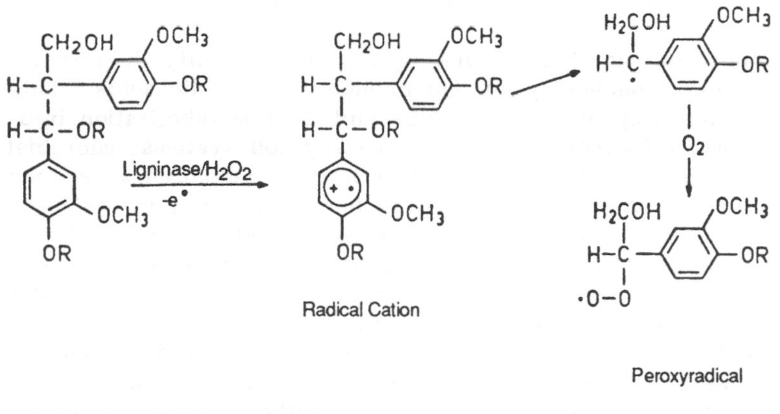

Figure 3   Action of the ligninase—H_2O_2 complex on lignin models: R, connection to lignin subunits or ethyl methyl substituents. (According to Ref. 50.)

The concept of a nonspecific enzyme-catalyzed radical mechanism provides, however, a better way in understanding the peculiarities of lignin degradation and its transformation into humus. It indicates that radical formation results in a random opening of the bond system, in an addition of nucleophilic groups and, in this connection, in the formation of more polar groups. In soils, these still high-molecular-mass fragments, containing additional functional groups and radicals, can react with humic compounds already present, with metal cations, or with clay surfaces saturated with transition metal ions [55]. By this mechanism of degradation, the lignin polymer undergoes gradual slow modifications during incorporation into humus. The earlier concept that lignin is first degraded into phenols, which then are either microbially metabolized or repolymerized into humus [14,15,56], is likely to be revised. It appears more probable that partly degraded lignin or lignin fragments become adsorbed or bonded by the already present organic and inorganic matrix.

C. Transformation of Phenolic Compounds

As the macromolecular lignin matrix is attacked only randomly, lignin-derived phenols are released in only small amounts during progressive stages of degradation. Phenols and phenolic acids, however, have been identified in soil extracts from many different soil systems [57,58]. They are derived from decomposing plant residues, microbial biosynthesis, or root exudation. Phenols have been implicated in various soil processes, including abiotic humus formation [59], dissolution of minerals [60], or as possible phytotoxic or allelopathic chemicals in field situations or hydroponic cultures [61, 62].

Indigenous soil microbial population have no difficulty degrading or transforming exogenously applied phenols or phenolic acids. This includes the splitting of aromatic rings and their metabolization into microbial biomass [63,64]. However, in many soil systems, microbial or physicochemical reactions with soil or humus particles may reduce the concentration of phenols in the free form and result in their adsorption by the bulk soil material. Low concentrations of ferulic acid, catechol, and catechol derivatives were more readily adsorbed and linked into soil humus, whereas higher concentrations were more readily microbially degraded [65]. Lehmann et al. [66] noted that phenolic acids (ferulic, caffeic, protocatechuic, or *p*-hydroxybenzoic acids) were oxidized in soil by Fe^{3+} or Mn^{4+}. They concluded that the reactions of phenols were predominantly chemical and were oxidatively coupled with soil humus upon reduction of Fe^{3+}- and Mn^{4+}-oxides. The activity of microbes associated with the sorption of phenolic acids in soil is also likely [65,67]. A stimulation of the microbial decomposition of certain aromatic and phenolic aldehydes in soil with added montmorillonite was also observed by Kunc and Stotzky [67a] and reviewed in more detail by Stotzky [67b]. In view of this rapid disappearance of certain phenolic acids in various soil types, it is obvious that, under field conditions, the free form of these compounds may not persist for long periods. Only in the vicinity of decaying plant residues or at root surfaces can they be present in appreciable concentrations. Diffusion into the soil environment rapidly reduces their dissolutive or allelopathic potentials.

The chemical structure of phenolic compounds greatly influences their sorption or degradation in soil. Several authors [63,66–68] observed increasing adsorption from p-hydroxybenzoic < vanillic < p-hydroxycoumaric < ferulic < protocatechuic < caffeic acids; but decreasing degradation rates. This suggests that sorption and degradation of the compounds is affected by their functional groups and, most probably, by their respective ease of being oxidized into semiquinones or quinones. Cheng et al. [69] observed that 80% of [^{14}C]catechol applied at 1 μg g^{-1} soil was bound in a neutral soil after 80 days of incubation, whereas about 75% of ^{14}C–ring-labeled vanillic acid was decomposed to ^{14}CO$_2$ during the same time [63]. Dalton et al. [67] discussed that the acrylic side chain (e.g., in ferulic, p-coumaric, or caffeic acids) could also account for the enhanced binding, as the α, β unsaturated bond in the side chain is more susceptible to electrophilic addition or to reactivity with polyvalent metal cations.

In a previous chapter in this series by Haider et al. [70], the microbial formation of phenols and quinones and their contribution to humus synthesis were described and, therefore, will not be addressed here.

D. Degradation and Transformation of Nitrogen-Containing Compounds

Proteins, amino acids, amino sugars, and nucleic acids, either in plant residues or in dead microbial cells, are generally readily degraded or transformed in soil. Natural grasslands and forest ecosystems normally develop a dynamic equilibrium between organic N inputs to the soil and mineral N uptake by roots. Agroecosystems, on the other hand, have developed from mixed- and multiple-cropping systems to intensively managed monocultures with large and pulsed N inputs in the form of commercial fertilizer. In these latter systems, the return of organic C and N in the form of plant residues is small, compared with the total input of N.

The turnover of nitrogen in soils is closely related to heterotrophic metabolism, energy generation, and its utilization for biosynthesis

[71,72]. Empirical research has demonstrated that decomposition of agricultural or other plant residues, in general, is influenced by their C:N ratio [73]. If the plant C:N ratio is greater than 25:1, N will be taken up from the mineral N pool or from simultaneously attacked soil organic matter, as the developing heterotrophic microbiota has a C:N ratio of about 10:1. Observations by Allison and Killham [74], however, suggest that with repeated application of straw to arable soils, the degradation becomes progressively more rapid. This may be because of a marked increase in the activity of fungal biomass that has a wider C:N ratio than the bacterial biomass or has the ability to attack lignin. Fungi may then be able to derive nitrogen that is associated with lignocelluloses; this was described for forest soils during the decomposition of tree litters, with a C:N ratio of about 50:1 and, also, for the decomposition of straw that was low in available N content [74,75].

The soil microflora is the prime decomposer of organic substrates and, therefore, is an important mediator in the metabolic turnover of C and N. Similar to the turnover of C, the heterogeneous availability of N in organic materials can be observed. Van Veen and Frissel [76] developed a mineralization—immobilization submodel for N and C. This model differentiated between N-containing readily decomposable materials, N in resistant active materials and lignin, and N in old organic matter. Similar to the model for carbon flow (see Fig. 1), a model for the N flow was presented by Parton et al. [23] that distinguishes between structural and metabolic N in plant residues and flows into an active, a slow, and a passive soil N pool. The value of defining and measuring soil organic fractions in the characterization of the quality and activity of organic matter was also demonstrated by Janssen [4]. By using a model to separate the decomposition rates of "young" and "old" SOM, he illustrated the effect of long-term amendments with mineral N, green manure, and animal manure on mineralizable soil N. This soil N fraction was more closely related to the amounts of young organic matter than to the total N contents of the soil. The young organic matter was defined here as the fraction that accumulated by addition of crop residues or farmyard manure during 25 years of various management practices.

As is now also becoming apparent, the soil fauna [77,78] and the plant rhizosphere [79] have an important role in N cycling. It is, however, difficult to quantify these effects, because in addition to their direct contribution to N fluxes, there is a wide range of indirect effects on the soil as an environment for microorganisms and plant roots.

III. NEWER CONCEPTS ABOUT THE STRUCTURE AND CHEMICAL COMPOSITION OF SOIL ORGANIC MATTER

Two approaches to the elucidation of humic structure are commonly utilized. These are the so-called degradative and nondegradative approaches. In the first approach, humic substances are chemically or physically broken down into various "subunits" that are subsequently isolated and identified. In the nondegradative approach, isolated humic substances are analyzed directly, utilizing techniques that allow structural inferences to be made. Nuclear magnetic resonance spectroscopy (NMR) is central to this approach.

It is beyond the scope of this chapter to discuss the chemical structure of humus and its fractions in detail, and it is described only to the extent that is directly related to humification and to soil functions. More detailed information can be found in books edited by Christman and Gjessing [80], Aiken et al. [81], Frimmel and Christman [82], and Hayes et al. [82a]. Details on degradative methods for studying humic compounds, particularly by permanganate oxidation, were published by Schnitzer [83]. The application of pyrolysis—mass spectrometry, including gas chromatography—mass spectrometry (GC-MS) analysis of the fragments, was recently reviewed by Schulten [84]. A review on the application of NMR techniques to soil chemistry was published by Wilson [85].

Integration of ^{13}C-NMR spectral areas can be used to estimate the relative concentrations of the various carbon signals. Usually, liquid or solid-state cross-polarization—magic-angle spinning (CP-MAS) ^{13}C-NMR spectra ranging from 5 to 200 ppm are divided into four ranges of chemical shifts: 5 to 46 ppm, designates the aliphatic region; 46 to 110 ppm, the C-O/C-N region; 110 to 160 ppm, the aromatic or olefinic region; and 160 to 200 ppm, the carboxylic or carbonyl region. Typical spectra of humic compounds and the average composition determined by integration of solution or CP-MAS spectra from humic acids from different soils are shown in Figure 4 and Table 4 [86].

Schnitzer and Preston [87] show that the relative intensities of areas, even from a set of similar humic materials, should be regarded as estimates and interpreted with caution. Moreover, Norwood [9] criticized that only few workers bother to ensure that their NMR acquisition parameters have been optimized for quantitative analysis. Fründ and Lüdemann [86] made an extensive comparative study of representative humic materials from Bavarian and northern German soils, to learn how much quantitative structural

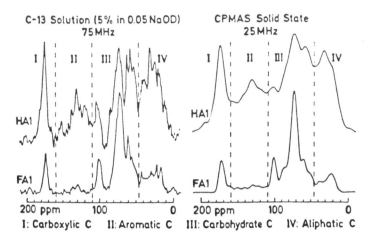

I: Carboxylic C   II: Aromatic C   III: Carbohydrate C   IV: Aliphatic C

Figure 4   Comparison of the solution (NAOD, deuterated sodium hydroxide) and CP—MAS (cross-polarization—magic-angle spinning) ^{13}C-NMR spectra of humic and fulvic acid fractions from a Mollisol Randoll [86].

information could be gathered from ^{13}C-NMR studies by optimizing NMR parameters.   In general, solution and CP-MAS spectra of the humic materials agreed within ±2% of the distinct spectral areas. However, the data obtained for soluble humic fractions from highly aromatic soils made it obvious that, by CP-MAS spectra, the aromatic region was grossly underrepresented.

A. Aromaticity of Humic Compounds

At the low field of ^{13}C-NMR spectra, signals of aromatic rings, 115 to 130 ppm, indicate highly protonated rings.   Resonances between 130 and 147 ppm suggest substitution of the aromatic rings by carbon functions [87].   Distinct peaks between 150 and 160 ppm appear to arise mainly from phenolic C- or from N-substituted aromatics.   In humic acids the latter signals are usually rather small. Humic acids from highly oxidized soils show the lowest level of phenolic carbons, whereas humic acids from peats are higher in phenolic and methoxyl (58 to 60 ppm) carbons.   Therefore, ^{13}C-NMR spectra of soil humic acids and, even more, of fulvic acids, show less polyphenolic character than formerly thought [8,9].   This theory of a significant contribution of phenols and other aromatic compounds to both humic and fulvic acids is based mainly on the former

Table 4 Average Composition of Humic Materials from Different Soils Determined by Quantification of ^{13}C-NMR Spectra in NaOD Solution[a] or of CP-MAS[b] Spectra

	Carboxyl-C (205—160 ppm) (%)	Aromatic-C (160—110 ppm) (%)	C-O/C-N-C (110—46 ppm) (%)	Aliphatic-C (46—5 ppm) (%)
Solution spectra[c] (75 MHz)	14.3%	14.4%	47.3%	23.8%
CP-MAS spectra[c] (75 MHz)	14.8%	11.3%	49.0%	24.9%
Solution spectra[d] (75 MHz)	12.0%	17.7%	49.0%	21.3%
CP-MAS spectra[d] (75 MHz)	12.5%	13.6%	51.9%	22.0%

[a]Deuterated sodium hydroxide.
[b]Cross-polarization—magic-angle spinning.
[c]From nine German soils (south and north Germany).
[d]From 20 German and Spanish soils.
Source: Data from Ref. 86.

theory of a close structural relationship between humic compounds and lignin. Furthermore, earlier degradative studies with permanganate indicated that humic and fulvic acids consist mainly of aromatic nuclei and highly substituted aromatics, with either crosslinking aliphatic side chains or functional groups, such as carboxyl, hydroxyl, or methoxyl [10,83]. From ^{13}C-NMR spectra of humic acids from Typic Borolls in Saskatchewan, Canada, Schnitzer and Preston [87] reported that the aromatic region contributed between 35 and 45% to the total carbon distribution. Fulvic acids from aerobic soils show many structural similarities to humic acids, but with a higher level of carboxyl and C-O groups and with a lower aromatic content [8,88].

The polyphenolic structure of lignin undergoes substantial modifications during degradation and humification, which include side chain and ring cleavage reactions (see under Sect. II.B). Almendros et al. [89] showed by ^{13}C-NMR spectroscopy of whole compost or humic acid extracts during a continuous aerobic composting process of straw and grape husks, that typical lignin structures

completely disappeared or decreased strongly after 4 months, whereas aromatic signals in the range from 100 to 130 ppm became more
prominent. There was also a large increase of signals in the aliphatic region.

If, however, woody plant tissues are degraded anaerobically, as
in anoxic sediments, swamps, or during peat formation, materials
directly related to lignin structures become relatively enriched and
appear in ^{13}C-NMR spectra as the result of a more rapid degradation of polysaccharides [90]. Fragments obtained by pyrolysis [91]
or fragments obtained by CuO oxidation of humic acid extracts from
buried woods or peat [92] confirmed the presence of typical lignin-
like materials. This indicates that the alterations of lignin during
decomposition under anaerobic or O_2-limiting conditions are less
complete than in well-aerated soils.

After CuO oxidation, most of the phenylpropanoid units of lignin in lignocelluloses appear in the form of aldehydes and very little
as acids. Fresh vascular plant tissues, after this oxidation, yield
ratios of vanillic acid/vanillin and syringic acid/syringaldehyde in
the range of 0.1 to 0.2. The acid/aldehyde ratios are significantly
elevated after CuO oxidation of microbially altered sedimentary plant
fragments [93] and in the profile of mull or moder raw humus (Mollisols) of forest soils with increased depth [94,95]. In humic or
fulvic acids from arable soils, acid/aldehyde ratios, if they can be
measured at all, are generally very high [92]. These consistent
patterns of increased acid/aldehyde ratios correspond to decreased
lignin and aromatic methoxyl contents and are characteristic of the
progressive microbial decay of lignin [92,94,95].

B. Aliphatic Derivatives in Humic Compounds

The ^{13}C-NMR spectra of humic compounds generally exhibit strong
signals between 17 and 33 ppm, which probably indicate CH_2 groups
in long paraffinic chains or in alicyclic or heterocyclic saturated
ring structures. This assumption is based on the low intensity of
signals in the region where CH_3 groups (around 16 ppm) are to be
expected [86,87]. By integration, the aliphatic region represents
from 16 to 20% or more of the entire carbon content (see Table 3).
In addition to the aliphatic carbon content of humic compounds, a
lipid fraction can be extracted from soils by simple Soxhlet extraction with organic solvents [96]. These lipidic compounds are accumulated in acidic soils, where they can represent up to 30% of
the total organic matter. By extraction with supercritical nonpolar
solvents, the aliphatic fraction can be obtained in even higher
yields [13,97—99].

Most of the ^{13}C-NMR signals of the extracted aliphatics and lipids appear in the zero to 40 ppm region and are mostly the result

of CH_2 groups with little contribution of CH_3 groups. Signals from carboxyl or ester groups are generally weak and appear around 180 ppm [99]. In n-pentane supercritical extracts from several soils, a number of n-C_{24} and n-C_{26} straight-chain hydrocarbons, several unsaturated branched or hydroxy fatty acids, and dicarboxylic acids were identified by gas chromatography [100,101].

The origin of the aliphatic and lipid fractions is not quite clear. An origin from plant materials, mainly from plant cuticles [102,103] or from microbes [101,104] has been suggested. Extensively degraded lignin also shows numerous signals in the aliphatic region, which may originate from ring carbons after cleavage of the aromatic nuclei and after microbial reduction of double bonds and oxygen functions [46,47].

The importance of the aliphatic and lipid fractions in the stability of soil structure, derives from observations that small amounts of lipids added to soil considerably increase aggregate stability [12]. It has been suggested that the lipidic substances form water-repellant films of oriented molecules on the surface of aggregates. Capriel et al. [13] showed a positive correlation between the quantity of an aliphatic fraction extracted from soils by supercritical n-pentane and the aggregate stability.

C. Fulvic Acids and Dissolved Organic Carbon in Soil

Another prominent region in the ^{13}C-NMR spectra of humic compounds from 46 to 110 ppm represents carbons linked with singly bonded O or N. This region is sometimes called the "carbohydrate-derived region," as it is suspected that in most humic materials, polysaccharides are the major parent compounds for these carbons. However, ether and amino derivatives also have shifts in this region.

Polysaccharides themselves can be separated from soil or soil organic matter by several extraction techniques [105,106]. The polysaccharides consist of relatively well-defined sugar units that can be isolated after hydrolysis [39]. By ^{13}C-NMR spectroscopy, the signals can be assigned mostly to sugar units by chemical shift tables [107].

This 46 to 110 ppm area is less well-defined in humic and fulvic acids. As the result of the broad peaks, an assignment to known sugar signals is difficult. Furthermore, aliphatic ethers, which also absorb in this region, could be of importance as structural units in humic compounds [108]. In fulvic acids, however, some better-defined signals at 105 ppm can be assigned to the anomeric C-1 of sugars. Hatcher et al. [8], as well as Preston and Ripmester [109], pointed out that polysaccharide-derived structures contribute substantially to fulvic acids. Saiz-Jimenez and de Leeuw

[110] also concluded this from results obtained by pyrolysis–gas
chromatography of fulvic acids from Typic Xerochrept and podzol
soils. They claimed that the fulvic acids consist mainly of polysaccharide units or remains of polysaccharides, in addition to varying contributions from lignin and fatty acids. Care must be exer-
cised, however, in interpreting evidence of a more-or-less polysaccharide-derived nature of fulvic acids, unless precautions have
been taken to separate associated from unassociated polysaccharides
[111].

As the result of a higher content of functional groups, fulvic
acids have a higher cation-exchange capacity than humic acids.
The well-known and frequently described function of fulvic acids in
forming water-soluble and water-insoluble metal complexes and their
activity in the weathering of minerals will not be discussed here.
The reader is referred to reviews by Schnitzer [112] as well as by
Stevenson and Fitch [113], which deal particularly with these topics. More important in the context of this chapter is the role of
soil fulvic acids in the formation of most of the dissolved organic
matter (DOC) in soil and ground waters and probably also in
streams and lakes. In surface waters, unique humification processes, including those of falling leaves, also may be of importance
[114,115].

The DOC can act as a carrier of trace metals, and in surface
waters, it also can bind organic chemicals, including pesticide residues or other pollutants. Gauthier et al. [116] provided evidence
that extremely hydrophobic pollutants (e.g., polycyclic hydrocarbons, DDT, mirex) have a strong tendency to associate with organic matter in water. Zepp [117] indicated that this association
results in a large enhancement in the rate of light-induced dechlor-
ination and degradation of such persistent pollutants. The mechanism of this enhanced light-induced degradation of pollutants bound
to DOC is unclear. It is possible that superoxide, organoperoxyl
radicals, or singlet oxygen, in concert with the direct photoreac-
tions of humic substances, may be involved [117; and literature
cited therein].

The DOC also has an important role in the formation of chlorinated organic compounds, which are sometimes toxic, as a result of
the chlorination of water for drinking purposes [118]. Some of the
chlorinated by-products are weakly mutagenic. However, a chloro-
furanone (Fig. 5) and several similar compounds with highly mutagenic potentials have been isolated and identified. They occur in
pulp effluents as by-products after chlorine bleaching or after
chlorination of DOC-containing waters for drinking purposes [118].

Concentrations of DOC in groundwaters or aquifers are generally
small and range from about 0.5 to 0.7 mg carbon per liter [119].
Sometimes, however, acquifiers that receive recharges from peat or

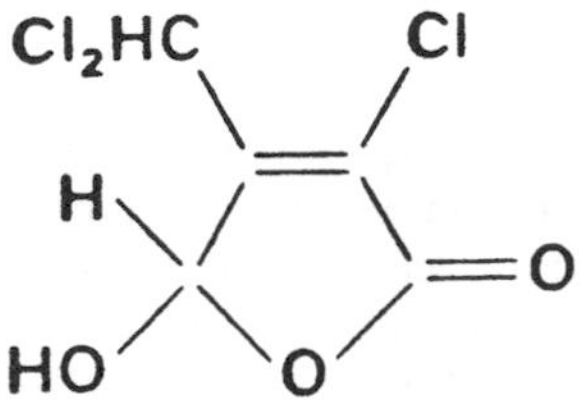

Cl₂HC, Cl, H, HO, O, O (chemical structure diagram)

Figure 5   3-Chloro-4(dichloromethyl)5-hydroxy-2(5*H*)-furanone, one of the mutagenic chlorofuranones identified in chlorinated pulp effluents or in chlorinated humic substances [118].

swamps are higher in groundwater DOC. Similar to soil fulvic acids, ^{13}C-NMR spectra indicate that DOC in groundwater contains little aromatic C, but is rich in aliphatic C, including C-O functions and carboxyl groups [114,119]. Only 1 to 4% of the organic carbon, however, can be accounted for by carbohydrate components. Apparently, humic substances present as DOC in groundwaters have lost their carbohydrate contents through microbial decay, probably as the result of their long residence times in groundwaters [119]. Only DOC from upper soil layers seem to contain more typical polysaccharides, as indicated by Guggenberger [120] for DOC from mull horizons of forest soils.

The microbial availability of DOC is important for its use as a carbon source for denitrifying organisms. Inasmuch as the nitrate content of groundwaters appears to be increasing because of the increasing use of N fertilizers, the question is whether DOC in groundwaters helps to diminish nitrate contents. Sometimes the denitrifying potential of DOC is calculated as if its carbon contents were equal to those of glucose and other readily available carbon sources and could be used completely by heterotrophic denitrifiers [121]. Given the limited knowledge about the structure of DOC, precaution is urged in assuming that DOC is an easily available carbon source, particularly at greater depths in the unsaturated and saturated zones of water catchment areas.

IV.   IMPACT OF THE ENVIRONMENT ON HUMUS FORMATION AND DEGRADATION

A decline in soil organic matter levels generally occurs when virgin soils under natural grasslands or forests are cleared and brought into agricultural production. This net mineralization of soil organic matter is accompanied by an enhanced release of mineral nutrients that are utilizable for plant growth. Once the net excess of readily decomposable organic matter is depleted, a new equilibrium becomes established between humus synthesis from the input of plant residues and humus decompositon [3,122]. Furthermore, changes in agricultural management, including changes in cropping and tillage, fertilizer, or water management and climatic changes, influence soil organic matter levels. These changes impinge on the crop productivity of soils, erosion potential of soils, and the storage and turnover of nutrients in soil [3,123].

A. Aggregation

As already described in the preceding sections, microbes decompose plant residues, and the resulting microbial products become substrates for humus formation. Polymerization and adsorption result in compounds with relatively high molecular mass that are susceptible to various types of chemical and physical bonding with clays and amorphous mineral colloids [67b].

Organic amendments have favorable effects on physical properties and improve soil structure. The aggregation of microaggregates into macroaggregates is typical for well-structured soils that have a greater resistance to erosion and improved air–water relations [124, 125]. Tisdall and Oades [11] suggested that microaggregates are combined into macroaggregates by transient (i.e., easily decomposed) microbial- or plant-derived polysaccharides and by roots and hyphae. Easily degradable lipids also seem to be of importance [12,13]. More persistent binding agents are responsible for the integrity of microaggregates and consist of humic materials in association with amorphous Fe and Al compounds and polyvalent metal cations [11,125].

The reduction of humus levels, as the result of cultivation, causes a decrease in macroaggregates and an increase of microaggregates. Elliot [126] indicated that the organic matter associated with macroaggregates is more readily mineralized than that associated with microaggregates. Tisdall and Oades [11], Elliot [126], and others [127] suggested that aggregation is an important factor in controlling organic matter levels and the nutrient flow in soil. Therefore, availability of organic materials as substrates for microorganisms is not determined only by chemical constitution, but also by their location in soil. This view has also been supported by Tiessen and Stewart [128], who demonstrated that organic matter associated with the fine clay (<0.2 μm) of a soil became more rapidly depleted during 60 years of continuous cultivation, whereas losses of fine silt (5 to 2 μm) and coarse clay associated organic materials were less. The proportion of residual soil organic matter in coarse clay- and fine silt-sized fractions increased with the time of cultivation.

B. Tillage

Changes in soil structure, as the result of tillage, have a strong influence on the decomposition and mineralization of organic materials. In undisturbed plant—soil environments, soil organic matter levels are fairly constant, but tillage alters this steady state, as the result of associated physical changes in the structure, aeration, water status, and availability of C and N as nutrient sources for microorganisms and plants [3]. A variety of models have been developed in recent years, to simulate soil organic matter development and the cycling of nutrients under long-term continuous management [22,23,129—131]. Parton et al. [122] simulated the dynamics of C, N, P, and S in uncultivated and cultivated grassland soils. After 60 years of cultivation, soil organic carbon levels had been reduced by 23%. Decomposition rates for the months when tillage occurred were increased by 25, 50, and 50% for active, slow, and passive soil organic matter pools, respectively (see Fig. 1). This reduction should be paralleled by a decrease in organic C, N, P, and S levels. The addition of N, P, and S fertilizers reduced the rate of decrease in soil carbon with cultivation and suggested that an equilibrium in soil C, N, P, and S levels should be reached after 150 to 200 years of cultivation, whereas these levels in the unfertilized soil should still be declining after this time. Other examples of continuously managed plots also show similar patterns [17, 131]. However, after reaching an equilibrium, organic matter levels do not decrease any further with regular additions of crop residues in the form of straw, together with intercropping of plants used as green manure [132]. The organic matter content of soil can be slowly increased by the regular addition of farmyard manure or by the alteration in tillage practices [1,131].

C. Effects of Growing Plants

The root systems of growing plants may disturb the association of humic compounds with inorganic materials or within aggregates [133], and they may also cause a "priming" effect (additional SOM degradation resulting from added undecomposed material) that results from root deposits that can lead to changes in the microbial activity in the rhizosphere [134]. Roots can also influence the soil that they contact by the adsorption of nutrients, particularly nitrogen. The uptake of water by roots may create localized rapid drying and rewetting effects that enhance soil organic matter degradation [135,136].

The literature on the effects of growing plants on organic carbon mineralization is, however, contradictory. Several authors [137,138] suggest that mineralization of native soil or plant residue carbon proceeds faster in a planted soil than in an unplanted soil under comparable conditions. On the other hand, it has also been reported [139,140] that growing plants tend to conserve carbon from added plant residues or from native soil. For example, Martin [140] and Reid and Goss [139], in experiments with ^{14}C-labeled plant residue or soils reported that the lower rate of [14]CO_2 evolution from soil as the result of cropping was partly compensated for by the uptake of labeled carbon into roots and transport into sprouts [140]. In a recent study [141], plants were grown in a soil that was either amended with ^{14}C-labeled plant residues or the soil organic matter was uniformly ^{14}C-labeled. Plant residues were either uniformly ^{14}C-labeled or specifically in the lignin portion. The experiments were conducted in a phytotron, wherein the temperature and moisture conditions were carefully controlled, and the [14]CO_2 released from planted and unplanted soils was monitored throughout the growth period. Similar to the observations of Reid and Goss [139] and Martin [140], growing corn plants did not accelerate the mineralization of plant residues, but had even a retarding effect on carbon mineralization from soil organic matter. The uptake of ^{14}C-labeled compounds from decomposing plant residues by growing plants was small (1 to 2% of the mineralized carbon) and consisted mainly of degradation products of lignin and, most probably, of phenolic compounds.

The results of these experiments seem to indicate that at medium and constant soil water tensions, growing plants have little retarding effect on the mineralization of organic residues or of soil organic matter. However, if cropping practices lead to enhanced drying and rewetting cycles, in comparison with an unplanted soil, growing plants probably may have a more critical effect on carbon mineralization.

D. Microbial Biomass

Although microbial biomass accounts only for 1 to 3% of the soil organic carbon, it is important for both decomposition of plant residues and the net energy flux in soil. It is also an important mediator of the turnover of nutrients and is more labile than the bulk of the soil organic matter [23,142—144].

This section will not address the measurement of biomass by various methods [142,143]. It only attempts to highlight some of the aspects about the role of the microbial biomass in humus formation and degradation and in functioning as a unique indicator of stress situations in soil.

As pointed out by Jenkinson [143], changes in the quantity of biomass reveal changes caused by soil management, long before such changes can be detected in the total carbon or nitrogen contents. Beck [6,7] combined the quantity of microbial biomass and several of its enzymatic activities into an index and correlated this index with the humus content of soils. By this procedure, he was able to predict long-term decreases or increases in humus levels, even before they were proved to be drastically altered. Gröblinghoff et al. [145] has used this method to characterize the microbial biomass and its activities in field plots, located near Donauwörth, Germany, that have been managed continuously for 15 years [146]. The first set of parallel plots was fertilized only with farmyard manure; the second set received manure and a low dose of mineral fertilizer; a third set was fertilized with minerals at a medium dose and also received straw and green manure; and the fourth set was heavily fertilized with inorganic fertilizer only, but most of the crop residues remained in the field. As shown in Table 5, the organic matter contents were only about 10% lower in the inorganic-fertilized than in the organic-fertilized plots. However, the decreases in microbial biomass and its enzymatic activities were in the inorganic-fertilized plots much more drastic and amounted to about 35%.

Newbould [25] reviewed the various environmental and management factors that influence humus formation and degradation. Among these, similarly to what Birch [136] first observed, the effects of alternate drying and rewetting were rather drastic in accelerating C and N mineralization in soil. Sörensen [136a] observed that repeated drying and rewetting caused a 16 to 120% increase in $^{14}CO_2$ evolution from a soil incubated for 1.5 to 8 years with ^{14}C-labeled straw, cellulose, or glucose, when compared with the same soil maintained continuously under moist conditions.

From a soil incubated with ^{14}C- and ^{15}N-labeled plant material, Amato and Ladd [32] observed that the labeled biomass decreased by 15% during the first drying and rewetting cycle, compared with a continuously moist soil. Similarly, Bottner [135] reported that about 25 to 35% of the microbial biomass was destroyed after drying and then restored upon subsequent rewetting.

Jenkinson and Powlson [147] explained the effect of drying and rewetting on biomass and on the turnover of C and N by "partial sterilization" of the microbial biomass during drying and by release of nonbiomass organic matter. After the soil is remoistened, these materials become available to the surviving microflora. Van Veen [148] suggested that the enhancement of CO_2 evolution and N mineralization after rewetting of a soil arose from increased availability of organic substrates as the result of chemical and physical reactions and the death of microbial cells during drying. He used these two factors in a model to simulate the effect of drying and

Table 5 Influence of Continuous Management[a] on Contents of Soil Organic Matter and Microbial Biomass and Its Enzymatic Activities [145]

Regular fertilization	$\%C_t$	$\%N_t$	Biomass[b] (mgC/ 100g)	Arginine ammonifi- cation	Cata- lase[c]	Avail- able N[d] (mgN/ 100g)
1. Fym[e] only 18 Mg ha^{-1} a^{-1}	1.3	0.13	51	4.6	11.4	8.5
2. Fym[e] + inorganic fert, 12 Mg + 40 kg N ha^{-1} a^{-1}	1.3	0.14	42	4.2	10.8	9.0
3. Inorganic fert. 124 kg N ha^{-1} a^{-1} + green manure	1.2	0.14	33	3.2	9.3	7.4
4. Inorganic fert. 180 kg N ha^{-1} a^{-1}	1.1	0.14	33	2.9	7.5	5.8

[a]An Inceptisol soil for 15 years under continuous cultivation.
[b]Biomass in spring before fertilization, determined by the respiration method [146].
[c]Ammonification of arginine and catalase activity according to Refs. 6, 7, 146.
[d]Available N determined according to Stanford and Smith (1976) Soil Sci. 122:71–76.
[e]Fym = farmyard manure.

rewetting on changes in both microbial biomass and soil organic matter.

Some microbes in soil are defined as "zymogenic" organisms, which can go through a rapid phase of growth and division in the presence of readily metabolizable organic substrates and, then, when the substrates have been depleted, shut down and wait for the next input of substrate. Other organisms have been characterized as "autochthonous" and are able to maintain low and relatively constant activities by using the more resistant components of soil organic matter [143,144]. However, it is not clear what kinds of organisms utilize humic compounds in the soil environment, wherein the compounds owe their resistance not only to their chemical structure, but mainly to the protective physical and chemical effects of interactions between the soil mineral matrix and humic compounds [14]. Once the compounds are released from this matrix as the result of alterations in management or climate, they can be slowly, but steadily, degraded. From data in the literature [149–152] and from in vitro studies with pure cultures, it is probable that organisms active in the degradation of humic compounds are also degraders of lignin. Similar to lignin, humic compounds are not readily degraded unless the media are supplemented with readily available C sources and contain relatively small concentrations of available N. Furthermore, degradation of humic compounds is largely enhanced at high oxygen tensions, which also enhance the degradation of lignin (see under Sect. II.B).

Well-aerated soils always have a marked ligninolytic activity with an appropriate microflora [153], and it is possible that this microflora is also active in the degradation of humic compounds, once their intimate association with an inorganic or a structural matrix has been disturbed.

ACKNOWLEDGMENT

During my career as a soil biochemist I was stimulated by many scientists, especially by the late J. P. Martin and by W. Flaig and F. E. Clark. I also thank Mrs. H. Lemke for helping collect the literature and Mrs. C. Seidel for typing the manuscript. Thanks are also given to Mrs. D. Massmann for her help in designing the figures and tables.

REFERENCES

1. Follett, R. H., S. C. Gupta, and P. G. Hunt. 1987. Conservation practices: relation to the management of plant nutrients for crop production, p. 19–51. *In* J. J. Mortvedt and D. R. Buxton (eds.), Soil fertility and organic matter as critical components of production systems. SSSA Spec. Publ. No. 19, American Society of Agronomy, Madison, Wisconsin.
2. Parton, W. J., D. W. Anderson, C. V. Cole, and J. W. B. Stewart. 1983. Simulation of soil organic matter formations and mineralization in semiarid agroecosystems, p. 533–550. *In* R. R. Lowrance et al. (eds.), Nutrient cycling in agricultural ecosystems. University of Georgia, Spec. Publ. No. 23p, Athens.

3. Doran, J. W., and M. S. Smith. 1987. Organic matter management and utilization of soil and fertilizer nutrients, p. 53–72. *In* J. J. Mortvedt and R. D. Buxton (eds.), Soil fertility and organic matter as critical components of production systems. SSSA Spec. Publ. No. 19. American Society of Agronomy, Madison, Wisconsin.

4. Janssen, B. H. 1984. A simple method for calculating decomposition and accumulation of "young" soil organic matter. Plant Soil 76:297–304.

5. Scerbakow, A., and E. Kislych. 1986. Effektive Fruchtbarkeit der Böden: Angewandte und theoretische Aspekte, p. 943–949. Trans. 13th Congr. Int. Soc. Soil Sci. Hamburg, Vol. 4.

6. Beck, T. 1984. Mikrobiologische und biochemische Charakterisierung landwirtschaftlich genutzter Böden. I. Mitt. Die Ermittlung einer bodenmikrobiologischen Kennzahl. Z. Pflanzernerähr. Bodenk. 147:456–466.

7. Beck, T. 1984. Der Einfluss unterschiedlicher Bewirtschaftungsmassnahmen auf die bodenmikrobiologischen Eigenschaften und die Stabilität der organischen Substanz im Boden. Kali-Briefe 17:331–340.

8. Hatcher, P. G., I. A. Breger, L. W. Dennis, and G. E. Maciel. 1983. Solid-state ^{13}C-NMR of sedimentary humic substances: new revelations on their chemical composition, p. 37–81. *In* R. F. Christman and E. T. Gjessing (eds.), Aquatic and terrestrial humic materials. Ann Arbor Science Publishing, Ann Arbor, Michigan.

9. Norwood, D. L. 1988. Critical comparison of structural implications from degradative and nondegradative approaches, p. 133–150. *In* F. H. Frimmel and R. F. Christman (eds.), Humic substances and their role in the environment. Dahlem workshop reports, No. 41. Wiley Interscience Publishers, New York.

10. Schnitzer, M. 1977. Recent findings on the characterization of humic substances extracted from soils from widely differing climatic zones, p. 117–132. Proc. Symp. IAEA and FAO, Soil organic matter studies. Braunschweig, 6–10, September 1976. Vol. 2.

11. Tisdall, J. M., and J. M. Oades. 1982. Organic matter and water-stable aggregates in soils. J. Soil Sci. 33:141–163.

12. Jambu, P., G. Coulibaly, P. Bilong, P. Magnoux, and A. Ambles. 1983. Influence of lipids on physical properties of soils. Studies about Humus. Humus Planta Prague, 1:46–50.

13. Capriel, P., T. Beck, H. Borchert, and P. Härter. 1990. Relationship between soil aliphatic fraction extracted with supercritical hexane, soil microbial biomass and soil aggregate stability. Soil Sci. Soc. Am. J. 54:415–420.

14. Stout, J. D., K. M. Goh, and T. A. Rafter. 1987. Chemistry and turnover of naturally occurring resistant organic compounds in soil, p. 1–73. *In* E. A. Paul and J. N. Ladd (eds.), Soil biochemistry, Vol. 5. Marcel Dekker, New York.

15. Tate, R. L. 1987. Soil organic matter, p. 1–291. John Wiley & Sons, New York

16. Jenkinson, D. S. 1977. Studies on the decomposition of plant materials in soil. V. The effects of plant cover and soil type on the loss of carbon from ^{14}C-labeled ryegrass decomposing under field conditions. J. Soil Sci. 28:417–423.

17. Jenkinson, D. S., and J. H. Rayner. 1977. The turnover of soil organic matter in some of the Rothamsted classical experiments. Soil Sci. 123:298–305.

18. Sauerbeck, D., and M. A. Gonzalez. 1977. Field decomposition of carbon-14-labeled plant residues in various soils of the Federal Republic of Germany and Costa Rica, p. 117–132. Proc. Symp. IAEA and FAO, Soil organic matter studies, Braunschweig 6–10 September 1976. Vol. 1.

19. Martin, J. P., and K. Haider. 1986. Influence of mineral colloids on turnover rates of soil organic carbon, p. 283–304. *In* P. M. Huang and M. Schnitzer (eds.), Interactions of soil minerals with natural organics and microbes. Spec. Publ. No. 17. American Society of Agronomy, Madison, Wisconsin.

20. Kolenbrander, G. J. 1974. Efficiency of organic manure in increasing soil organic matter content, p. 129–136. Trans. 10th Int. Congr. Soil Sci., Vol. 2.

21. Herman, W. A., W. B. McGill, and J. F. Dormaar. 1977. Effects of initial chemical composition on decomposition of roots of three grass species. Can. J. Soil Sci. 57:205–215.

22. Van Veen, J. A., J. N. Ladd, and M. J. Frissel. 1984. Modelling C and N turnover through the microbial biomass in soil. Plant Soil 76:257–274.

23. Parton, W. J., D. S. Schimel, C. V. Cole, and D. S. Ojima. 1987. Analysis of factors controlling soil organic matter levels in Great Plains grasslands. Soil Sci. Soc. Am. J. 51:1173–1179.

24. Stott, D. E., G. Kassim, W. M. Jarrell, J. P. Martin, and K. Haider. 1983. Stabilization and incorporation into biomass of specific plant carbon during biodegradation in soil. Plant Soil 70:15–26.

25. Newbould, P. 1982. Losses and accumulation of organic matter in soils, p. 107–131. *In* D. Boels, D. B. Davies, and A. E. Johnston (eds.), Soil degradation. A. A. Balkema, Rotterdam.

26. Van Dijk, H. 1982. Survey of Dutch soil organic matter research with regard to humification and degradation rates in

arable land, p. 133—144. *In* D. Boels, D. B. Davies, and A. E. Johnston (eds.), Soil degradation. A. A. Balkema, Rotterdam.

27. Umbreit, W. W. 1962. Modern microbiology. Freeman Publications, San Francisco.

28. Allison, F. E. 1973. Soil organic matter and its role in crop production, p. 1—637. Elsevier Scientific Publishing, Amsterdam.

29. Sörensen, L. H. 1975. The influence of clay on the rate of decay of amino acid metabolites synthesized in soils during decomposition of cellulose. Soil Biol. Biochem. 7:171—177.

30. Haider, K., and J. P. Martin. 1981. Decomposition in soil of specifically ^{14}C-labeled model and cornstalk lignins and coniferylalcohol over two years as influenced by drying, rewetting, and additions of an available C substrate. Soil Biol. Biochem. 13:447—452.

31. Haider, K., and F. Azam. 1983. Umsetzung ^{14}C-markierter Pflanzeninhaltsstoffe im Boden in Gegenwart von ^{15}N-Ammonium. Z. Pflanzenernähr. Bodenk. 146:151—159.

32. Amato, M., and J. N. Ladd. 1980. Studies of nitrogen immobilization and mineralization in calcareous soils. V. Formation and distribution of isotope-labelled biomass during decomposition of ^{14}C- and ^{15}N-labelled plant material. Siol Biol. Biochem. 12:405—411.

33. Sörensen, L. H. 1987. Organic matter and microbial biomass in a soil incubated in the field for 20 years with ^{14}C-labeled barley straw. Soil Biol. Biochem. 19:39—42.

34. Wagner, F., and P. Sistig. 1979. Verwertung von Cellulose durch Mikroorganismen. Forum Mikrobiol. 2:74—80.

35. Eriksson, K.-E. 1981. Microbial degradation of cellulose and lignin, p. 60—65. *In* The Ekman Days, Vol. 3. Int. Symp. Wood Pulping Chem., Stockholm, 9—21 June.

36. Fengel, D., and G. Wegener. 1984. Wood: Chemistry, ultrastructure, reactions, p. 1—613. Gruyter, Berlin.

36a. Butler, G. W., and R. W. Bayley. 1973. Chemistry and biochemistry of herbage, Vol. 1, p. 1—639. Academic Press, London.

37. Benoit, R. E., and R. L. Starkey. 1968. Inhibition of decomposition of cellulose and some other carbohydrates by tannin. Soil Sci. 105:291—296.

38. Sörensen, H. 1963. Studies on the decomposition of C^{14}-labelled barley straw in soil. Soil Sci. 95:45—51.

39. Cheshire, M. V. 1979. Nature and origin of carbohydrates in soils, p. 1—216. Academic Press, London.

40. Kirk, T. K., and R. L. Farrell. 1987. Enzymatic combustion: the microbial degradation of lignin. Annu. Rev. Microbiol. 41:465—505.

41. Haider, K. 1988. Der mikrobielle Abbau des Lignins und seine Bedeutung für den Kreislauf des Kohlenstoffs. Forum Mikrobiol. 11:477–483.

42. Hackett, W. F., W. J. Connors, T. K. Kirk, and J. G. Zeikus. 1977. Microbial decomposition of synthetic ^{14}C-labeled lignin in nature: lignin biodegradation in a variety of natural materials. Appl. Environ. Microbiol. 33:43–51.

43. Benner, R., and R. E. Hodson. 1984. Thermophilic anaerobic biodegradation of ^{14}C-lignin, ^{14}C-cellulose, and ^{14}C-lignocellulose preparations. Appl. Environ. Microbiol. 50:971–976.

44. Colberg, P. J., and L. Y. Young. 1985. Aromatic and volatile acid intermediates observed during anaerobic metabolism of lignin-derived oligomers. Appl. Environ. Microbiol. 49:350–358.

45. Chua, M. G. S., C. L. Chen, H. M. Chang, and T. K. Kirk. 1982. ^{13}C NMR spectroscopic study of spruce lignin degradated by *Phanerochaete chrysosporium*. Holzforsch. 36:165–172.

46. Ellwardt, P.-C., K. Haider, and L. Ernst. 1981. Untersuchungen des mikrobiellen Ligninabbaues durch ^{13}C-NMR-Spektroskopie an spezifisch ^{13}C-angereichertem DHP-Lignin aus Conferylalkohol. Holzforschung 35:103–109.

47. Haider, K., H. Kern, and L. Ernst. 1988. Chemical synthesis of lignin alcohols and model lignins enriched with carbon isotopes. Meth. Enzymol. 161:47–56.

48. Chen, C. L., H. M. Chang, and T. K. Kirk. 1982. Aromatic acids produced during degradation of lignin in spruce wood by *Phanerochaete chrysosporium*. Holzforschung 36:3–9.

49. Kirk, T. K., and M. Shimada. 1985. Lignin biodegradation: the microorganisms involved in biochemistry of degradation by white-rot fungi, p. 579–605. *In* T. Higuchi (ed.), Biosynthesis and biodegradation of wood components. Academic Press, San Diego.

50. Schoemaker, H. E., P. J. Harvey, R. M. Bowen, and J. M. Palmer. 1985. On the mechanism of enzymatic lignin breakdown. Fed. Eur. Biochem. Soc. 183:7–12.

51. Kirk, T. K. 1987. Lignin-degrading enzymes. Phil. Trans. R. Soc. Lond. 321:461–474.

52. Umezawa, T., and F. Higuchi. 1986. Aromatic ring cleavage of β-O-4-lignin model dimers without prior demeth(ox)ylation by lignin peroxidase. FEBS Lett. 205:293–298.

53. Kern, H. W., K. Haider, W. Pool, J. W. de Leeuw, and L. Ernst. 1989. Comparison of the action of *Phanerochaete chrysosporium* and its extracellular enzymes (lignin peroxidases) on lignin preparations. Holzforschung 43:375–384.

54. Chua, M. G. S., S. Choi, and T. K. Kirk. 1983. Mycelium binding and depolymerization of synthetic ^{14}C-labeled lignin during decomposition by *Phanerochaete chrysosporium*. Holzforschung 37:55—61.

55. Schwertmann, U., H. Kodama, and W. R. Fischer. 1986. Mutual interactions between organics and iron oxides, p. 223—250. *In* P. M. Huang and M. Schnitzer (eds.), Interaction of soil minerals with organics and microbes. SSSA Spec. Publ. No. 17. American Society of Agronomy, Madison, Wisconsin.

56. Stevenson, F. J. 1982. Humus chemistry, p. 1—443. John Wiley & Sons. New York.

57. Whitehead, D. C. 1964. Identification of *p*-hydroxybenzoic, vanillic, *p*-coumaric and ferulic acids: Nature 202:417—418.

58. Whitehead, D. C., H. Dibb, and R. D. Hartley. 1983. Bound phenolic compounds in water extracts of soils, plant roots and leaf litter. Soil Biol. Biochem. 15:133—136.

59. Wang, T. S. C., P. M. Huang, C.-H. Chou, and J.-H. Chen. 1986. The role of soil minerals in the abiotic polymerization of phenolic compounds and formation of humic substances, p. 251—282. *In* P. M. Huang and M. Schnitzer (eds.), Interaction of soil minerals with organics and microbes. SSSA Spec. Publ. No. 17. American Society of Agronomy, Madison, Wisconsin.

60. Huang, P. M., and A. Violante. 1986. Influence of organic acids on crystallization and surface properties of precipitation products of aluminum, p. 159—222. *In* P. M. Huang and M. Schnitzer (eds.), Interaction of soil minerals with organics and microbes. SSSA Spec. Publ. No. 17. American Society of Agronomy, Madison, Wisconsin.

61. Yakle, G. A., and R. M. Cruse. 1984. Effects of fresh and decomposing corn plant residue extracts on corn seedling development. Soil Sci. Soc. Am. J. 48:1143—1146.

62. Blum, U., B. R. Dalton, and J. O. Rawlings. 1984. Effect of ferulic acid and some of its microbial products on radicle growth of cucumber. J. Chem. Ecol. 10:1169—1191.

63. Haider, K., and J. P. Martin. 1975. Decomposition of specifically carbon-14 labeled benzoic and cinnamic acid derivatives in soil. Soil Sci. Soc. Am. Proc. 39:657—662.

64. Dao, T. H. 1987. Sorption and mineralization of plant phenolic acids in soil, p. 358—370. *In* G. R. Waller (ed.), Allelochemicals: role in agriculture and forestry. American Chemical Society, Washington, D.C.

65. Martin, J. P., and K. Haider. 1979. Effect of concentration on decomposition of some ^{14}C-labelled phenolic compounds, benzoic acid, glucose, cellulose, wheat straw, and *Chlorella* protein in soil. Soil Sci. Soc. Am. J. 43:917—920.

66. Lehmann, R. G., H. H. Cheng, and J. B. Harsh. 1987. Oxidation of phenolic acids by soil iron and manganese oxides. Soil Sci. Soc. Am. J. 51:352–356.

67. Dalton, B. R., U. Blum, and B. W. Sterling. 1989. Differential sorption of exogenously applied ferulic, p-coumaric, p-hydroxybenzoic, and vanillic acids in soil. Soil Sci. Soc. Am. J. 53:757–762.

67a. Kunc, F., and G. Stotzky. 1977. Acceleration of aldehyde decomposition in soil by montmorillonite. Soil Sci. 124:167–172.

67b. Stotzky, G. 1986. Influences of soil mineral colloids on metabolic processes, growth, adhesion, and ecology of microbes and viruses, p. 305–428. *In* P. M. Huang and M. Schnitzer (eds.), Interactions of soil minerals with organics and microbes. SSSA Spec. Publ. No. 17. American Society of Agronomy, Madison, Wisconsin.

68. Dalton, B. R., S. B. Weed, and U. Blum. 1987. Plant phenolic acids in soils: A comparison of extraction procedures. Soil Sci. Soc. Am. J. 51:1515–1521.

69. Cheng, H. H., K. Haider, and S. S. Harper. 1983. Catechol and chlorocatechols in soil. Degradation and extractability. Soil Biol. Biochem. 15:311–317.

70. Haider, K., J. P. Martin, and Z. Filip. 1975. Humus biochemistry, p. 195–244. *In* E. A. Paul and A. D. McLaren (eds.), Soil biochemistry, Vol. 4. Marcel Dekker, New York.

71. Jansson, S. L. 1958. Tracer studies on nitrogen transformations in soil. Ann. R. Agric. Coll. Sweden 24:101–361.

72. Jansson, S. L., and J. Persson. 1982. Mineralization and immobilization of soil nitrogen, p. 229–252. *In* F. J. Stevenson (ed.), Nitrogen in agricultural soils. American Society of Agronomy, Madison, Wisconsin.

73. Bartholomew, W. V. 1965. Mineralization and immobilization of nitrogen in the decomposition of plant and animal residues, p. 285–306. *In* W. V. Bartholomew and F. E. Clark (eds.), Soil nitrogen. American Society of Agronomy, Madison, Wisconsin.

74. Allison, M. F., and K. Killham. 1988. Response of soil microbial biomass to straw incorporation. J. Soil Sci. 39:237–242.

75. Paul, E. A., and F. E. Clark. 1989. Soil microbiology and biochemistry, p. 1–273. Academic Press, San Diego.

76. Van Veen, J. A., and M. J. Frissel. 1981. Simulation model of the behavior of N in soil, p. 126–144. *In* M. J. Frissel and J. A. Van Veen (eds.), Simulation of nitrogen behavior of soil-plant systems. PUDOC, Wageningen.

77.   Clarholm, M. 1985. Interactions of bacteria, protozoa and plants leading to mineralization of soil nitrogen. Soil Biol. Biochem. 17:181–187.

78.   Anderson, J. M. 1987. The role of soil fauna in agricultural systems, p. 89–112. *In* J. R. Wilson (ed.), Advances in nitrogen cycling in agricultural ecosystems. CAB International, Wallingford, Connecticut.

79.   Whipps, J. M., and J. M. Lynch. 1986. The influence of the rhizosphere on crop productivity. Adv. Microb. Ecol. 9:187–244.

80.   Christman, R. F., and E. T. Gjessing (eds.). 1983. Aquatic and terrestrial humic materials, p. 1–538. Ann Arbor Science, Ann Arbor, Michigan.

81.   Aiken, G. R., D. M. McKnight, R. L. Wershaw, and P. MacCarthy (eds.). 1985. Humic substances in soil, sediment and water, p. 1–692. John Wiley & Sons, New York.

82.   Frimmel, F. H., and R. F. Christman (eds.). 1988. Humic substances and their role in the environment, p. 1–271. Dahlem Workshop Reports No. 41. John Wiley & Sons, Chichester.

82a. Hayes, M. H. B., P. MacCarthy, R. L. Malcolm, and R. S. Swift (eds.). 1989. Humic substances II — in search of structure, p. 1–764. John Wiley & Sons, Chichester.

83.   Schnitzer, M. 1978. Humic substances: chemistry and reactions, p. 1–64. *In* M. Schnitzer and S. U. Khan (eds.), Soil organic matter. Elsevier Scientific Publishers, New York.

84.   Schulten, H.-R. 1987. Pyrolysis and soft ionization mass spectrometry of aquatic/terrestrial humic substances and soils. J. Anal. Appl. Pyrolysis 12:149–187.

85.   Wilson, M. A. 1987. NMR techniques and application in geochemistry and soil chemistry. Pergamon Press, Oxford.

86.   Fründ, R., and H.-D. Lüdemann. 1989. The quantitative analysis of solution and CPMAS-C-13 NMR spectra of humic material. Sci. Total Environ. 81/82:157–168.

87.   Schnitzer, M., and C. M. Preston. 1986. Analysis of humic acids by solution and solid-state carbon-13 nuclear magnetic resonance. Soil Sci. Soc. Am. J. 50:326–331.

88.   Hatcher, P. G., and E. C. Spiker. 1988. Selective degradation of plant biomolecules, p. 59–74. *In* F. H. Frimmel and R. F. Christman (eds.), Humic substances and their role in the environment, Dahlem Workshop Reports No. 41. John Wiley & Sons, Chichester.

89.   Almendros, G., R. Fründ, F. J. Gonzalez-Vila, H.-D. Lüdemann, and F. Martin. 1987. NMR and ESR investigation of the humification processes in defined vegetable starting materials. Z. Pflanzenernähr. Bodenkd. 150:201–207.

90. Hedges, J. I., G. L. Cowie, J. R. Ertel, R. J. Barbour, and P. G. Hatcher. 1985. Degradation of carbohydrates and lignins in buried woods. Geochim. Cosmochim. Acta 49:701—711.

91. Saiz-Jimenez, C., J. J. Boon, J. I. Hedges, J. K. C. Hessels, and J. W. de Leeuw. 1987. Chemical characterization of recent and buried woods by analytical pyrolysis. Comparison of pyrolysis data with ^{13}C NMR and wet chemical data. J. Anal. Appl. Pyrolysis 11:437—450.

92. Ertel, J. R., and J. I. Hedges. 1984. The lignin component of humic substances: distribution among soil and sedimentary humic, fulvic, and base-insoluble fraction. Geochim. Cosmochim. Acta 48:2065—2074.

93. Ertel, J. R., and J. I. Hedges. 1985. Sources of sedimentary humic substances: vascular plant debris. Geochim. Cosmochim. Acta 49:2097—2107.

94. Kögel, I. 1986. Estimation and decomposition pattern of the lignin component in forest humus layers. Soil Biol. Biochem. 18:589—594.

95. Kögel, I. 1987. Organische Stoffgruppen in Waldhumusformen und ihr Verhalten während der Streuzersetzung und Humifizierung. Bayreuther Bodenkundl. Ber. 1:1—131.

96. Fustec-Mathon, E., P. Jambu, G. Joly, and R. Jacquesi. 1977. Analyse et rôle des bitumes dans les sols sableux acides, p. 105—114. *In* Proc. Symp. IAEA and FAO, Soil organic matter studies. Braunschweig, 6—10 September, 1976, Vol. 2.

97. Spiteller, M. 1982. Ein neues Verfahren zur Extraktion von organischen Stoffen aus Boden mit überkritischen Gasen. I. Z. Pflanzenernähr. Bodenkd. 145:483—492.

98. Spiteller, M. 1985. Extraction of soil organic matter by supercritical fluids. Org. Geochem. 8:111—113.

99. Schnitzer, M., and C. M. Preston. 1987. Supercritical gas extraction of a soil with solvents of increasing polarities. Soil Sci. Soc. Am. J. 51:639—646.

100. Spiteller, M., and A. Ashauer. 1982. Ein neues Verfahren zur Extraktion von organischen Stoffen aus Boden mit überkritischen Gasen. II. Z. Pflanzenernähr. Bodenkd. 145: 567—575.

101. Schnitzer, M., D. A. Hindle, and M. Meglic. 1986. Supercritical gas extraction of alkanes and alkanoic acids from soils and humic materials. Soil Sci. Soc. Am. J. 50:913—919.

102. Kolattukudy, P. E. 1980. Cutin, suberin and waxes, p. 571—645. *In* P. K. Stumpf (ed.), The biochemistry of plants, Vol. 4: Lipids, structure and function. Springer-Verlag, New York.

103. Nip, M., E. W. Tegelaar, J. W. de Leeuw, P. A. Schenck, and P. J. Holloway. 1986. Analysis of modern and fossil plant cuticles by Curie point Py-GC and Curie point Py-GC-MS: recognition of a new, highly aliphatic and resistant biopolymer. Org. Geochem. 10:769–778.

104. Sevenson, F. J. 1966. Lipids in soils. Am. Oil Chem. Soc. 43:203–210.

105. Acton, C. J., E. A. Paul, and D. A. Rennie. 1963. Measurement of the polysaccharide content of soils. Can. J. Soil Sci. 43:141–150.

106. Saiz-Jimenez, C., and J. W. de Leeuw. 1984. Pyrolysis-gas chromatography–mass spectrometry of soil polysaccharides, soil fulvic acids and polymaleic acid. Org. Geochem. 6:287–293.

107. Bremser, W., L. Ernst, B. Franke, R. Gerhards, and A. Hardt. 1981. Carbon-13 NMR spectral data. Verlag Chemie, Weinheim.

108. Bayer, E., K. Albert, W. Bergmann, K. Jahns, W. Eisener, and K.-H. Peters. 1984. Aliphatische Polyether, Grundbausteine von natürlichen Huminstoffen. Angew. Chem. 96: 151–153.

109. Preston, C. M., and J. A. Ripmester. 1982. Application of solution and solid-state ^{13}C NMR to four organic soils, their humic acids, fulvic acids, humins, and hydrolysis residues. Can. J. Spectrosc. 27:99–105.

110. Saiz-Jimenez, C., and J. W. de Leeuw. 1986. Chemical characterization of soil organic matter fractions by analytical pyrolysis-gas chromatography–mass spectrometry. J. Anal. Appl. Pyrolysis 9:99–119.

111. Wershaw, R. L. 1985. Application of nuclear magnetic resonance spectroscopy for determining functionality in humic substances, p. 561–584. *In* G. R. Aiken, et al. (eds.), Humic substances in soil, sediment and water. John Wiley & Sons, New York.

112. Schnitzer, M. 1986. Binding of humic substances by soil mineral colloids, p. 77–101. *In* P. M. Huang and M. Schnitzer (eds.), Interactions of soil minerals with organics and microbes. SSSA Spec. Publ. No. 17. American Society of Agronomy, Madison, Wisconsin.

113. Stevenson, F. J., and A. Fitch. 1986. Chemistry of complexation of metal ions with soil solution organics, p. 29–58. *In* P. M. Huang and M. Schnitzer (eds.), Interactions of soil minerals with organics and microbes. SSSA Spec. Publ. No. 17. American Society of Agronomy, Madison, Wisconsin.

114. Malcolm, R. L. 1985. Geochemistry of stream fulvic and humic substances, p. 181–210. *In* G. R. Aiken, et al. (eds.),

Humic substances in soil, sediment and water. John Wiley & Sons, New York.

115. Frimmel, F., and H. Bauer. 1987. Influence of photochemical reactions on the optical properties of aquatic humic substances gained from fall leaves. Sci. Total Environ. 62:139—148.

116. Gauthier, T. D., E. C. Shane, W. F. Guerin, W. R. Seitz, and C. L. Grant. 1986. Fluorescence quenching method for determining equilibrium constants for polycyclic aromatic hydrocarbons binding to dissolved humic materials. Environ. Sci. Technol. 20:1162—1166.

117. Zepp, R. G. 1988. Environmental photoprocesses involving natural organic matter, p. 193—214. *In* F. H. Frimmel and R. F. Christem (eds.), Humic substances and their role in the environment, Dahlem Workshop Reports No. 41, John Wiley & Sons, New York.

118. Hemming, J., B. Holmbom, M. Reunanen, and L. Kronberg. 1986. Determination of the strong mutagen 3-chloro-4-(dichlormethyl)-5-hydroxy-2(5H)-furanone in chlorinated drinking and humic waters. Chemosphere 15:549—556.

119. Thurman, E. M. 1985. Humic substances in groundwater, p. 87—103. *In* G. R. Aiken, et al. (eds.), Humic substances in soil, sediment and water. John Wiley & Sons, New York.

120. Guggenberger, G. 1988. Chromatographische (GPC, RP-HPLC) und spektroskopische (IR, ^{13}C-NMR) Kennzeichnung wasserlöslicher organischer Bodensubstanzen aus Waldhumus. Bayreuther Bodenkundl. Ber. 10:1—110.

121. Isermann, K., and G. Henjes. 1990. Potentials for biological denitrification in the (un-)saturated zone with different soil management. Mitt. Deutsch. Bodenk. Ges. 60:271—276.

122. Parton, W. J., J. W. B. Stewart, and C. V. Cole. 1988. Dynamics of C, N, P and S in grassland soils: a model. Biogeochemistry 5:109—131.

123. Fenster, C. R., and G. A. Peterson. 1979. Effects of no-tillage fallow as compared to conventional tillage in a wheat-fallow system. Nebr. Agric. Studies Bull. 289. University of Nebraska, Lincoln.

124. Chaney, K., and R. S. Swift. 1984. The influence of organic matter on aggregate stability in some British soils. J. Soil Sci. 35:223—230.

125. Dormaar, J. F. 1983. Chemical properties of soil and water-stable aggregates after sixty-seven years of cropping to spring wheat. Plant Soil 75:51—61.

126. Elliott, E. T. 1986. Aggregate structure and carbon, nitrogen, and phosphorus in native and cultivated soils. Soil Sci. Soc. Am. J. 50:627—633.

127. Skjemstad, J. O., I. Vallis, and R. J. K. Myers. 1988. Decomposition of soil organic nitrogen, p. 134–144. *In* J. R. Wilson (ed.), Advances in nitrogen cycling in agricultural ecosystems. CAB International, Wallingford.

128. Tiessen, H., and J. W. B. Stewart. 1983. Particle size fractions and their use in studies of soil organic matter II. Cultivation effects on organic matter composition in size fractions. Soil Sci. Soc. Am. J. 47:509–514.

129. Van Veen, J. A., and E. A. Paul. 1981. Organic carbon dynamics in grassland soils. I. Background information and computer simulation. Can. J. Soil Sci. 61:185–201.

130. Voroney, R. P., J. A. Van Veen, and E. A. Paul. 1981. Organic C dynamics in grassland soils. 2. Model validation and simulation of the long-term effects of cultivation and rainfall erosion. Can. J. Soil Sci. 61:211–224.

131. Van der Linden, A. M. A., J. A. Van Veen, and M. J. Frissel. 1987. Modelling soil organic matter levels after long-term applications of crop residues, and farmyard and green manures. Plant Soil 101:21–28.

132. Franken, H. 1985. Einfluss der Landbewirtschaftung auf die Humusversorgung, p. 19–29. VDLUFA-Schriftenreihe, 16, Kongressband 1985.

133. Reid, J. B., M. J. Goss, and P. D. Robertson. 1982. Relationship between the decreases in soil stability effected by the growth of maize roots and the changes in organically bound iron and aluminum. J. Soil Sci. 33:397–410.

134. Keith, H., J. M. Oades, and J. K. Martin. 1986. Input of carbon to soil from wheat plants. Soil Biol. Biochem. 18:445–449.

135. Bottner, P. 1985. Response of microbial biomass to alternate moist and dry conditions in a soil incubated with ^{14}C- and ^{15}N-labelled plant material. Soil Biol. Biochem. 17:329–337.

136. Birch, H. F. 1958. The effect of soil drying on humus decomposition and nitrogen availability. Plant Soil 10:9–31.

136a. Sörensen, L. N. 1974. Rate of decomposition of organic matter in soil as influenced by repeated air drying-rewetting and repeated additions of organic material. Soil Biol. Biochem. 6:287–292.

137. Billes, G., and P. Bottner. 1981. Effet des racines vivantes sur la décomposition d'une litiere racinaire marquée au ^{14}C. Plant Soil 62:193–208.

138. Helal, H. M., and D. Sauerbeck. 1986. Influence of plant roots on the stability of soil organic matter and of soil aggregates, p. 776–777. Trans. 13th Congr. Int. Soc. Soil Sci. Vol. III.

139. Reid, J. B., and M. J. Goss. 1983. Growing crops and transformations of ^{14}C-labelled soil organic matter. Soil Biol. Biochem. 15:687–691.

140. Martin, J. K. 1987. Effect of plants on the decomposition of ^{14}C-labelled soil organic matter. Soil Biol. Biochem. 19: 473–474.

141. Haider, K., O. Heinemeyer, and A. R. Mosier. 1989. Effects of growing plants on humus and plant residue decomposition in soil; uptake of decomposition products by plants. Sci. Total Environ. 81/82:661–670.

142. Jenkinson, D. S., and J. N. Ladd. 1981. Microbial biomass in soil: measurement and turnover, p. 415–471. *In* E. A. Paul and J. N. Ladd (eds.), Soil biochemistry, Vol. 5, Marcel Dekker, New York.

143. Jenkinson, D. S. 1988. Determination of microbial biomass carbon and nitrogen in soil, p. 368–386. *In* J. R. Wilson (ed.), Advances in nitrogen cycling in agricultural ecosystems. CAB International, Wallingford, Connecticut.

144. Ladd, J. N., and R. C. Foster. 1988. Role of soil microflora in nitrogen turnover, p. 113–133. *In* J. R. Wilson (ed.), Advances in nitrogen cycling in agricultural ecosystems. CAB International, Wallingford, Connecticut.

145. Gröblinghoff, F.-F., K. Haider, and T. Beck. 1989. Abbau von Pflanzenrückständen und Humusbildung in Böden unterschiedlicher Bewirtschaftung. Mitt. Dtsch. Bodenk. Ges. 59: 563–568.

146. Beck, T. 1984. Einfluss unterschiedlicher organischer und mineralischer Düngeintensitäten auf die bodenmikrobiologischen Eigenschaften. Bayer. Landw. Jahrb. 61:57–64.

147. Jenkinson, D. S., and D. S. Powlson. 1976. The effects of biocidal treatments on metabolism in soil. V. A method for measuring soil biomass. Soil Biol. Biochem. 8:209–213.

148. Van Veen, J. A. 1986. The use of simulation models of the turnover of soil organic matter: an intermediate report, p. 626–643. *In* Trans. 13th Congr. Int. Soc. Soil Sci., Vol. 6.

149. Hurst, H. M., and N. A. Burges. 1967. Lignin and humic acids, p. 260–286. *In* A. D. McLaren and G. H. Peterson (eds.), Soil biochemistry, Vol. 1, Marcel Dekker, New York.

150. Mathur, S. P. 1970. Degradation of soil humus by the fairy ring mushroom. Plant Soil 33:717–720.

151. Monib, M., I. Hosny, L. Zohdy, and M. Khalafallah. 1981. Studies on humic-acid decomposing streptomycetes. 2. Efficiency of different species in decomposition. Zentralbl. Bakteriol. Parasitenk. Infektionskr. Hyg. Abt. II 136:15–25.

152.   Haider, K., and J. P. Martin. 1988. Mineralization of [14]C-
       labelled humic acids and of humic-acid bound ^{14}C-xenobiotics
       by *Phanerochaete chrysosporium*. Soil Biol. Biochem. 20:
       425—429.
153.   Martin, J. P., K. Haider, and G. Kassim. 1980. Biodegra-
       dation and stabilization after 2 years of specific crop, lignin,
       and polysaccharide carbons in soils. Soil Sci. Soc. Am. J.
       44:1250—1255.

3

Nematophagous Fungi and Their Activities in Soil

CARIN DACKMAN, HANS-BÖRJE JANSSON, and
BIRGIT NORDBRING-HERTZ *Lund University, Lund, Sweden*

I. INTRODUCTION

During recent decades, intensive research has been conducted on nematode-destroying fungi at several levels: biological, physiological, biochemical—molecular, and ecological. This research effort has encompassed both basic and applied approaches and has resulted in data sufficient to enable a more unified view of these diverse fungi and their activities in soil, including their survival strategies and potential as regulators of nematode populations. Several reviews have summarized these research activities [1—18].

The diversity within this group of fungi is reflected both in the range of strategies used to attack nematodes and in the different developmental stages of nematodes attacked by the fungi. Three main strategies of attack are used: (1) the nematode-trapping fungi attack free-living nematodes, with adhesive or nonadhesive trapping devices formed by the vegetative mycelium; (2) the endoparasites use adhesive or nonadhesive spores for the same purpose; and (3) the fungal parasites of cyst and root-knot nematodes attack the eggs and females of these nematodes with their hyphal tips.

The continued interest in these fungi has been partly the result of an increasing nematode problem in agricultural and horticultural soils and the continued need for alternative methods for the control of nematodes. Knowledge of the role that these fungi have in soil is still incomplete, as is knowledge concerning the importance of environmental factors in triggering changes in their activities at the developmental, physiological, and ecological levels. However, the data compiled from applied and basic research on the aforementioned groups of fungi provide a sound basis for an evaluation of their activities in such a complex system as soil. The intention of this review is to contribute to such an evaluation.

II. NEMATODES

Nematodes are found in large numbers in marine, freshwater, and soil environments, from the tropics to the polar regions [19]. The nematodes contain animal-parasitic, plant-parasitic, bacteria-feeding, and predatory (cannibalistic) species. The vast majority of soil-dwelling nematodes feed on bacteria, fungi, and small organic particles, but many species are also severe plant pathogens. One characteristic that distinguishes the different types of nematodes is the construction of the buccal cavity: the bacteria-feeding nematodes simply have an open tube leading to the esophagus (Fig. 1A); the predatory nematodes have a much more complicated mouth, often including teeth and mandibles; and the plant-parasitic (including fungus-feeding) nematodes are equipped with a stylet with which they pierce plant cells (see Fig. 1B,C).

During their life cycle, nematodes shed their cuticles in four moults, producing four juvenile stages (J_1–J_4) before becoming adults. The juveniles are often the infective stages. In animal-parasitic nematodes, the third-stage juveniles (J_3) are infective, and in the root-knot and cyst nematodes, the J_2 stage infects plants.

The plant-parasitic nematodes can be divided into sedentary and free-living species (see Fig. 1). The sedentary nematodes [e.g., the root-knot nematodes (*Meloidogyne* spp.) and the cyst nematodes (*Heterodera* spp. and *Globodera* spp.)], in contrast with the migratory nematodes, spend a major part of their life cycles within infected plant roots. In these nematodes, the second-stage juveniles (J_2) penetrate the roots and continue their development to adults inside the infected roots. When the cyst nematode females grow, the root cortex cells rupture and the females become exposed on the root surface, where they develop into cysts containing several hundred eggs (see Fig. 1C). The cyst wall is resistant, and the eggs inside can remain viable in soil for many years before exposure to a suitable plant host. In the root-knot nematodes, the females produce egg masses that may contain more than 1000 eggs. Under suitable conditions, the eggs hatch to produce new J_2 stages, which can begin a new infection cycle.

The migratory plant-parasitic nematodes attack roots, stems, or leaves, piercing the cells and sucking out the cytoplasm, which may cause severe damage (see Fig. 1B). Examples of such parasites are found with the genera *Pratylenchus*, *Ditylenchus*, and *Aphelenchoides*.

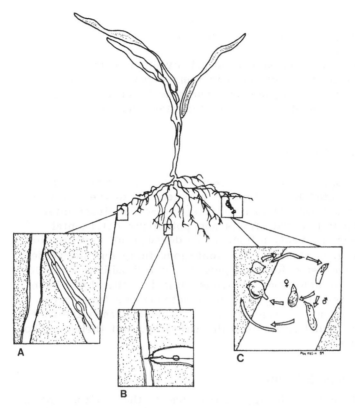

Figure 1 Behavior of different types of nematodes in the vicinity
of plant roots. (A) Bacteria-feeding nematode searching the root
surface for bacteria. Note the simple outline of the buccal cavity,
which generally consists of a tube leading to the esophagous. (B)
Free-living plant-parasitic nematode piercing the root tissue with its
stylet. Apart from feeding, the life cycle of these nematodes occurs
outside the root. (C) Life cycle of a cyst nematode. The second
stage juvenile (top) penetrates the root and starts feeding. It moults
through the third and fourth juvenile stages and develops into male
or female. The female, still attached to the root, produces eggs
that later will fill the entire female body. The cyst (i.e., the dead
female) filled with eggs can later be found in the soil.

Soilborne nematodes affect plants in different ways. Plant-parasitic nematodes can decrease the growth of plants by piercing and
destroying root tissue, seeds, and aboveground parts. Bacteria-
feeding nematodes may be important in nutrient cycling in the

rhizosphere, as the result of their bacterial grazing. Insect-parasitic nematodes may function as biocontrol agents by infecting and killing plant-parasitic insects.

The nematodes are vulnerable to attack by other organisms, notably by nematophagous fungi, both in their vermiform, or wormlike, stages by nematode-trapping and endoparasitic fungi and as eggs and females by the fungal parasites of sedentary nematodes.

III. NEMATOPHAGOUS FUNGI

Numerous morphological adaptations occur in nematophagous fungi, each of which is suited to the type of nematode or to the developmental stage of the nematode to be attacked. Soil environmental factors, including nutritional effects and potentially detrimental physical conditions, probably influence the development and function of these morphological stages. Nematophagous fungi and nematodes have apparently coexisted for long periods, and 25 million-year-old fossil records of possible nematophagous fungi in amber have been found [20]. Representative genera of nematophagous fungi are found in all the major taxonomic groups. Table 1 summarizes these genera, their functional relations with nematodes, and their morphological adaptations.

A. Nematode-Trapping Fungi

The ability to change morphology in response to the environment is a valuable property of many nematode-trapping (or predatory) fungi that increases their chances of survival. Although several of the nematode-trapping fungi can survive in soil as saprophytes, their ability to use nematodes as a food source is a transition from a saprophytic to a predaceous phase, with a concomitant morphological change in which nematode-trapping structures are formed. This transition may be influenced by many environmental factors, both biotic and abiotic, and is typical for those species of the genus *Arthrobotrys*, which form adhesive network traps (Fig. 2A). Other predatory fungi that form adhesive knobs or branches (see Fig. 2B,C) or those with constricting or nonconstricting rings are "more predaceous," as they form trapping structures spontaneously and seem to be more dependent on nematodes as a food source. *Dactylaria candida, Monacrosporium cionopagum, M. ellipsosporum,* and *Dactylella brochopaga* are examples of this group.

Both biotic and abiotic factors affect hyphal differentiation in nematode-trapping fungi. Exposure to nematodes or proteinaceous material induces the formation of trapping structures [21]. For example, in *Arthrobotrys oligospora*, small peptides that have a high

Table 1 Taxonomic Classification of Nematophagous Fungi;
Representative Genera and Functional Classes

Division	Genus	Interaction[a]	Infection structure
Chytridiomycetes	*Catenaria*	EP, FP	Zoospores
Oomycetes	*Myzocytium*	EP	Zoospores
	Nematophthora	FP	Zoospores
Zygomycetes	*Stylopage*	NT	Adhesive hyphae
	Cystopage	NT	Adhesive hyphae
Deuteromycetes	*Arthrobotrys*	NT	(Non-)adhesive traps
	Monacrosporium	NT	Adhesive traps
	Dactylaria	NT	(Non-)adhesive traps
	Dactylella	NT, EggP	Adhesive traps, hyphal tips
	Nematoctonus	NT, EP	Adhesive traps, adhesive conidia
	Harposporium	EP	Nonadhesive conidia
	Drechmeria	EP	Adhesive conidia
	Hirsutella	EP	Adhesive conidia
	Verticillium	EP, EggP, FP	Adhesive conidia, hyphal tips
	Paecilomyces	EggP	Hyphal tips
	Cylindrocarpon	EggP, FP	Hyphal tips
	Fusarium	EggP, FP	Hyphal tips
Basidiomycetes	*Hohenbuhelia* (teleomorph of *Nematoctonus*)	NT, EP	
	Pleurotus	NT, toxic	Adhesive traps, toxic droplets
Ascomycetes	*Atricordyceps* (teleomorph of *Harposporium*)	EP	

[a]EP, endoparasitic; NT, nematode-trapping; FP, female parasite; EggP, egg parasite; toxic, fungus producing toxins on special hyphal devices.

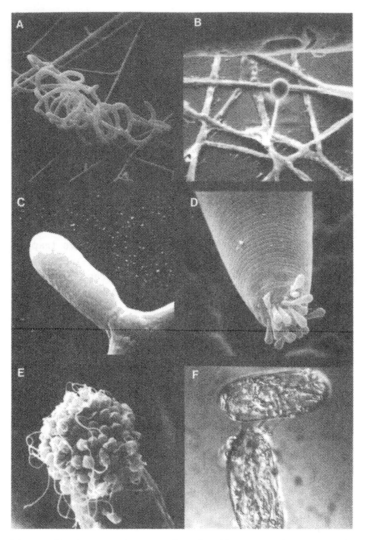

Figure 2 Nematophagous fungi. (A) Adhesive network trap of *Arthrobotrys oligospora* (from Nordbring-Hertz, *Appl. Environ. Microbiol.* 45:290–293, 1983). (B) Adhesive knob trap of *Dactylaria candida*. Top of the figure shows vulvar region of a female nematode. (C) Adhesive branch trap of *Monacrosporium cionopagum*. (D) Conidia of the endoparasitic fungus, *Drechmeria coniospora*, adhering to the sensory structures of a nematode [44]. (E) Zoospores of *Catenaria anguillulae* adhering to the head of a nematode. (F) Mycelium of *Verticillium suchlasporium* proliferating inside eggs of a cyst nematode [36].

proportion of nonpolar and aromatic amino acids or their amino acid constituents in combination with a low nutrient status induce trap formation on solid [22,23] or in liquid [24] laboratory media. Both peptide- and nematode-induced trap formation are also influenced by such environmental factors as nutrient level, moisture content, pH, light, O_2 and CO_2, and volatile compounds [25,26].

The ability to change morphology in response to environmental factors is especially pronounced among the more saprophytic species of nematode-trapping fungi. In some species of *Arthrobotrys*, nematodes may be captured not only in fully developed traps, but also on the basal cells of conidiophores [27] or even on undifferentiated hyphae [28]. There is a similarity here with the zygomycetous nematode-destroying fungi, which capture nematodes with nonseptate adhesive hyphae (see Table 1). Nematode trapping in nature might not be restricted to fully developed traps, but also may be accomplished by other environmentally regulated structures, which increases the importance of the role of these fungi as regulators of nematode populations. The delicate balance between saprophytic and predaceous development is especially evident when fungal germ tubes are subjected to trap-inducing factors. Traps may be formed immediately upon germination of conidia in both *Arthrobotrys dactyloides* [1] and *A. oligospora* (in cow dung; B. Nordbring-Hertz, unpublished). The sensitivity of young hyphae to environmental factors, detrimental as well as beneficial, may be of great importance in the natural environment. For example, germ tubes and young mycelium have been shown to respond to low concentrations of polyoxin D, a chitin synthetase inhibitor, resulting in increased trap formation [29]. This increased trap formation is believed to be due to weakening of cell walls that have not yet reached their final stability, thus making the mycelium susceptible to trap-inducing factors.

B. Endoparasitic Fungi

In contrast to nematode-trapping fungi, endoparasitic fungi do not produce specific traps, but use their spores to infect nematodes (see Table 1). Most of these fungi are obligate parasites of nematodes and spend the vegetative part of their life cycles within infected nematodes. In zoosporic fungi (e.g., *Catenaria anguillulae*; see Fig. 2D), motile zoospores are released from zoosporangia, produced inside digested nematodes. In other fungi, mostly Deuteromycetes, conidia are produced on conidiophores that protrude through the cuticles of infected hosts. These conidia can be either adhesive and stick to the nematode cuticle, as in *Drechmeria coniospora* (see Fig. 2E), or be ingested mainly by bacteria-feeding nematodes (e.g., conidia of *Harposporium* spp.). The conidia of the first type produce germ tubes that penetrate the cuticle and trophic hyphae that digest the nematode contents, whereas the second type continues the infection from the esophagus of their hosts.

The conidia of *Harposporium* spp., which are ingested and lodge in the buccal cavity of the nematodes, often have remarkable shapes [1] and appear to be just as complex as the traps of predatory fungi. Some of the conidia of the endoparasitic fungi also exhibit morphological changes in response to the environment. Conidia of *D. coniospora*, for instance, go through a maturation phase and produce an adhesive bud that is responsible for the attachment of the conidia to the nematode cuticle [30]. The details of the maturation process have not been clarified, but the water content of the substrate and the presence of nematodes may be, at least partly, involved. The conidia of *D. coniospora* show a high degree of specificity as, in most nematode species, adhesion occurs close to the sensory structures in the head (see Fig. 2E) and tail regions [31,32].

The spores of the zoosporic fungi, which are common soil inhabitants, also show morphological changes during their life cycle. After release from the zoosporangia, the zoospores may swim around for several hours until they find a suitable host on which some, but not all, species encyst. The encysted zoospores produce germ tubes within 30 minutes, under suitable conditions, that penetrate the nematode cuticle and, thus, infect the animals. Inside the host, chytridiomycetes (e.g., *C. anguillulae*) produce zoosporangia and resting sporangia, whereas oomycetes (e.g., *Myzocytium lenticulare*) may also produce thick-walled resting oospores, the function of which has been suggested to be mainly for persistence during harsh conditions, rather than for transfer of genetic material [1].

C. Fungal Parasites of Sedentary Nematodes

Nematode-trapping and endoparasitic fungi also attack the free-living stages of cyst and root-knot nematodes. The second-stage juveniles that hatch from the eggs and enter the soil searching for a root of a host plant, may be trapped or infected by spores of these fungi. The vermiform males may also become trapped or infected in the same manner, but fungi from these groups have not been found infecting sedentary stages (i.e., females and eggs in cysts and egg masses). Fungi that infect these structures are common soil inhabitants (e.g., species of *Verticillium*, *Cylindrocarpon*, *Fusarium*, *Paecilomyces*, *Gliocladium*, and *Dactylella*; see Table 1). These fungi infect their hosts using hyphal tips (see Fig. 2F). No trapping structures are needed, as only sedentary stages are infected.

Similar fungal species have been isolated from several nematode species in different parts of the world, suggesting a restricted flora of soil fungi that parasitize or are otherwise associated with eggs and females of cyst and root-knot nematodes [33,34]. A common feature of fungi capable of parasitizing eggs and females might be the production of specific enzymes involved in the penetration process. There is evidence that chitinase and protease are involved in the penetration of the rigid egg shell, which consists of a chitin–protein complex [35,36].

The ability to utilize nematode eggs and females as a food source is probably an advantage to these fungi, but little is known about their dependence on nematodes for propagation and survival. There is some indirect evidence for the importance of nematodes for the propagation of *Verticillium chlamydosporium*. Considerably higher numbers of dichtyochlamydospores of this species have been found in fields where population densities of the cereal cyst nematode were declining, indicating parasitic activity of this and other fungi [37].

Obligate parasites are completely dependent on their nematode hosts for survival. A few such fungi are known (e.g., *Nematophthora gynophila* and *Catenaria auxiliaris*). Females are infected with zoospores, and trophic hyphae subsequently develop in the female body. Zoospores and resting hyphae are then formed in the hyphal fragments. It has not been possible to cultivate *N. gynophila* on common laboratory media, which points to the very intimate relationship between this fungus and its host.

IV. POSSIBLE MECHANISMS INVOLVED IN FUNGUS–NEMATODE INTERACTIONS

Many different mechanisms may be involved in the antagonistic action of nematophagous fungi toward nematodes. The growth of the fungi in the natural environment may be subject to competition and antibiosis from other organisms, which may influence hyphal lengths and the number of infection units in a specific environment. In addition, nematodes in soil may have an important role in inducing morphogenesis in some fungi and in functioning as hosts of endoparasites, thereby increasing the fungal biomass.

More specifically, our approach to the study of the interaction mechanisms has been to elucidate how the fungi recognize their hosts (Table 2). The question of host specificity is interesting for two main reasons: (1) host-specific antagonists are often desirable when biological control is contemplated; and (2) the question of recognition of prey is often approached from the point of view of host specificity. Chemoattraction of nematodes is one recognition mechanism used by the fungi, which might be considered as a form of "long-range" communication. This is followed by a "short-range"

Table 2 Possible Recognition Mechanisms
in Fungus—Nematode Relations

Host specificity

Chemoattraction of nematodes

Adhesion of nematodes
  Lectin—carbohydrate interaction
  Presence of adhesive
  Enzymatic activity

---

or contact communication: the adhesion process. The adhesion may
include a lectin—carbohydrate interaction, presence of an adhesive,
and activity of extracellular fungal enzymes.

A. Host Specificity

Although nematophagous fungi rarely attack any animals other than
nematodes, no distinct host specificity has been detected among any
of the fungi. However, the zoosporic fungus *N. gynophila*, is an
obligate parasite, with a very restricted host range. Only cyst
nematode females of the genus *Heterodera*, are parasitized, whereas
members of *Globodera* are not affected [38]. Among egg-parasitic
fungi, there is a tendency toward host specificity, as *Heterodera*
and *Meloidogyne* species are frequently parasitized in the field,
whereas cyst nematodes of the genus *Globodera*, are never infected
to the same extent, if at all. There also seem to be differences in
the species composition of fungi that infect these nematodes, although comparatively few investigations have been published involving *Globodera*.
    Some host specificity also occurs among the endoparasitic fungi.
A restricted host range has been found for *D. coniospora* [32,39]
and *Verticillium balanoides* [40]. This is especially evident when
not only adhesion of *D. coniospora* conidia to the nematode surface
but also the entire infection process is considered. Jansson et al.
[32] found that 10 out of 17 nematode species could be considered
as hosts. The predatory fungi attack most free-living nematodes,
but some differences in infectivity and parasitic development have
been reported [4]. The reasons for these differences are unclear,
but morphological and physiological differences of the cuticle have
been suggested to influence the activity of fungal enzymes or toxins [4].

B. Chemoattraction of Nematodes

Chemotactic signals from various sources are likely to affect both fungi and nematodes. Accumulation of zoospores at the natural openings of nematodes, possibly owing to chemotaxis, is well known [1], and chemotropism of germ tubes and hyphae to specific nutrient sources or to immobilized hosts is likely to occur [41]. Chemoattraction of nematodes to fungi has been extensively studied in laboratory systems [reviewed in Ref. 8] and may be considered as one recognition mechanism in this relationship [9,14].

In laboratory studies, nematodes are attracted to most of the nematophagous fungi tested [42]. Although the more saprophytic species show a relatively weak attraction, the endoparasites exhibit increased ability to attract, which may be associated with their increased dependence on nematodes as a food source. Furthermore, an increased attraction of nematodes has been detected when infection structures are present on the mycelium, compared with the attraction to hyphae without these structures. This was true for *A. oligospora* with adhesive network traps, as well as for *D. coniospora* and other endoparasitic species in the presence of their adhesive conidia [43,44]. These examples reflect the possibility that chemotaxis of nematodes may be important to the survival of these fungi in soil. This was demonstrated in microcosm experiments with sterilized soil, during which a high correlation between chemotaxis and predacity was observed [43]. Although random movement of the nematodes would be the most probable mechanism for movement over long distances, the behavior of the nematodes in the vicinity of the fungi might be explained by the chemoattractance to specific sites on the mycelium.

C. Adhesion of Nematodes

Adhesion to surfaces is a fundamental feature in microbial ecology and not least so in the interactions between microorganisms and higher organisms. In many host—parasite interactions in which fungi are involved, an adhesive phase of the fungal partner is a prerequisite for penetration of the host and the subsequent development of the relationship. In many fungus—nematode interactions, an adhesive with specific properties is necessary for the firm anchoring of a moving nematode to specific mycelial trapping structures or to conidia [9,30,45,46]. The consequences of the adhesion are penetration of the cuticle, development of trophic hyphae, and digestion of the nematode [9,47,48]. An adhesive phase in the cyst nematode parasites is not so obvious, but some fungi develop appressorialike swellings that attach to the female or egg [49].

Lectin—Carbohydrate Interactions

A lectin—carbohydrate interaction has been suggested to operate in initial stages of the capture of nematodes by *A. oligospora* [50], *D. coniospora* [31], and other fungi. These studies have been recently reviewed in detail [9,14,15]. The results summarized in Table 3 use *A. oligospora* as a model organism. The background to these studies was that many predatory and endoparasitic fungi produce structures that have unique adhesive properties. We postulated that the firmness of the attachment of the nematode to the adhering structure, despite the vigorous movements of the nematode, results from a series of events, beginning with an interaction

Table 3 Steps of Evidence for a Trap Lectin in *Arthrobotrys oligospora*

Step	Ref.
1. Inhibition of capture of nematodes by N-acetyl-D-galactosamine (GalNAc)	50,58
2. Binding of red blood cells to traps	50
3. Demonstration of GalNAc on nematode surface	50
4. Inhibition of capture by trypsin and glutaraldehyde treatment of traps	51
5. Binding of ^{125}I-labeled fungal homogenate to GalNAc-Sepharose	55
6. Inhibition of binding of homogenate by trypsin treatment	55
7. Isolation and characterization of a carbohydrate-binding GalNAc-specific and Ca^{2+}-dependent protein (subunit M_r 20,000)	56,58
8. Isolation of one major lectin-binding glycoprotein (M_r 65,000) from nematode cuticle	57
9. Localization of the lectin in the trap cell wall by the immunogold technique	Nordbring-Hertz et al. 1988 (Abstr)

between complementary molecular configurations on the nematode and fungal surfaces. If a lectin–carbohydrate interaction is involved, preexposure of the lectin-carrying structure to the specific carbohydrate would lead to inhibition of the adhesion. Biological inhibition experiments using intact organisms of *A. oligospora* and the nematode, *Panagrellus redivivus* (see Table 3, points 1–4) showed that capture was inhibited principally by N-acetyl-D-galactosamine (GalNAc), and that this carbohydrate was present on the nematode surface. A model prey in the form of red blood cells also bound to the traps and then lysed under the influence of the trap. Capture was further inhibited by trypsin and glutaraldehyde, indicating the proteinaceous nature of the binding molecule. From these biological interaction studies, it was suggested that a lectin with primary specificity for GalNAc was present on the fungus and that the lectin binding to surface structures of the nematode initiated further events in the interaction [50]. Similar techniques have been used to demonstrate possible carbohydrate-binding proteins in other nematophagous fungi (e.g., in nematode-trapping species [51-53]) and in the endoparasite *D. coniospora* [31,54]. In the latter, sialic acid inhibited adhesion of conidia to the sensory organs of the nematodes, and treatment of nematodes with sialidase and a sialic acid-specific lectin reduced attraction of the nematodes to the fungus [54].

In *A. oligospora*, the GalNAc specificity was used to isolate and partially characterize the *A. oligospora* lectin (see Table 3, points 5–7) [55,56]. The lectin was GalNAc-specific, Ca^{2+}-binding and had a molecular mass of approximately 20 kilodaltons (kd) [56,57]. Furthermore, these studies clearly showed the presence of the lectin in only trap-containing cultures, indicating its role in the recognition of prey. Further studies were conducted to isolate the possible receptor of this lectin on the nematode surface (point 8) and to localize the lectin on the trap using immunogold–protein A (point 9). These results showed that the lectin was located in cell walls of the traps, but not in cell walls of hyphae, and that the adhesive outside the cell wall was not the site of the lectin [B. Nordbring-Hertz, M. Veenhuis, B. Mattiasson and C. A. K. Borrebaeck, Abstr., 5th Int. Congr. Plant Pathol., 1988, P.IV-1-12, p. 155].

Adhesive Properties

Thus, although the lectin is present on the structure responsible for the trapping of nematodes and its role is probably in recognition of nematode surface structures, the lectin might not explain fully the attachment of nematodes. The tenacity of the adhesive is an outstanding feature in nematode-trapping fungi [58,59], and increased secretion of adhesive after initial attachment [45,56,60–62] seems to be a prerequisite for invasion of the host, as appears to be true for many other host–parasite interactions. During the initial phases of adhesion, before invasion of the nematode, the random fibrils of the *A. oligospora* adhesive become oriented in one direction, perpendicular to the surface of the nematode [46]. Directed fibrils are also present on the adhesive bud of *D. coniospora* conidia [9]. In *A. oligospora*, at least, directed fibrils may lead to a necessary strengthening of the adhesive properties.

Enzymatic Activity

The penetration of nematode cuticles and eggshells may be either mechanical or enzymatic, or both. Inasmuch as collagen is a major component of nematode cuticles, fungal collagenase has been suggested to operate and has been detected in cultures of several species of nematode-trapping fungi [63,64]. When cytochemical methods and electron microscopy were used to study penetration of the *P. redivivus* cuticle, penetration was considered to be largely mechanical, although acid phosphatase activity (as a marker enzyme for hydrolytic activity) was detected in the adhesive layer and within the nematode at a later stage [46]. Peroxidase activity was similarly detected in the adhesive layer after nematode attachment (B. Nordbring-Hertz, unpublished).

To infect nematode eggs, the fungus has to penetrate the eggshell, which is a very rigid structure consisting of three layers: an outer protein layer, a chitin layer, and a lipid layer [65]. Lytic activity, in particular that of chitinase, seems to be important for successful infection [35]. Reinfection experiments with isolated fungi showed that those which infected eggs in vitro produced chitinase, whereas those that did not infect eggs, seldom showed chitinase activity. Protease was produced by most of the fungi studied, irrespective of reinfection ability [36; C. Dackman, unpublished].

V. ACTIVITIES IN SOIL

Nematophagous fungi occur frequently in soils in the tropics as well as in the Antarctic. In most soil samples studied, at least one species of nematophagous fungi is present. The nematophagous fungi are generally poorly competitive saprophytes, and they will not appear when ordinary methods for isolation of soil fungi are used. A step that uses nematodes as bait for the fungi is usually necessary. Some of the techniques used for the detection and isolation of nematophagous fungi will be briefly described in the following sections.

A. Techniques to Detect Nematophagous Fungi in Soil

Nematode-Trapping and Endoparasitic Fungi

One of the oldest and most frequently used techniques is the sprinkled-plate method. In this method, a petri dish with a dilute nutrient medium or water agar is sprinkled with 0.5 to 1.0 g of soil. Observation of the plates at regular intervals for up to 2 months will reveal trapped and infected nematodes, trapping organs, and the characteristic conidia of the nematophagous fungi. In this way, the naturally occurring nematodes are used as bait for the fungi. The efficiency of the method can be improved by the addition of 500 to 1000 nematodes per plate [66]. The conidia of the nematophagous fungi present on the plates can be picked and easily isolated in pure cultures on such media as corn meal or malt agar. The sprinkled-plate technique will detect all types of nematophagous fungi, but it will underestimate the endoparasitic fungi and is, thus, mainly suitable for detection of nematode-trapping species.

Other methods have been devised for the isolation of endoparasitic fungi. One of these, the Baermann funnel technique [1], uses a method for the extraction of nematodes from soil. Adding the extracted nematodes to a water agar plate and observing signs of fungal infection will usually reveal the presence of endoparasitic fungi. The possibility of detecting zoosporic nematophagous fungi is increased by maintaining a moist surface on the plate.

A differential centrifugation technique has also been developed to detect endoparasitic fungi [1]. In this method, two centrifugation steps are used. The first low-speed centrifugation will spin down heavy particles and the large conidia of nematode-trapping fungi. The supernatant is then centrifuged again at a higher speed, and the resulting pellet, containing the smaller conidia of endoparasitic fungi, is spread on a water agar plate. Nematodes added to this plate subsequently become infected and the endoparasites can be detected and isolated. Techniques for the detection and isolation of nematophagous fungi are fully described elsewhere [1,2,5].

In addition to detecting nematophagous fungi, there is a need for techniques to quantify these fungi in soil. Because of the laborious and time-consuming nature of such methods, only a few techniques have been developed [66-70]. We have devised a method that uses a combination of the differential centrifugation technique and most probable number (MPN) estimations for the quantification of the number of propagules of nematophagous fungi. This technique has been used to estimate the level of nematophagous fungi in soils that have received different treatments [70; Y. Persson, C. Dackman, H.-B. Jansson, and B. Nordbring-Hertz, Abstr., 5th Int. Congr. Plant Pathol., 1988, P.IV-1-15, p. 156].

Cyst and Root-Knot Nematode Parasites

Fungal parasitism of females of cyst nematodes can be observed by removing intact roots from soil and examining the females on the root surface. Females infected with zoosporic fungi (e.g., *N. gynophila* and *C. auxiliaris*) will soon be filled with resting spores, which can be used for species identification. Females infected with hyphomycetes (e.g., *V. chlamydosporium* and *Cylindrocarpon destructans*) will have to be subcultured for development of fungal colonies. Observation chamber techniques have been developed that make continuous observations of fungal parasitism of females possible [71,72] and enable a more precise estimation of the true level of parasitism.

The presence of parasites of female nematodes (e.g., *N. gynophila* and *V. chlamydosporium*) in soil can also be detected through spore extraction of soil samples. These two fungi have persistent resting spores (oospores and dichtyochlamydospores, respectively) that can be extracted by a soil-washing and centrifugation technique [37]. The number of spores per gram of soil seems to reflect the level of infection in soil.

The egg-parasitic fungi are easy to detect and isolate. Techniques used to extract nematode cysts from soil samples are well developed and simple to use [73]. After the cysts have been extracted with water and collected in a test tube, the cyst wall is ruptured and the eggs are released into the suspension. Aliquots of this suspension are spread on water agar plates, fungi develop from infected eggs, and the percentage of infected eggs is calculated. For identification of the fungal species involved, eggs with developing mycelium are transferred to nutrient agar, whereon the fungi will develop a characteristic colony [74]. Figure 3 shows an overview of three different procedures used to assess the number of fungi parasitic on nematode eggs or otherwise associated with cysts.

B. Role of Soil Characteristics and Environmental Pollution

Although nematophagous fungi exist in a wide variety of habitats and can be isolated from almost any part of the world, the distribution of individual species is influenced by biotic as well as abiotic factors. Information on the influence of soil characteristics on growth and trapping ability of these fungi is of great importance when the use of nematophagous fungi in biological control is considered.

It is generally accepted that nematophagous fungi are favored by a high organic matter content in soil. Gray analyzed the distribution of nematophagous fungi in different types of soil in relation to pH and the contents of moisture, organic matter, and nematodes [75]. He found endoparasites to be more frequent in soils with higher contents of moisture and organic matter and lower pH. Endoparasites were also isolated more often than nematode-trapping fungi from soils with higher densities of nematodes [76], reflecting the dependence of endoparasites on nematodes for propagation and survival. The influence of soil moisture on endoparasitic fungi belonging to the Mastigomycotina is obvious, as the infection process is mediated by zoospores. This was also shown for *N. gynophila*, an obligate parasite of cyst nematode females [77]. Infection of nematode females ceased almost completely under dry conditions, whereas watering the soil increased the number of infected females. Among the nematode-trapping fungi, network formers were more common in soils with lower contents of organic matter and moisture.

Individual species of nematophagous fungi have most probably adapted to distinct habitats. This was indicated by Gray and Bailey [78], who studied the vertical distribution of these fungi and found a degree of separation of species between the humus layer and the mineral soil. Differences in vertical distribution were also found for nematode-trapping fungi that occur in the rhizosphere of peach trees [79]. The significance of these results with respect to the distribution of nematophagous fungi in soil is not fully understood.

In addition to general soil characteristics, the distribution and activity of nematophagous fungi may also be influenced by xenobiotic compounds, such as pesticides and heavy metals, which are commonly present in the environment. A toxic compound can affect nematophagous fungi at two levels: vegetative growth and morphogenesis. The effect of zinc, cadmium, and lead on growth of seven species of nematode-trapping fungi showed individual variations in sensitivity, although the overall sensitivity was comparable with that of other fungi [64]. Usually, inhibition of growth was directly correlated with a decreased capacity to form traps. In a few cases, trap formation and growth were differently influenced. The growth of *Monacrosporium eudermatum* was not affected by 50 µg Zn per milliliter, but trap formation was completely inhibited, whereas 300 µg Pb per milliliter reduced growth of *A. oligospora* to 64% of the control, but trap formation was not affected [64]. In another study, *A. oligospora*, *A. superba*, and *D. candida*

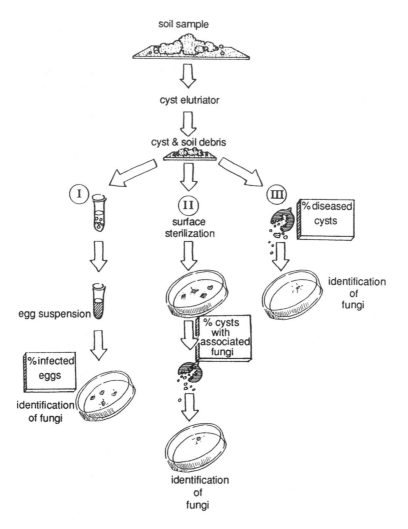

soil sample

cyst elutriator

cyst & soil debris

(I)

(II)
surface
sterilization

(III)

%diseased
cysts

egg suspension

%infected
eggs

identification
of fungi

% cysts
with
associated
fungi

identification
of
fungi

identification
of
fungi

Figure 3 Different methods for assessing the degree of fungal parasitism and fungal association with cyst nematodes. Cysts are recovered from soil samples with a cyst elutriator, and cysts are picked
out of the soil debris. (I) Cysts are collected in a test tube with
water, and eggs are released in the water by crushing the cysts.
Eggs are washed in sterile water with antibiotics, and aliquots of
the egg suspension are transferred to agar media. The proportion
of infected eggs is determined, and fungal species are identified.
(II) Cysts are surface sterilized, put on agar media and fungi allowed to develop. Cysts with associated fungi are cut open, and

were very sensitive to cadmium, and trap formation was inhibited
at lower concentrations (<1 μg/ml) than was growth [M. Grip and
B. Nordbring-Hertz, unpublished).

The effect of pesticides on nematophagous fungi is not well understood, and the studies that have been conducted have evaluated only the effects on growth and not on trap formation, which
is of obvious importance to the activity of these fungi in soil. The
effect of several fungicides and herbicides on growth of *A. irregularis* was studied in vitro [80]. When doses comparable with those
remaining in soil after a treatment were used, only the fungicide,
Benomyl, retarded the growth of the fungus. Benomyl inhibited
or greatly reduced the growth of the egg- and female-parasitic
fungi, *C. destructans* and *V. chlamydosporium*, when tested in
vitro [81]. The insecticides and herbicides used in the same
study had no significant effect on either fungus at 1 ppm and,
generally, had only a small effect at 10 and 100 ppm. However,
most of the 15 fungicides tested retarded growth at concentrations
as low as 1 ppm, and only one nematicide (Oxamyl) inhibited fungal growth. As fungi differ in their susceptibility to pesticides,
the balance of species may be altered in soil for longer or shorter
periods after pesticide application [82].

C. Reduction in Nematode Numbers

One of the major questions concerning the ecological aspects of
nematophagous fungi is whether these organisms are capable of
reducing nematode populations substantially or whether their effect is only marginal. To study these effects, a few methods have
been devised to measure the predacity of nematophagous fungi.
Predacity is defined as the ability of the fungi to destroy nematodes in soil. The methods used vary from the determination of
predatory indices on agar plates [83], the "agar disk method" in
nonsterile soil [84] and sterile sand with nutrient solution [85] to
sterilized soil microcosms [43]. Recently, a dilution plate assay
has been developed to quantify specifically parasitism of the plant-

---

Figure 3 (continued)
the eggs are examined for internal hyphae. (III) Cysts are cut
open, and the egg content is analyzed under a dissecting microscope. The proportion of diseased cysts is determined. Samples of
infected eggs are subcultured for identification of fungal species.
(From C. Dackman, PhD thesis, Lund University, Lund, Sweden,
1988.)

parasitic nematode, *Criconemella xenoplax*, by the endoparasite, *Hirsutella rhossiliensis* [86]. None of the methods is ideal, but all have made small contributions to our understanding of the ecology of nematophagous fungi.

In the simplest system, the agar plate, most nematode-trapping fungi destroy the nematodes within a few days after their addition, as do the endoparasitic fungi when a suitable host is added. In sterile soil microcosm experiments, for which one fungus and one nematode species are added at a time, varying predaceous abilities of the different fungi tested have been observed. Nematode-trapping fungi (e.g., *A. oligospora*) in which trap formation has to be induced by the nematodes show a lower ability to destroy nematodes than both fungi displaying spontaneous trap formation (e.g., *D. candida*) and endoparasitic fungi (e.g., *D. coniospora*) [43]. In none of the experiments were the nematodes completely destroyed, and after 30 days in the microcosms, the nematode population usually started to recover, indicating a variability in the predaceous ability with time, which may be a reflection of the succession of nematophagous fungi that was described by Cooke [87].

In other microcosm experiments that used nonsterile cow dung inoculated with conidia of *A. oligospora*, the reduction in infective juveniles of the animal-parasitic nematode, *Ostertagia ostertagi*, was 90% after 4 weeks of incubation. When the same experiment was conducted in the field, the reduction in nematode numbers was much lower [88].

D. Activity in the Rhizosphere

The rhizosphere of plant roots is the site of accumulation of such organisms as bacteria, including actinomycetes, algae, fungi, and soil microfauna. Root exudations and sloughed dead and dying root fragments are believed to be the reasons for this increased biological activity. In early rhizosphere studies by Katznelson and co-workers [89,90], the influence of actinomycetes and fungi on the accumulation of nematodes and the incidence of nematode-trapping fungi in the vicinity of plant roots were estimated in the rhizospheres of different plants. *Arthrobotrys oligospora*, for instance, was consistently more common in the rhizosphere of soybean than of wheat. The observed attraction of nematodes to the vicinity of the roots was attributed not only to root diffusates, but also to the abundance and activity of soil microorganisms close to the roots. Nematode-trapping fungi, although considered as soil-inhabiting fungi, in accordance with Garrett's terminology [91], were present in the rhizosphere, but were almost completely absent from the surface of the roots (rhizoplane).

Other than these early rhizosphere studies, few similar studies have been performed. The detection of parasites of female cyst nematodes, in particular, has made rhizosphere observations necessary. Crump [72] found that some fungi that infect females (e.g., *C. destructans*) colonized the root cortex, whereas others (e.g., *V. chlamydosporium*) did not, although the latter species had the ability to colonize the rhizosphere [92]. In recent studies of citrus rhizosphere [93], nematode-trapping fungi were isolated from 43 of 58 orchards that contained high populations of the citrus nematode, *Tylenchulus semipenetrans*. Nematode-trapping fungi were the principal fungal antagonists recovered, but no correlation with suppressiveness was evident in these studies. In contrast with the studies by Katznelson and co-workers [89, 90], propagule densities of nematode-trapping fungi were much higher in the rhizoplane than in the rhizosphere.

The relatively few rhizosphere studies of nematophagous fungi is surprising, as occurrence and high activity of nematophagous fungi of all groups would be expected as the result of the release of organic nutrients and the subsequent accumulation of microorganisms and nematodes. An increased level of nematophagous fungi in the rhizosphere would lead to a degree of natural control in sites where both sedentary and free-living plant-parasitic nematodes prevail. It is tempting to speculate that nematode-trapping fungi are true rhizosphere fungi [11,90,93]. For example, in nitrogen stress, fungi might be more advantageous than bacteria in the competition for root carbon, as they may translocate nutrients through their mycelium from outside the rhizosphere. Nematode-trapping fungi would be even more advantageous, since they can use nematodes as an additional nitrogen source. A prerequisite for such a rhizosphere effect then is a degree of nitrogen stress and, consequently, high levels of inorganic nitrogen would be expected to reduce the amount and activity of nematode-trapping fungi. In contrast, endoparasitic fungi would not be directly influenced by root exudates, but would be expected to be present in large amounts in soils with high nematode densities [70,75].

Apart from the aforementioned rhizosphere effect, both the abundance of nematodes and proteinaceous root exudates would, additionally, increase trap formation significantly.

E. Survival Strategies

The ability of nematophagous fungi to gain access to an additional food source, such as nematodes, increases the chances of survival of the fungi and may, in specific environments, give them a selective advantage. A shift from a saprophytic mode of life to a predaceous one, with the concomitant formation of specialized trapping

structures, is an aid in this survival strategy. Moreover, it has
been shown in laboratory cultures that traps of *A. oligospora* remain intact, even after prolonged cultivation (40 to 50 days) and
that vegetative cells break down before trap cells [61,94]. Traps
might, therefore, be considered as survival structures, similar to,
for example, conidia. One reason for the survival of traps might
be their high contents of so-called dense bodies that may be considered as storage organelles to supply carbon and energy or nitrogen required for hyphal development [47,61]. In addition, the
adhesive layer outside the trap may have a role in protecting the
structure from drying out [94]. Thus, the great variety of morphological adaptations of both traps and conidia may aid the survival of nematode-trapping and endoparasitic species. The obvious
need for nematodes of the endoparasites and the obligate parasites
of cyst nematodes further stresses this point. In addition, the
ability of the fungi to attract nematodes probably also gives them
a strong selective advantage.

Typical fungal survival structures, such as chlamydospores, are
found both among nematode-trapping and endoparasitic fungi and
are common among fungal parasites of sedentary nematodes. In the
latter group, both chlamydospores and oospores are common in soil.
The thick-walled oospores of the female cyst-nematode parasite, *N.
gynophila*, can survive more than 5 years in soil [7], and the number of oospores extracted from soil reflects the level of infection.
The importance of chlamydospores of nematode-trapping and endoparasitic fungi for survival in the field has not been evaluated, but
it is known that several species readily form chlamydospores under
laboratory conditions [95; M. Peloille, personal communicaton]. Investigations of the conditions necessary for both formation and germination of chlamydospores, now under way, will add to our understanding of the survival strategies of these fungi.

In addition to their ability to capture nematodes by special trapping structures, several of the predatory fungi show quite a different type of morphogenesis, in that their hyphal tips may coil around
hyphae of other fungi in their environment and, thus, act as myco-
parasites [96,97]. In laboratory studies with *A. oligospora*, a rather broad host range has been found [97]. Although hyphal coiling by nematode-trapping fungi has not been detected in natural
soil, the possibility of mycoparasitism as a survival strategy cannot
be neglected. The mechanism behind this antagonism is not fully
understood, but the interaction may be interpreted in terms of competition for nutrients with or without toxic substances involved.

The use of a toxin in the survival of fungi has been detected in
a quite different system: wood-decaying basidiomycetes, such as
Pleurotus spp., produce a potent toxin by which they paralyze and
kill nematodes, which then are penetrated by hyphae through their

natural openings and digested. The behavior is interpreted as a strategy of these fungi to supplement the low level of nitrogen available in wood by "capturing" nematodes in this way [98,99].

VI. BIOLOGICAL CONTROL

Nematodes are parasites on both plants and animals and may cause severe damage to agricultural crops and livestock. The increased restriction of the use of nematicides in many countries indicates an urgent need for alternative control practices. Such methods can involve interference with chemical communication systems, whether these affect food-finding, mate-finding, or adhesion factors [8]. More traditional biological control using antagonists (e.g., invertebrate predators and predatory nematodes [100] or microbial parasites and predators [101]) have long been discussed. In this chapter, we will restrict our discussion to the possibility of using nematophagous fungi for the biocontrol of plant- as well as animal-parasitic nematodes.

Possible methods with which to study the role of nematophagous fungi in biological control can be divided into three categories: natural control, addition of fungi to soil, and stimulation of existing fungal antagonists in soil.

A. Natural Control

Observations that crop damage caused by nematodes may be less severe in certain soils than in other comparable soils have resulted in a search for the biological causes of these phenomena. In a few cases, nematophagous fungi have been suggested to be the cause of the decline of populations of plant-parasitic nematodes.

It was found in Great Britain that the cereal cyst nematode showed a heavy decline in soils, despite intensive cultivation of cereals. The reason for this decline was shown to be the presence of antagonistic fungi, mainly *V. chlamydosporium* and *N. gynophila* [74]. Several methods were used to show the importance of the fungal antagonists, including soil drenches with formalin, which selectively killed the antagonists, but did not affect the nematodes. The results showed that the population of the cereal cyst nematode increased in formalin-treated soils compared with untreated controls [6]. The mechanisms involved in the infection of nematodes, as well as the potential use of these fungal antagonists for biocontrol purposes, are now being studied in several laboratories around the world.

Other examples of natural control have been found on nonresistant root stocks of peach in the United States. The damage caused

by root-knot nematodes was found to decrease in soils containing
the fungus, *Dactylella oviparasitica*, which attacks not only the
eggs of *Meloidogyne incognita*, but also the females of this species
[102]. This fungus was shown to have a high ability to infect
small egg masses of *M. incognita* in laboratory and greenhouse experiments. Later experiments in vineyards showed that the fungus
was unable to infect successfully the large egg masses found on
grapes [6,101]. Similar observations were made in other peach
orchards where a sudden decline of the plant-parasitic nematode
C. xenoplax, was observed. Here, the endoparasitic fungus, *H.
rhossiliensis*, was suggested to be responsible for the decline in the
nematode population. Under laboratory conditions, the fungus had
a high ability to parasitize *C. xenoplax* [103], although recent research has suggested that *H. rhossiliensis* may be only a weak parasite of *C. xenoplax* under field conditions [104].

B. Addition of Antagonists

The earliest attempts to control plant-parasitic nematodes were made
in the 1930s by applying nematode-trapping fungi to soil. Linford
and Yap [105] added five nematode-trapping fungi to potted soils
containing root-knot nematodes and found that one of these, *Monacrosporium ellipsosporum*, reduced the number of nematodes significantly, compared with the other nematode-trapping fungi and control soils. These early experiments have been followed by others,
wherein different nematodes and nematode-trapping fungi have been
tested in various combinations, with a widely varying degree of success [11,101]. A number of commercial products have been developed (e.g., the French preparations, Royal 300 and Royal 350, containing *Arthrobotrys* spp., for biological control of myceliophagous
nematodes in mushroom cultivations and of root-knot nematodes in
greenhouse cultivations of tomato, respectively) [106,107].

The efficacy of the use of nematode-trapping fungi for biological control of plant-parasitic nematodes was questioned by Cooke
[108], and he suggested that endoparasitic fungi might be more
efficient. A few experiments have been conducted in which endoparasites have been utilized for biocontrol purposes (e.g., *H. rhossiliensis* mentioned earlier). In greenhouse experiments, this fungus
was able to suppress populations of *C. xenoplax* in sterilized soil
[109], but in nonsterile soil, the fungus was less successful and
showed signs of malformed germ tubes suggesting that fungistatic
effects in natural soil may limit its potential use for biological control [110]. Another endoparasitic fungus, *D. coniospora*, has been
tested for biological control of root-knot nematodes on tomato in
greenhouse experiments [111]. In these tests, the fungus was successful in controlling root-knots in both sterile and nonsterile soils

when added as a spore suspension, or with living nonparasitic nematodes as a carrier. A problem with this and other endoparasitic fungi is that they are obligate parasites and, thus, are difficult to cultivate to obtain the large quantities of inoculum necessary for field application. Similar problems also exist with the procaryotic parasite, *Pasteuria penetrans*, which is an obligate parasite of a number of plant-parasitic nematodes [112].

A different and very interesting approach to biological control using nematophagous fungi is presently being studied by groups in France and Denmark [113,114]. In these experiments, attempts are made to prevent the spread of animal-parasitic nematodes in feces and grass in pastures by using different types of nematode-trapping fungi: *A. oligospora* reduced the nematode burden in cattle pastures, as did *D. candida, Candilabrella musiformis*, and *A. tortor* when tested for control of nematodes parasitic in sheep. This type of biocontrol is promising, but much research is necessary before a successful product is produced that can control nematode infections in cattle and other livestock.

C. Stimulation of Resident Antagonists

Nematophagous fungi are ubiquitous and are frequently found in most soils studied. The possibility of stimulating these fungi to an increased level of effectiveness has been studied by several groups. Early experiments in this direction were again performed by Linford and co-workers in Hawaii [115]. They added fresh organic matter to soils, and the decomposition initiated a series of events, starting with an increase in bacterial populations, which increased the number of bacteria-feeding nematodes. These nematodes stimulated growth and trap formation in the nematode-trapping fungi, and the fungi then captured nonparasitic as well as parasitic nematodes. This type of experiment has been followed by many others [101]. The most serious criticism of these experiments has been the difficulty in correlating the decrease in plant-parasitic nematodes with nematophagous fungi, while excluding the effect of toxic products produced by the decomposition of the organic material. We have preliminary results that indicate that a derease in nematode numbers is the result of an increase in the population of nematophagous fungi [Y. Persson et al., Abstr., 5th Int. Congr. Plant Pathol. 1988, P.IV-1-15, p. 156], but further studies are needed before conclusive evidence is obtained.

In addition to the aforementioned nonspecific means of stimullating a higher activity of nematophagous fungi, experiments using selective stimulants should be encouraged. One possible way to accomplish a reduction in root-knot nematodes by selectively stimulating chitinolytic microorganisms has been discussed by Rodriguez-

Kabana et al. [116]. The nematode eggshell contains large amounts of chitin, and, the addition of chitin-containing materials to soil stimulates the activities of chitinolytic microorganisms, mainly actinomycetes and other bacteria, which will attack the nematode eggs and result in a lower number of infective juveniles. These interesting results show that it is possible to stimulate selectively certain antagonists, and further means to activate nematophagous fungi by selective stimulation should be investigated.

VII. FUTURE OUTLOOK

In the past, most investigations of nematophagous fungi have focused on the interrelations between the fungi and nematodes in controlled systems. Although the question of how to use the fungi in biological control has been the driving force, it has been in the background. In the future, information from both laboratory and field investigations should be collected and combined in an effort to elucidate the potential of these fungi for biological control.

Research should be performed in the laboratory, as well as in the field, with intermediate experiments in microcosms and greenhouses, to learn as much as possible about the activities of the nematophagous fungi at all levels. The difficulties of interpreting data from microcosms and extrapolating these to natural systems are obvious, and the present lack of suitable methods limits our knowledge of the predaceous ability of these fungi in field situations. Perhaps the methods outlined by Jaffe and co-workers [104], by which population dynamics studies with single nematode-fungus associations, may lead to a better understanding of the natural system. There is also an urgent need for investigations designed to evaluate the effects of environmental factors (e.g., inorganic nutrients as well as heavy metals, pesticides, and other pollutants) not only on growth of these fungi, but also on their morphogenesis in the natural environment.

Combined efforts in the future should focus available knowledge of all three groups of fungi on in-depth studies of the following main features:

1. Development of accurate but simple methods
2. Continued search for examples of natural control and isolation of the antagonists
3. Studies of the population dynamics of both nematodes and fungi
4. Studies of the survival strategy of the fungi in soil
5. Studies of rhizosphere biology
6. Continued studies of the basic mechanisms of interaction

As described in the foregoing, biocontrol may be approached from different points of view. The observations of disease decline and detection or isolation of antagonists from suppressive soils may result in further clarification of the mechanisms behind the activity. This was how the activity of female and egg parasites was detected and studied [7]. Basic knowledge of the biology of an organism may encourage investigations into its use in a particular disease complex. This was the case with *D. coniospora*, which was applied in greenhouse experiments to control *Meloidogyne* infection in tomato [111], and with *A. oligospora*, which was applied in field studies designed to reduce infection by *Ostertagia ostertagi* in cattle [88].

The possibility of manipulating attraction behavior of the nematodes or the adhesion mechanisms of the fungi has opened up new ways to design experiments to elucidate the factors controlling the fungi in a competitive environment. Nematode attraction to different sources has already been used to explain the presence and behavior of nematodes and nematophagous fungi in various investigations [5,8,18]. The role of adhesive factors in a competitive environment has not yet been explored.

Further in-depth studies of the adhesive of traps of predatory fungi and of spores of endoparasites and the role of the adhesive in soil are needed. Besides its function in nematode capture, the adhesive coat might aid in the survival of the fungi [47] by protecting the surface from drying effects. Remarkable protective properties are known for the extracellular mucilage of fungal spores, such as those of *Colletotrichum* spp., suggesting a role of adhesive in fungal survival [117]. Furthermore, the influence of soil particles on the performance of the adhesive has not been investigated. Stotzky [118] reviewed the effects of clay minerals on microbes at different levels of complexity, from field observations to pure culture studies. More specifically, Lavie and Stotzky [119] observed a significant reduction of the respiration of the dimorphic yeast, *Histoplasma capsulatum*, in the presence of clay. It was concluded that the reduced respiration rate was caused, in part, by clay particles adhering to mycelial surfaces and, thereby, interfering with the movement of nutrients, metabolites, and gases across the mycelial wall. Research on the adhesive properties of nematophagous fungi at different levels of experimental complexity is, therefore, desirable, along with continued analysis of the chemical nature of the adhesives.

The mechanisms of interaction between nematodes and nematophagous fungi, and the signals involved in these interactions, are questions of high priority also from the point of view of general biology. The relevance of data from laboratory investigations should be tested under field conditions. The techniques of molecular

biology might be useful tools in understanding the behavior of these
organisms in the laboratory as well as in the field.

REFERENCES

1. Barron, G. L. 1977. The nematode-destroying fungi. Topics
    in mycobiology No. 1. Canadian Biological Publications Ltd.,
    Guelph, Ontario, Canada.
2. Barron, G. L. 1981. Predators and parasites of microscopic
    animals, p. 167–200. *In* G. T. Cole and B. Kendrick (eds.),
    Biology of conidial fungi, Vol. 2. Academic Press, New York.
3. Barron, G. L. 1982. Nematode-destroying fungi, p. 533–552.
    In R. G. Burns and J. H. Slater (eds.), Experimental micro-
    bial ecology. Blackwell Scientific Publications, Oxford.
4. Dowe, A. 1987. Räuberische Pilze. Die Neue Brehm-Bücherei.
    A. Ziemsen Verlag, Wittenburg Lutherstadt.
5. Gray, N. F. 1988. Fungi attacking vermiform nematodes, p.
    3–38. *In* G. O. Poinar, Jr. and H.-B. Jansson (eds.), Dis-
    eases of nematodes, Vol. 2. CRC Press, Boca Raton, Florida.
6. Kerry, B. R. 1980. Biocontrol: fungal parasites of female
    cyst nematodes. J. Nematol. 12:253–259.
7. Kerry, B. R. 1984. Nematophagous fungi and the regulation
    of nematode populations in soil. Helminthol. Abstr. Ser. B
    Plant Nematol. 53:1–14.
8. Jansson, H.-B. 1987. Receptors and recognition in nematodes,
    p. 153–158. *In* J. A. Veech and D. W. Dickson (eds.), Vis-
    tas on nematology. Society of Nematologists, Hyattsville.
9. Jansson, H.-B., and B. Nordbring-Hertz. 1988. Infection
    events in the fungus-nematode system, p. 59–72. *In* G. O.
    Poinar, Jr. and H.-B. Jansson (eds.), Diseases of nematodes,
    Vol. 2. CRC Press, Boca Raton, Florida.
10. Lysek, G., and B. Nordbring-Hertz. 1983. Die Biologie der
    Nematoden-fangender Pilze. Forum Mikrobiol. 6:201–208.
11. Mankau, R. 1980. Biological control of nematode pests by nat-
    ural enemies. Annu. Rev. Phytopathol. 18:415–440.
12. Mankau, R. 1981. Microbial control of nematodes, p. 475–494.
    In B. M. Zuckerman and R. A. Rohde (eds.), Plant parasitic
    nematodes, Vol. 3. Academic Press, New York.
13. Nordbring-Hertz, B. 1984. Mycelial development and lectin–
    carbohydrate interactions in nematode-trapping fungi, p. 419–
    432. *In* D. H. Jennings and A. D. M. Rayner (eds.), The
    ecology and physiology of the fungal mycelium. Cambridge
    University Press, Cambridge.
14. Nordbring-Hertz, B. 1988. Ecology and recognition in the
    nematode–nematophagous fungus system, p. 81–114. In

K. C. Marshall (ed.), Advances Microbial Ecology. Plenum Publishing, New York.

15. Nordbring-Hertz, B. 1988. Nematophagous fungi: strategies for nematode exploitation and for survival. Microbiol. Sci. 5:108–116.

16. Nordbring-Hertz, B., and I. Chet. 1986. Fungal lectins and agglutinins, p. 393–408. *In* D. Mirelman (ed.), Microbial lectins and agglutinins: properties and biological activity. John Wiley & Sons, New York.

17. Nordbring-Hertz, B., and H.-B. Jansson. 1984. Fungal development, predacity, and recognition of prey in nematode-destroying fungi, p. 327–333. *In* M. J. Klug and C. A. Reddy (eds.), Current perspectives in microbial ecology. Proc. Third Int. Symp. Microb. Ecol., August 1983. American Society for Microbiology, Washington, D.C.

18. Zuckerman, B. M., and H.-B. Jansson. 1984. Nematode chemotaxis and possible mechanism of host/prey recognition. Annu. Rev. Phytopathol. 22:95–113.

19. Nicholas, W. L. 1975. The biology of free-living nematodes. Clarendon Press, Oxford.

20. Jansson, H.-B., and G. O. Poinar, Jr. 1986. Possible fossil nematophagous fungi. Trans. Br. Mycol. Soc. 87:471–474.

21. Pramer, D., and S. Kuyama. 1963. Symposium on biochemical bases of morphogenesis in fungi: II. Nemin and the nematode-trapping fungi. Bacteriol. Rev. 27:282–292.

22. Nordbring-Hertz, B. 1973. Peptide-induced morphogenesis in the nematode-trapping fungus *Arthrobotrys oligospora*. Physiol. Plant. 29:223–233.

23. Nordbring-Hertz, B., and C. Brinck. 1974. Qualitative characterization of some peptides inducing morphogenesis in the nematode-trapping fungus *Arthrobotrys oligospora*. Physiol. Plant 31:59–63.

24. Friman, E., S. Olsson, and B. Nordbring-Hertz. 1985. Heavy trap formation by *Arthrobotrys oligospora* in liquid culture. FEMS Microbiol. Ecol. 31:17–21.

25. Nordbring-Hertz, B. 1977. Nematode-induced morphogenesis in the predacious fungus *Arthrobotrys oligospora*. Nematologica 23:443–451.

26. Nordbring-Hertz, B., and G. Odham. 1980. Determination of volatile nematode exudates and their effects on a nematode-trapping fungus. Microb. Ecol. 6:241–251.

27. Jansson, H.-B., and B. Nordbring-Hertz. 1981. Trap and conidiophore formation in *Arthrobotrys superba*. Trans. Br. Mycol. Soc. 77:205–207.

28. Barron, G. L. 1979. Nematophagous fungi: a new *Arthrobotrys* with nonseptate conidia. Can. J. Bot. 57:1371–1373.

29. Persson, Y., B. Nordbring-Hertz, and I. Chet. 1990. The
     effect of polyoxin D on morphogenesis of the nematode-trap-
     ping fungus *Arthrobotrys oligospora*. Mycol. Res. 94:196–200.
30. Jansson, H.-B., A. von Hofsten, and C. von Mecklenburg.
     1984. Life cycle of the endoparasitic nematophagous fungus
     Meria coniospora: a light and electron microscopic study.
     Antonie Leeuwenhoek J. Microbiol. 50:321–327.
31. Jansson, H.-B., and B. Nordbring-Hertz. 1983. The endo-
     parasitic fungus *Meria coniospora* infects nematodes specifically
     at the chemosensory organs. J. Gen. Microbiol. 129:1121–
     1126.
32. Jansson, H.-B., A. Jeyaprakash, and B. M. Zuckerman. 1985.
     Differential adhesion and infection of nematodes by the endo-
     parasitic fungus *Meria coniospora* (Deuteromycetes). Appl.
     Environ. Microbiol. 49:552–555.
33. Morgan-Jones, G., B. O. Gintis, and R. Rodriguez-Kabana.
     1981. Fungal colonisation of *Heterodera glycines* cysts in
     Arkansas, Florida, Mississippi and Missouri soils. Nema-
     tropica 11:155–164.
34. Rodriguez-Kabana, R., and G. Morgan-Jones. 1988. Poten-
     tial for nematode control by mycofloras endemic in the trop-
     ics. J. Nematol. 20:191–203.
35. Lysek, H., and D. Krajci. 1987. Penetration of ovicidal fun-
     gus *Verticillium chlamydosporium* through the *Ascaris lumbri-
     coides* egg-shells. Folia Parasitol. 34:57–60.
36. Dackman, C., I. Chet, and B. Nordbring-Hertz, 1989. Fun-
     gal parasitism of the cyst nematode *Heterodera schachtii*: in-
     fection and enzymatic activity. FEMS Microbiol. Ecol. 62:201–
     208.
37. Crump, D. H., and B. R. Kerry. 1981. A quantitative
     method for extracting resting spores of two nematode para-
     sitic fungi, *Nematophthora gynophila* and *Verticillium chla-
     mydosporium*, from soil. Nematologica 27:330–339.
38. Kerry, B. R., and D. H. Crump. 1980. Two fungi parasitic
     on females of cyst nematodes (*Heterodera* spp.). Trans. Br.
     Mycol. Soc. 74:119–125.
39. Dürschner, U. 1983. Pilzliche Endoparasiten an beweglichen
     Nematodenstadien. Mitt. Biol. Bundesanst. Land Forstwirtsch.
     217:1–83.
40. Dürschner-Pelz, U. U., and H. J. Atkinson. 1988. Recog-
     nition of *Ditylenchus* and other nematodes by spores of the
     endoparasitic fungus *Verticillium balanoides*. J. Invert.
     Pathol. 51:96–106.
41. Zachariah, K. 1981. Chemotropism by isolated ring traps of
     Dactylella dodycoides. Protoplasma 106:173–182.

42. Jansson, H.-B., and B. Nordbring-Hertz. 1979. Attraction of nematodes to living mycelium of nematophagous fungi. J. Gen. Microbiol. 112:89—93.

43. Jansson, H.-B. 1982. Predacity by nematophagous fungi and its relation to the attraction of nematodes. Microb. Ecol. 8: 233—240.

44. Jansson, H.-B. 1982. Attraction of nematodes to endoparasitic nematophagous fungi. Trans. Br. Mycol. Soc. 79:25—29.

45. Nordbring-Hertz, B., and M. Stålhammar-Carlemalm. 1978. Capture of nematodes by *Arthrobotrys oligospora*, an electron microscope study. Can. J. Bot. 56:1297—1307.

46. Veenhuis, M., B. Nordbring-Hertz, and W. Harder. 1985. An electron microscopical analysis of capture and initial stages of penetration of nematodes by *Arthrobotrys oligospora*. Antonie Leeuwenhoek J. Microbiol. 51:385—398.

47. Veenhuis, M., C. van Wijk, U. Wyss, B. Nordbring-Hertz, and W. Harder. 1989. Significance of electron dense microbodies in trap cells of the nematophagous fungus *Arthrobotrys oligospora*. Antonie Leeuwenhoek J. Microbiol. 56:251—261.

48. Veenhuis, M., W. Harder, and B. Nordbring-Hertz. 1989. Occurrence and metabolic significance of microbodies in trophic hyphae of the nematophagous fungus *Arthrobotrys oligospora*. Antonie Leeuwenhoek J. Microbiol. 56:241—249.

49. Lopez-Llorca, L. V., and G. H. Duncan. 1988. A study of fungal endoparasitism of the cereal cyst nematode (*Heterodera avenae*) by scanning electron microscopy. Can. J. Microbiol. 34:613—619.

50. Nordbring-Hertz, B., and B. Mattiasson. 1979. Action of a nematode-trapping fungus shows lectin-mediated host—microorganism interaction. Nature 281:477—479.

51. Nordbring-Hertz, B., E. Friman, and B. Mattiasson. 1982. A recognition mechanism in the adhesion of nematodes to nematode-trapping fungi, p. 83—90. *In* T. C. Bøg-Hansen (ed.), Lectins, biology, biochemistry and clinical biochemistry, Vol. 2. W. de Gruyter, Berlin.

52. Rosenzweig, W. D., and D. Ackroyd. 1983. Binding characteristics of lectins involved in the trapping of nematodes by fungi. Appl. Environ. Microbiol. 46:1093—1096.

53. Rosenzweig, W. D., D. Premachandran, and D. Pramer. 1985. Role of trap lectins in the specificity of nematode capture by fungi. Can. J. Microbiol. 31:693—695.

54. Jansson, H.-B., and B. Nordbring-Hertz. 1984. Involvement of sialic acid in nematode chemotaxis and infection by an endoparasitic nematophagous fungus. J. Gen. Microbiol. 130:39—43.

55. Mattiasson, B., P. A. Johansson, and B. Nordbring-Hertz. 1980. Host—microorganism interaction: studies on the molecular mechanisms behind the capture of nematodes by nematophagous fungi. Acta Chem. Scand. B34:539—540.

56. Borrebaeck, C. A. K., B. Mattiasson, and B. Nordbring-Hertz. 1984. Isolation and partial characterization of a carbohydrate-binding protein from a nematode-trapping fungus. J. Bacteriol. 159:53—56.

57. Borrebaeck, C. A. K., B. Mattiasson, and B. Nordbring-Hertz. 1985. A fungal lectin and its apparent receptors on a nematode surface. FEMS Microbiol. Lett. 27:35—39.

58. Premachandran, D., and D. Pramer. 1984. Role of *N*-acetylgalactosamine-specific protein in trapping of nematodes by *Arthrobotrys oligospora*. Appl. Environ. Microbiol. 47:1358—1359.

59. Nordbring-Hertz, B., U. Zunke, U. Wyss, and M. Veenhuis. 1986. Trap formation and capture of nematodes by *Arthrobotrys oligospora*. Film, C 1622, Inst. Wiss. Film, Göttingen, Germany.

60. Nordbring-Hertz, B., M. Veenhuis, and W. Harder. 1984. Dialysis membrane technique for ultrastructural studies of microbial interactions. Appl. Environ. Microbiol. 47:195—197.

61. Veenhuis, M., B. Nordbring-Hertz, and W. Harder. 1985. Development and fate of electron-dense microbodies in trap cells of the nematophagous fungus *Arthrobotrys oligospora*. Antonie Leeuwenhoek J. Microbiol. 51:399—407.

62. Dowsett, J. A., and J. Reid. 1979. Observations on the trapping of nematodes by *Dactylaria scaphoides* using optical, transmission and scanning-electron-microscopic techniques. Mycologia 71:379—391.

63. Schenck, S., T. Chase, Jr., W. D. Rosenzweig, and D. Pramer. 1980. Collagenase production by nematode-trapping fungi. Appl. Environ. Microbiol. 40:567—570.

64. Rosenzweig, W. D., and D. Pramer. 1980. Influence of cadmium, zinc and lead on growth, trap formation, and collagenase activity of nematode-trapping fungi. Appl. Environ. Microbiol. 40:694—696.

65. Bird, A. F., and M. A. McClure. 1976. The tylenchid (Nematoda) egg shell: structure, composition and permeability. Parasitology 72:19—28.

66. Wyborn, C. H. E., D. Priest, and C. L. Duddington, 1969. Selective techniques for determination of nematophagous fungi in soil. Soil Biol. Biochem. 1:101—102.

67. Eren, J., and D. Pramer. 1965. The most probable number of nematode-trapping fungi in soil. Science 99:285.

68. Klemmer, H. W., and R. Y. Nakano. 1964. A semiquantitative method of counting nematode-trapping fungi in soil. Nature 203:1085.

69. Mankau, R. 1975. A semiquantitative method for enumerating and observing parasites and predators of soil nematodes. J. Nematol. 7:119–122.

70. Dackman, C., S. Olsson, H.-B. Jansson, B. Lundgren, and B. Nordbring-Hertz. 1987. Quantification of predatory and endoparasitic fungi in soil. Microb. Ecol. 13:89–93.

71. La Mondia, J. A., and B. B. Brodie. 1984. An observation chamber technique for evaluating potential biocontrol agents of *Globodera rostochiensis*. J. Nematol. 16:112–115.

72. Crump, D. H. 1987. A method for assessing the natural control of cyst nematode populations. Nematologica 33:232–243.

73. Seinhorst, J. W. 1964. Methods for extraction of *Heterodera* cysts from not previously dried soil samples. Nematologica 10:87–94.

74. Kerry, B. R., and D. H. Crump. 1977. Observations on fungal parasites of females and eggs of the cereal cyst-nematode, *Heterodera avenae*, and other cyst nematodes. Nematologica 23:193–201.

75. Gray, N. F. 1985. Ecology of nematophagous fungi: effect of soil moisture, organic matter, pH and nematode density on distribution. Soil Biol. Biochem. 17:499–507.

76. Gray, N. F. 1985. Nematophagous fungi from the maritime antarctic: factors affecting distribution. Mycopathologia 90:165–176.

77. Kerry, B. R., D. H. Crump, and L. A. Mullen. 1980. Parasitic fungi, soil moisture and multiplication of the cereal cyst nematode, *Heterodera avenae*. Nematologica 26:57–68.

78. Gray, N. F., and F. Bailey. 1985. Ecology of nematophagous fungi: vertical distribution in a decidious woodland. Plant Soil 86:217–223.

79. Mankau, R., and M. V. McKenry. 1976. Spatial distribution of nematophagous fungi associated with *Meloidogyne incognita* on peach. J. Nematol. 8:294–295.

80. Cayrol, J. C. 1983. Biological control of *Meloidogyne* by *Arthrobotrys irregularis*. Rev. Nematol. 6:265–274.

81. Crump, D. H., and B. R. Kerry. 1986. Effects of some agrochemicals on the growth of two nematophagous fungi, *Verticillium chlamydosporium* and *Cylindrocarpon destructans*. Nematologica 32:363–366.

82. Kapur, S., W. Belfield, and B. H. S. Gibson. 1981. The effect of fungicides on soil fungi with special reference to nematophagous species. Pedobiologia 21:172–181.

83. Heintz, C. E. 1978. Assessing the predacity of nematode-trapping fungi in vitro. Mycologia 70:1086—1100.

84. Cooke, R. C. 1962. The ecology of nematode-trapping fungi in the soil. Ann. Appl. Biol. 50:507—513.

85. Hayes, W. A., and F. Blackburn. 1966. Studies on nutrition of *Arthrobotrys oligospora* Fres. and *A. robusta* Dudd. II. The predaceous phase. Ann. Appl. Biol. 58:51—60.

86. Jaffee, B. A., J. T. Gaspard, H. Ferris, and A. E. Muldoon. 1988. Quantification of parasitism of the soil-borne nematode *Criconemella xenoplax* by the nematophagous fungus *Hirsutella rhossiliensis*. Soil. Biol. Biochem. 20:631—636.

87. Cooke, R. C. 1963. Succession of nematophagous fungi during the decomposition of organic matter in the soil. Nature 197:205.

88. Grønvold, J., P. Nansen, S. A. Henriksen, J. Thylin, and J. Wolstrup. 1988. The capability of the predacious fungus *Arthrobotrys oligospora* (Hyphomycetales) to reduce numbers of infective larvae of *Ostertagia ostertagi* (Trichostrongylidae) in cow pats and herbage during the grazing season in Denmark. J. Helminthol. 62:271—280.

89. Katznelson, H., and V. E. Henderson. 1962. Studies on the relationships between nematodes and other soil microorganisms. I. The influence of actinomycetes and fungi on *Rhabditis* (*Cephaloboides*) *oxycera* de Man. Can. J. Microbiol. 8:875—882.

90. Peterson, E. A., and H. Katznelson. 1965. Studies on the relationships between nematodes and other soil microorganisms. IV. Incidence of nematode-trapping fungi in the vicinity of plant roots. Can. J. Microbiol. 11:491—495.

91. Garrett, S. D. 1963. Soil fungi and soil fertility. Pergamon Press. Oxford.

92. Kerry, B. R., A. Simon, and A. D. Rovira. 1984. Observations on the introduction of *Verticillium chlamydosporium* and other parasitic fungi into soil for control of the cereal cyst-nematode *Heterodera avenae*. Ann. Appl. Biol. 105:509—516.

93. Gaspard, J. T., and R. Mankau. 1986. Nematophagous fungi associated with *Tylenchus semipenetrans* and the citrus rhizosphere. Nematologica 32:359—363.

94. Grønvold, J. 1989. Induction of nematode-trapping organs in the predacious fungus *Arthrobotrys oligospora* (Hyphomycetales) by infective larvae of *Ostertagia ostertagi* (Trichostrongylidae). Acta Vet. Scand. 30:1—11.

95. Mankau, R., and E. W. Bartnicki. 1987. Studies on the biology of nematode-trapping fungi associated with the citrus

nematode *Tylenchulus semipenetrans*. J. Nematol. 19:540. Abstract.

96. Tzean, S. S., and R. H. Estey. 1978. Nematode-trapping fungi as mycopathogens. Phytopathology 68:1266–1270.

97. Persson, Y., M. Veenhuis, and B. Nordbring-Hertz. 1985. Morphogenesis and significance of hyphal coiling by nematode-trapping fungi in mycoparasitic relationships. FEMS Microbiol. Ecol. 31:283–291.

98. Thorn, R. G., and G. L. Barron. 1984. Carnivorous mushrooms. Science 224:76–78.

99. Barron, G. L., and R. G. Thorn. 1987. Destruction of nematodes by species of *Pleurotus*. Can. J. Bot. 65:774–778.

100. Small, R. W. 1988. Invertebrate predators, p. 73–92. *In* G. O. Poinar, Jr. and H.-B. Jansson (eds.), Diseases of nematodes, Vol. 2. CRC Press, Boca Raton, Florida.

101. Stirling, G. R. 1988. Biological control of plant-parasitic nematodes, p. 93–139. *In* G. O. Poinar, Jr. and H.-B. Jansson (eds.), Diseases of nematodes, Vol. 2. CRC Press, Boca Raton, Florida.

102. Stirling, G. R., and R. Mankau. 1978. Parasitism of *Meloidogyne* eggs by a new fungal parasite. J. Nematol. 10:236–240.

103. Jaffee, B. A., and E. I. Zehr. 1982. Parasitism of the nematode *Criconemella xenoplax* by the fungus *Hirsutella rhossiliensis*. Phytopathology 72:1378–1381.

104. Jaffee, B. A., J. T. Gaspard, and H. Ferris. 1989. Density-dependent parasitism of the soil-borne nematode *Criconemella xenoplax* by the nematophagous fungus *Hirsutella rhossiliensis*. Microb. Ecol. 17:193–200.

105. Linford, M. B., and F. Yap. 1939. Root-knot nematode injury restricted by a fungus. Phytopathology 29:596–609.

106. Cayrol, J.-C., J.-P. Frankowski, A. Laniece, G. d'Hardemare, and J.-P. Talon. 1978. Contre les nématodes en champignonniere. Mise au point d'un méthode de lutte biologique à l'aide d'un Hyphomycète prédateur: *Arthrobotrys robusta* Soucha "Antipolis" (Royal 300). Rev. Hortic. 193:23–30.

107. Cayrol, J.-C., and J.-P. Frankowski. 1979. Une méthode de lutte biologique contre les nématode à galles des racines appartenant au genre méloidogyne. Rev. Hortic. 193:15–23.

108. Cooke, R. C. 1968. Relationships between nematode-destroying fungi and soil-borne phytonematodes. Phytopathology 58:909–913.

109. Eayre, C. G., B. A. Jaffee, and E. I. Zehr. 1987. Suppression of *Criconemella xenoplax* by the nematophagous fungus *Hirsutella rhossiliensis*. Plant Dis. 71:832–834.

110. Jaffee, B. A., and E. I. Zehr. 1985. Parasitic and sapro-
     phytic abilities of the nematode-attacking fungus *Hirsutella
     rhossiliensis*. J. Nematol. 17:341–345.
111. Jansson, H.-B., A. Jeyaprakash, and B. M. Zuckerman.
     1985. Control of root-knot nematodes on tomato by the en-
     doparasitic fungus *Meria coniospora*. J. Nematol. 17:327–329.
112. Sayre, R. M., and M. P. Starr. 1988. Bacterial diseases
     and antagonisms of nematodes, p. 69–101. *In* G. O. Poinar,
     Jr. and H.-B. Jansson (eds.), Diseases of nematodes, Vol. 1.
     CRC Press, Boca Raton, Florida.
113. Gruner, L., M. Peloille, C. Sauvé, and J. Cortet. 1985.
     Parasitologie animale. — Survie et conservation de l'activité
     prédatrice vis-à-vis de nématodes trichostrongylides après
     ingestion par des ovins de trois hyphomycètes prédateurs.
     C. R. Acad. Sci. Paris 300:525–528.
114. Grønvold, J., J. Wolstrup, S. A. Henriksen, and P. Nansen.
     1987. Field experiments on the ability of *Arthrobotrys oligo-
     spora* (Hyphomycetales) to reduce the number of larvae of
     Cooperia onchophora (Trichostrongylidae) in cow pats and
     surrounding grass. J. Helminthol. 61:65–71.
115. Linford, M. B., F. Yap, and J. M. Oliveira. 1938. Reduc-
     tion of soil populations of the root-knot nematode during de-
     composition of organic matter. Soil Sci. 45:127–141.
116. Rodriguez-Kabana, R., G. Morgan-Jones, and I. Chet. 1987.
     Biological control of nematodes: soil amendments and microbial
     antagonists. Plant Soil 100:237–247.
117. Nicholson, R. L., and L. Epstein. 1989. Adhesion of fungi
     to the plant surface: prerequisite for pathogenesis. *In* G. T.
     Cole and H. C. Hoch (eds.), The fungal spore and disease
     initiation in plants and animals. Plenum Press, New York.
118. Stotzky, G. 1980. Surface interactions between clay minerals
     and microbes, viruses, and soluble organics, and the prob-
     able importance of these interactions to the ecology of mi-
     crobes in soil, p. 231–247. *In* R. C. Berkeley, J. M. Lynch,
     J. Melling, P. R. Rutter, and B. Vincent (eds.), Microbial
     adhesion to surfaces. Ellis Horwood, Chichester, England.
119. Lavie, S., and G. Stotzky. 1986. Adhesion of the clay min-
     erals, montmorillonite, kaolinite, and attapulgite, reduces res-
     piration of *Histoplasma capsulatum*. Appl. Environ. Microbiol.
     51:65–73.

4

Applications of Molecular Techniques
to Soil Biochemistry

GARY S. SAYLER, KAVE NIKBAKHT, JAMES T. FLEMING, and
JANET PACKARD   *University of Tennessee, Knoxville,*
Tennessee

I. INTRODUCTION

During the past two decades, the extensive research efforts in biotechnology, especially in the area of molecular biology, have resulted in important discoveries, some of which carry the potential
for valuable contributions to the health and welfare of man. Techniques of genetic engineering are now being applied to the development of drugs, improvements in the quality and quantity of agricultural products, pollution control, and basic research in biology and
medicine.

The recent incorporation of molecular biological approaches into
studies in microbial ecology promises a new level of information concerning the structure and function of microbial communities as they
mediate degradation of organic compounds, cycle nutrients and trace
elements, and interact with other plant and animal communities.
Traditional methods of environmental analysis, which are mainly
based on cultural, physiological, and biochemical techniques, are
readily complemented by qualitatively and quantitatively specific
molecular techniques. For instance, nucleic acid hybridization
technology can be used to identify and quantitate individual bacterial species within a natural microbial community. Such enumeration can be achieved with specificity and speed to aid in the
isolation and characterization of desirable microorganisms.

The focus of this chapter is to provide an overview of the use
of molecular techniques in the analysis of the structure and activity of microbial community in soils. A major emphasis is placed on

analytical approaches to nucleic acid analysis in microbial communities. Although the use of genetically engineered microbes (GEMs) or recombinant DNA (rDNA) continues as an important consideration in environmental risk assessment, this topic is secondary and is considered as inclusive in the overall applications of molecular techniques to the analysis of the soil microbiota.

The principal starting material or target for many of the molecular analytic techniques are the nucleic acids of the microbial cell, both DNA and RNA. The comparative structure of these bipolymers is shown in Figures 1 and 2. Chromosomal or plasmid DNA from the microbial cell exists as a duplex, double-stranded, molecule, often exceeding 3×10^6 base pairs (bp) in length, which is ordered (folded or coiled) in its native state (see Fig. 2).

RNA exists in three predominant forms. Messenger RNA (mRNA) is enzymatically transcribed from the DNA template by RNA polymerase as a single-stranded polymer (see Fig. 1) up to several thousand nucleotides long. Ribosomal RNA (rRNA) is transcribed in sizes ranging from a few hundred to approximately 2000 nucleotides and is a primary structural component of the ribosome. Transfer RNAs (tRNA), 70 to 90 nucleotides in length, are amino acid carriers for protein synthesis at the ribosome. Both tRNA and rRNA are folded into three-dimensional structures. Of the RNA species, the mRNAs have a short half-life (generally less than a few minutes) and are rapidly degraded by ribonucleases (RNases).

Ribosomal RNA and mRNA are, along with DNA, primary targets for nucleic acid hybridization. In general, DNA is more stable under environmental conditions and somewhat less suceptible to enzymatic degradation than is RNA. Chromosomal DNA, because of its size, is subject to much shearing during isolation procedures. However, DNA is stable in alkali, although it denatures to the single-stranded form, whereas RNA degrades substantially above pH 8.5. The isolation and applications of these nucleic acid species to microbial analysis are described in detail in the following sections.

II. NUCLEIC ACID EXTRACTION FROM SOIL AND SEDIMENT

A. DNA Extraction

Total DNA extraction from soil or sediment and its subsequent analysis using specific gene probes is becoming an accepted method in structural and functional characterization of natural microbial communities. The genetic composition of bacteria is relatively uniform [1,2]; therefore, the amount of extracted DNA from an environmental sample should yield reliable information on microbial biomass. Inasmuch as numerous bacterial species present in soil or sediment

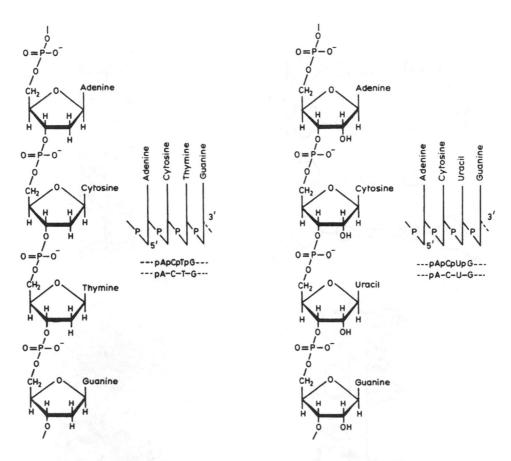

Figure 1 Comparative polymer structure of DNA and RNA: (left, single strand of DNA; right, RNA). The molecules are composed of a sugar phosphate backbone and four nucleic acid bases that hydrogen bond to form duplex, hybrid, or folded structures. From Ref. 136.

are not readily culturable (perhaps fewer than 10% at any time of sampling), DNA extraction of an environmental sample yields accurate information about a microbial community. Once extracted, the DNA may be targeted with specific gene probes to measure the abundance of an individual population or specific functional genes.

DNA extraction from soil or sediment requires that a larger amount of sample be processed when compared with the isolation of DNA from pure bacterial cultures. The biomass in sediments

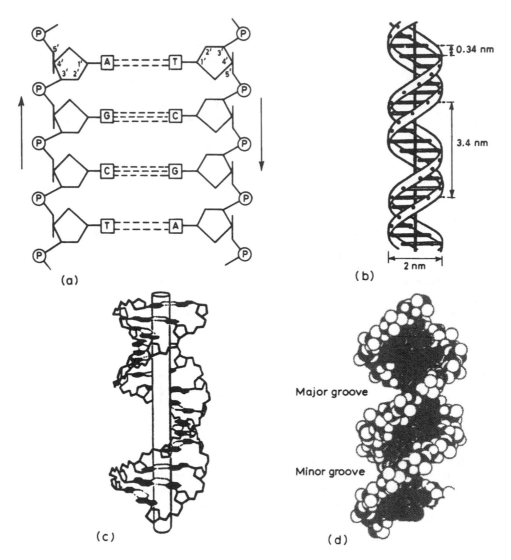

Figure 2   Comparative molecular representation of DNA (a, base pairing; b, helical twisting; c, helical twisting with base pairing; d, a space-filled model).   From Ref. 136.

and soils, however, is generally sufficient for nucleic acid extraction to be achieved without major difficulties.   There are currently two approaches to DNA extraction: the cell separation and the direct lysis methods. These methods are described in Figure 3 and are discussed in detail in the following.

Cell Separation Method

A procedure to separate bacterial cells from soil was first described by Faegri et al. [3]. With three natural soils, they showed that by

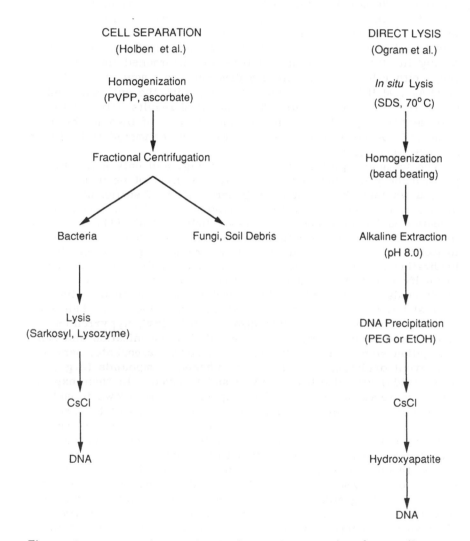

Figure 3 Comparative protocols for DNA extraction from soils.

homogenization of soil in Winogradsky's salt solution (1:20) in a
Waring blender and subsequent fractional centrifugation of the ho-
mogenate, two fractions could be obtained. One fraction contained 50
to 80% of the bacteria; the other contained the fungi, together with
soil debris and the rest of the bacterial population. To isolate DNA
from the bacterial fraction, Torsvik and Goksoyr [4] modified the
procedure by preextracting the bacterial suspension before DNA
isolation. In their method, acid-washed diatomaceous earth was
layered on 20-mm diameter Sartorius filters (0.45 µm) or Whatman
glass fiber filters. The bacterial suspension was filtered, washed
with 0.1 M sodium pyrophosphate, pH 7.0, and then extracted with
cold 5% trichloroacetic acid, followed by ethanol:ether (3:1). After
drying the filter, the amount of DNA was determined fluorometri-
cally, either by reaction with 3,5-diaminobenzoic acid, 2 M HCl
(DABA, 2 M HCl) [5–7] or after coupling with the antibiotic, plica-
mycin (mithramycin) [8–10]. The analytical results obtained by
both methods agreed well, and the total amount of DNA in the organic soil sample used in this experiment was estimated to be 90 µg
g^{-1} dry weight soil.

The foregoing protocol is time-consuming, and only a few samples can be processed each week. Also, it has not been demonstrated whether this method can generate DNA of sufficiently high
quality to be used in experiments involving restriction endonuclease
digestion and gene probe techniques. Holben et al. [11] have modified the fractionate centrifugation technique of Faegri et al. [3],
to isolate DNA from soil suitable for restriction digestion and hybridization reactions. A major consideration in their method was
to purify the DNA by removing soil contaminants, in particular,
humic acids, which render DNA refractory to complete digestion
by restriction endonucleases and hybridization analyses. This was
partially achieved by including insoluble polyvinylpolypyrrolidone
(PVPP) and sodium ascorbate in the initial homogenization and centrifugation steps. The reducing agent, sodium ascorbate, serves
to prevent oxidation of phenols, and phenolic compounds (e.g.,
humic acids) are adsorbed on PVPP and removed. In their experiments, cultures of *Bradyrhizobium japonicum* were grown to midlog
phase (optical density of 0.5), appropriate dilutions of the bacterial
suspensions were mixed with soil samples, and DNA was extracted
from the inoculated soil. For efficient DNA isolation, bacterial cells
collected at the end of the fractional centrifugation steps were disrupted through a comprehensive lytic procedure that included Sar-
kosyl and lysozyme. By using CsCl–ethidum bromide (EtBr) equilibrium density gradient centrifugation, bacterial DNA was purified
from the cell lysate and remaining soil contaminants. Ultraviolet
(UV) spectral analysis of the extracted DNA indicated that it was

relatively pure, and agarose gel electrophoresis showed no detectable contaminating nuclease activity. Furthermore, when DNAs isolated from soil and from a pure culture of the bacterium were incubated separately with the restriction endonuclease, *Eco*RI, both appeared to be completely digested. The soil DNA was successfully analyzed by slot blot and Southern blot hybridization and, by using specific probes, particular microbial species within a natural soil community were detected and quantitated. In general, the addition of PVPP and sodium ascorbate to the homogenization medium resulted in a twofold increase in DNA recovery and in significant improvement in DNA quality.

Although separation of bacteria from soil, and subsequent nucleic acid extraction of isolated cells, may yield a fairly pure DNA product, it is probable that this procedure may selectively separate bacteria that are easily dislodged, leaving behind those that are more strongly attached to the organic—mineral matrix of the soil. After three rounds of homogenization and low-speed centrifugation of a soil sample, the efficiency of recovery of the bacterial fraction was estimated to be roughly 34% [11]. This method, therefore, can result not only in a bias in the composition of the extracted DNA, but also may result in a low yield of the purified sample.

Direct Lysis Method

To circumvent the foregoing problems, Ogram et al. [12] developed a direct lysis method that enables isolation of bacterial DNA from sediment without prior cell separation. In this procedure, cells are first lysed by incubation of the sediment with sodium dodecyl sulfate (SDS) (or lysozyme—SDS) at 70°C and homogenization (5 min) in a Bead-Beater (Bio-Spec Products) with a 1:1 ratio of sediment to glass beads. As determined by acridine orange direct counts [13], a lysis efficiency of greater than 90% was obtained [12]. Following lysis, sequential alkaline extractions with 0.12 M sodium phosphate buffer (pH 8) were performed to recover adsorbed DNA. The combination of heat and SDS liberated a significant amount of humic materials, imparting a brown color to the extracts. The number of alkaline extractions necessary was dependent on the clay mineralogy and organic content of the sample; the amount of DNA adsorbed increased as the clay content increased [14].

A novel feature of this protocol is that it enables distinguishing between extracellular and intracellular DNA. Before cell lysis, sequential alkaline extraction of the soil or sediment results in isolation of extracellular DNA [14]; the DNA extracted from the soil or sediment after cell lysis is presumed to be completely intracellular. To concentrate DNA from the alkaline extracts, two methods were examined: precipitation by ethanol (EtOH) and by polyethylene glycol (PEG). During EtOH precipitation, a large quantity of humic materials was also precipitated along with DNA. To avoid this, the extract was brought to 0.5 M potassium acetate, stored on ice for 2 h, and centrifuged before EtOH addition; most of the humic materials were precipitated with negligible loss of DNA. Precipitation with PEG has the advantage that the sample volume remains small and manageable, in contrast with EtOH precipitation, which requires a large amount of alcohol before DNA precipitation and, subsequently, the manipulation of large volumes during centrifugation. Furthermore, phenol extraction of the resuspended PEG precipitate resulted in a substantial removal of contaminating humic materials without affecting the recovery of DNA. The purity of the concentrated DNA sample so obtained can be improved appreciably if CsCl—EtBr gradient centrifugation or hydroxyapatite column chromatography purification is employed. These steps are especially necessary if the DNA is to be purified from soil or sediments with high organic carbon contents; however, it was found that the CsCl—EtBr gradient centrifugation step could be omitted when the sample contained only a small amount of organic carbon. Depending on the soil or sediment sample, the amount of intracellular DNA recovered ranged from 8 to 26 μg g^{-1} of wet soil or sediment. Although some shearing of DNA occurred as a result of bead-beating, the recovered DNA was still of high quality, as judged by UV-visible spectroscopy, and was suitable for hybridization probing and nick-translation.

Comparative Effectiveness of Methods in DNA Recovery

Recently, Steffan et al. have compared and evaluated the effectiveness of cell separation and direct lysis methods for recovering DNA from bacterial communities in soils and sediments [15]. They have also examined modifications of individual steps within each protocol. In the cell separation method, it was found that an acidic 0.1 M phosphate buffer (pH 4.5) with SDS gave higher recovery of bacteria from soils and sediments than Winogradsky salt solution or other phosphate buffers that have been used for extraction from soils and sediments. Three repetitive extractions with such a buffer system containing PVPP resulted in the recovery of 32.2 and 27.8% of the total bacterial cells, as determined by the acridine orange direct count method, from sediment and soil, respectively. The efficiency of recovery was similar to that reported by Holben et al. [11], who were able to separate 34.9% of the bacterial cells from soil after three extractions with Winogradsky solution plus sodium ascorbate and PVPP, and by Bakken [16], who used four to seven extraction steps followed by density gradient centrifugation of a soil sample with high clay content. The use of only three extraction steps and removal of humic materials by PVPP, as performed by Steffan et al. [15], should, therefore, represent a compromise between a reasonable effort to obtain the greatest yield of DNA and the necessity to ensure that the purified DNA is free from contaminating material. It should be noted that in the cell separation method, a large proportion of bacterial cells usually remains attached to clay particles, even after extensive extraction [15–17]. According to Steffan et al. [15], a 50% recovery of bacterial cells was achieved when centrifugation speeds were lowered, although more clay was associated with the cells in the supernatant. Low-speed centrifugation may result in a higher recovery of microorganisms from soil or sediment and a greater yield of DNA, but the additional impurities incurred with low centrifugation speed may offset the actual increase in DNA recovery. The presence of clay (kaolinite or montmorillonite) does not appear to affect the lysis of bacteria by lysozyme and SDS. The lysis efficiency, however, may be reduced when the soil contains a higher proportion of polysaccharide-producing bacteria, in which case a more exhaustive lytic procedure should be used [13,15].

Problems associated with the direct lysis method [12] are as follows: (1) DNA shearing, as a result of physical disruption of the cells by bead-beating and (2) large amounts of humic material extracted with DNA. Although phenol extraction of the PEG–DNA pellet and subsequent purification steps through CsCl–EtBr density gradient centrifugation and hydroxyapatite column chromatography should rid the DNA of substantial amounts of humic material, a trace amount of this humic material seems to persist, as evidenced by the presence of a 230-nm absorbing component in the final purified DNA product. Steffan et al. [15] examined the applicability of PVPP treatment [11] to the direct lysis method [12] and found that DNA recovered from soil with PVPP treatment had A260/280 and A260/230 ratios of 1.5 and 2.1, respectively, indicating a highly purified DNA product. In the original work of Ogram et al. [12], DNA recovered from sediment typically had an A260/280 ratio of 1.8. Relative to the amount of DNA recovered, milligram quantities were obtained from 100-g samples of both soils and sediments by the direct lysis method, which was a tenfold improvement in yield over the cell extraction method.

The two major environmental variables affecting nucleic acid recovery are clay mineralogy and the organic content of soils [14]. The organic content of soils appears to interfere primarily through the copurification of humic material associated with the soil matrix. Humic acid is a polar, high-molecular-weight, aromatic polymer sharing physicochemical properties with nucleic acids. Humic substancs are alkali-stable and are often extracted in DNA preparation. Their UV absorption spectra also tends to mask that of DNA, thereby making quantitation of nucleic acids difficult. Although moderate contamination of DNA with humic substances does not exclude hybridization with probes, more extensive analysis and enzymatic digestion of DNA contaminated with humic material is usually impossible.

The efficiency of DNA recovery from soils and sediments is dependent upon the extent to which the DNA is adsorbed to the soil mineral and organic components [14]. As described in Figure 4, nucleic acids may adsorb directly on mineral surfaces or be adsorbed as a DNA—humic acid complex. Ogram et al. calculated sheared and unsheared herring sperm DNA adsorption isotherms for a variety of soils and sediments, with varying clay and organic matter contents [14]. They found that DNA adsorption was affected by the length of the DNA polymer and by the mineralogy, ionic strength, and pH of the soil matrix. In general, there was a direct correlation between the amount of DNA sorbed and the montmorillonite content of the soil or sediment at low to neutral pH values. As the pH was increased by the addition of NaOH, less DNA was adsorbed. This finding suggests that DNA is adsorbed to soil or sediment in the unchanged form at pH values near or below the pK_a of DNA (pH_a = ~5). Above pH 7.0 DNA is in the ionic form and is not adsorbed [14].

B. RNA Extraction

For RNA isolation techniques, the greatest effort has been applied to the isolation of rRNA as a potential target for phylogenetic group- or population-specific probes. There has been relatively little work on the isolation of mRNA, because of its short half-life and limited stability.

The direct isolation of RNA from natural microbial populations to characterize rRNAs as a means for the determination of microbial diversity has been examined by Stahl et al. [18,19]. The intent of direct rRNA isolation is the same as that for direct DNA isolation; that is, to avoid bias sampling errors associated with the culturing of environmental isolates. Once isolated, rRNA may be used for hybridization studies that are analogous to DNA hybridization analysis. The rRNA target, however, offers an increase in sensitivity by several orders of magnitude over DNA sequences, as a result of the great number of ribosomes in the cell (up to thousands per cell). Moreover, the occurrence in the rRNA of conserved sequences of disparate bacteria permits the determination of phylogenetic relationships in a population on the basis of rRNA sequencing, by using synthetic probes to the conserved or variable regions of the rRNA molecule.

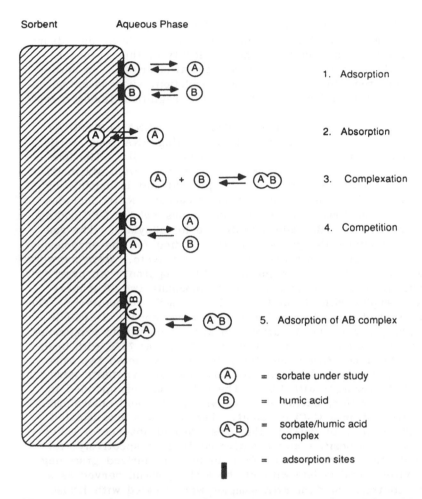

Figure 4 Conceptual model for the interaction of nucleic acids and humic acids with adsorption sites on clay minerals. (A, sorbate can be either DNA, RNA, or ssDNA.) Model adapted from Ref. 138.

Recently, Sayler et al. [20] tested the feasibility of directly obtaining mRNA from microorganisms in soils and sediments. The intent was to relate the quantitative abundance of genes, as determined by DNA—DNA hybridization, to the instantaneous expression of gene transcripts. In particular, this work has applied direct mRNA isolation to the study of bacterial catabolic genes, to relate the catabolic genetic potential of a population to the biodegradative activity. A major problem with this procedure has been the adsorption of RNA by humic acids, as the sensitivity of this method is limited by the absolute number of targeted transcripts. In addition, the inherent difficulties with RNA isolation, such as extracellular and intracellular ribonucleases, are magnified during direct soil extraction.

The feasibility of obtaining total RNA (both rRNA and mRNA) from soils and sediments was tested by isolating RNA from sterile soil inoculated with a known number of bacterial cells. The bacterium used in this experiment was *Pseudomonas putida* strain G7 (PpG7) which hosts a NAH7 catabolic plasmid and is capable of utilizing naphthalene as the sole source of carbon. NAH7 includes two operons, the first of which includes genes *nahABCDEF* coding for the enzymes that degrade naphthalene to salicylate. The second operon includes the genes *nahGHIJK* coding for the enzymes that oxidize salicylate to acetaldehyde and pyruvate. Both naphthalene and salicylate induce the *nahABCDEF* operon. An overnight culture of PpG7 cells was used to inoculate two 10 g sterilized soil samples with 10^8 and 10^9 cells g^{-1} soil, respectively. The soil was washed three times with diethylpyrocarbonate (DEPC) in sodium phosphate (NaPO$_4$) buffer (pH 8.0) to inactivate exogenous ribonucleases, followed by in situ lysis with SDS at 70°C with bead-beating. After centrifugation, the pellet was extracted two more times with the same buffer, and the combined supernatants were precipitated overnight with PEG. The pellet was resuspended in DEPC-treated buffer, extracted twice with phenol and chloroform, and centrifuged on a CsCl gradient. The RNA was also isolated from pure cultures of PpG7 in the presence and absence of salicylate, to serve as positive and negative controls, respectively, for mRNA induction. RNA isolated from an uncharacterized gram-negative environmental isolate without the NAH7 plasmid served as a negative control. Several RNA samples were treated with RNase-free DNase to eliminate any DNA—DNA hybridization artifacts. The RNA samples were electrophoresed in duplicate in 1.2% agarose formaldehyde gels and transferred to nylon membranes for Northern blot analysis. Blots were probed with nick-translated ^{32}P-labeled *NahA-D* in pBR322, washed, and placed on X-ray film for autoradiography.

The RNA obtained from the inoculated soil samples was electrophoresed on a 1.2% agarose denaturing gel (Fig. 5), in duplicate, along with the RNA isolated from induced (lanes 2,6) and uninduced (lanes 1,3) PpG7 cultures. When one-half of the gel was stained with EtBr, all preparations displayed distinct 16S and 23S ribosomal subunits, although the subunits from the soil sample inoculated with 10^8 cells g^{-1} soil was barely visible. There was little, if any, degradation of the ribosomal subunits from the inoculated

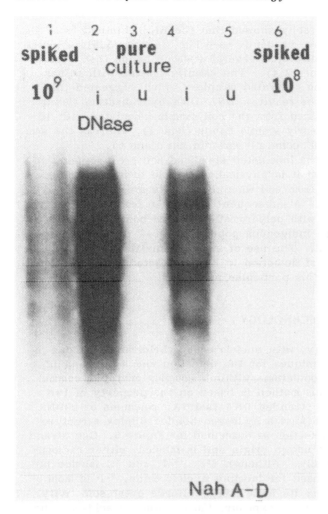

Figure 5 Northern blot analysis of total RNA isolated from sterile soils amended with *P. putida* G7 cells hybridized with the nick-translated. ^{32}P-labeled *nah A-D* genes from the NAH7 plasmid. Lanes 1,6: total RNA from 10 g soil inoculated with 10^9 and 10^8 cells per gram of soil, respectively. Lanes 2,4: 10 μg of salicylate induced PpG7 RNA with and without DNase treatment, respectively. Lanes 3,5: 10 μg of uninduced PpG7 RNA with and without DNase treatment, respectively.

samples. Autoradiography showed that the RNA obtained from the uninduced culture gave no bands (see Fig. 5, lanes 3,5), whereas the induced culture displayed several smeared bands that were indicative of mRNA (lanes 2,4). The samples treated with DNase looked identical to the untreated samples, which suggested that the bands were not the result of DNA—DNA hybridization (lanes 2,5). The RNA obtained from the soil sample inoculated with 10^9 cells g^{-1} soil gave clearly visible bands (lane 1), whereas the sample inoculated with 10^8 cells g^{-1} soil did not (lane 6).

The results with the inoculated sterilized soil are promising in that they suggest that it is physically possible to isolate undegraded NAH7 mRNA from soil when PpG7 cells are present in a high enough number. A subsequent attempt to isolate mRNA from nonsoil contaminated with polyaromatic hydrocarbons (including naphthalene) with an indigenous population of 10^5 to 10^6 cells g^{-1} soil was unsuccessful. The use of a nick-translated DNA probe here limits the level of detection to RNA extracted from >10^8 cells g^{-1} sterile soil with this particular protocol.

III. GENE PROBE TECHNOLOGY

Gene probe technology, with nucleic acid hybridization, is one of the most reliable techniques for the detection and monitoring of genetically defined populations within a complex microbial community [11,21—26]. This method is based on the property of two complementary single-stranded DNA (ssDNA) molecules or ssDNA and RNA molecules to form a hydrogen-bonded duplex structure and is the basis of detection as described for Figure 5. One strand serves as a probe of known origin and is labeled, either radioactively or nonisotopically. Although ^{32}P, ^{125}I, and ^{3}H-labeled nucleotides have been used for labeling nucleic acids, ^{32}P is most commonly used because its high energy affords great sensitivity and short autoradiographic exposure times. Nucleic acids can be nonradioactively labeled by the incorporation of biotinylated or digoxigenin nucleotide analogues into DNA followed by detection using enzymatic colorimetric techniques. The labeled nucleic acid strand is used in hybridization experiments to detect and identify a specific nucleic acid sequence or gene in the other or target strand. A basic protocol for nucleic acid hybridization is described in Figure 6. The nucleic acid target strand is usually transferred and immobilized on a nylon or nitrocellulose membrane before adding the DNA probe. Hybridization is allowed to proceed for several hours or overnight in a solution containing 50% formamide. The presence of formamide and the hybridization temperature prevent nonspecific binding of the probe to the target DNA.

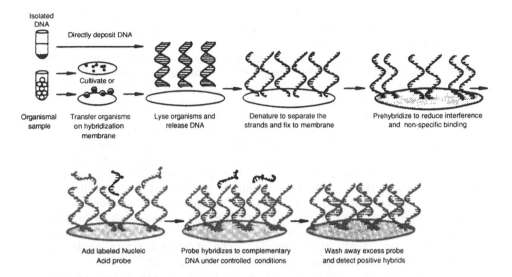

Figure 6 Fundamental protocol for the hybridization of a labeled nucleic acid probe with target DNA immobilized on a filter surface. From Ref. 21.

After completion of hybridization, the unbound probe is washed from the filter to minimize the background. When the probe used is radio-isotopically labeled, X-ray film is laid over the filter membrane, and after adequate exposure the presence of a specific DNA sequence complementary to the probe DNA is determined by the appearance of spot(s) on the film. With nonisotopically labeled probes, the presence of a specific sequence is determined by the appearance of colored spots on the filter membrane. Methods used in nucleic acid transfer and hybridization are treated in more detail in the following discussion.

A. Preparation of Gene Probes

Isotopic Labeling

In isotopic labeling, molecular probes are prepared by the labeling of nucleic acids with ^{32}P-deoxynucleotide triphosphate (dNTP). Any or all four of the most common ^{32}P-dNTPs may be used, and the specific activity of the labeled probe usually ranges from 3×10^8 to 1×10^9 dpm μg^{-1} DNA depending if one or all four ^{32}P-dNTPs are used.   The most commonly used methods for isotopic labeling of DNA are nick-translation [27–29] and the random primer method [30, 31], as will be described.

Probe preparation by nick-translation may be conducted on both intact DNA molecules or DNA fragments.   The nick-translation reaction mixture includes DNA polymerase I (DNA pol I), deoxynuclease I (DNase I), and the four dNTPs. DNase I generates random nicks on both strands of DNA.   By virtue of its 5'-3' exonuclease activity, DNA pol I excises nucleotides from one side of the nick and, subsequently, adds nucleotides to the 3' hydroxyl-terminus on the 5' side of the nick.   As nucleotides are removed from the 5' end and added to the 3' end, the nick appears to move or be translated along the DNA.   Labeled as well as nonlabeled dNTPs will be incorporated into DNA as the nicks are translated. Inasmuch as nicks are generated randomly on both strands of the DNA molecule, labeled strands of varying sizes (usually 80 to 200 nucleotides) are synthesized from both DNA strands.   The specific activity achieved by nick-translation ranges from 1 to 3 $\times 10^8$ dpm μg^{-1} DNA.

In random-primed labeling, a mixture of nucleotide hexamers that contain all four bases in all possible positions are hybridized to the heat-denatured DNA molecule to be labeled.   Because the hexamers lack sequence bias, hybrids are formed at many positions on the DNA, ensuring that the template will be copied with equal frequency at all positions.   With the hexamers as primers, the complementary strand is synthesized by the polymerization action of the large fragment of *Escherichia coli* DNA polymerase I (Klenow fragment) which lacks 5'-3' exonuclease activity.   Labeled and nonlabeled dNTPs are incorporated into the newly synthesized complementary strand.   As the input DNA is not nicked or degraded during the reaction, the amount of DNA needed for labeling (a few nanograms) is much smaller than that for nick-translation labeling.   Furthermore, the reaction is not influenced by the length of the DNA molecule, and the label is incorporated uniformly and equally along the length of the DNA.   As with nick-translationed probes, random-primed labeled probes are synthesized from both strands of the DNA molecule.   The specific activity obtained with the random primer method may reach as high as 1×10^9 dpm μg^{-1} DNA, which is approximately one order of magnitude higher than that achieved by the nick-translation technique.

Nonisotopic Labeling

Nonisotopic probes are becoming more common because of several advantages they offer over radioactive probes [32–46].   First, probe preparation does not involve handling of radioisotopes such as ^{32}P. Second, as most of the probes are chemically stable, they can be stored and used over long periods without appreciable loss of sensitivity or reproducibility. Third, at excessive probe concentrations, nonisotopic probes usually exhibit lower nonspecific binding to membranes than radiolabeled probes. Lastly, detection of a specific DNA sequence that is based on colorimetric visualization provides superior resolution over autoradiography with ^{32}P. The resolution of a radioactive probe is determined by the emission energy of the radionucleotide. For example, ^{32}P has a high-emission energy and, therefore, gives a very low resolution image. Ideally, the resolution of a probe for in situ hybridization should approach 0.2 μm (the resolving power of a light microscope). Radiolabels fall far below this ideal; ^{3}H and ^{125}I have resolutions of 1 μm, and ^{35}S has a resolution of 10 to 15 μm. Nonisotopic methods provide superior resolution because visualization is dependent on a localized color change, not on radiactive emission.

Preparation of biotin-labeled DNA probes may be carried out by using biotinylated analogues of dNTPs [36,37,43,45]. Biotinylated nucleotides may be incorporated into nucleic acid probes by enzymatic methods, including nick-translation and the random-primed technique. The biotin-labeled probes are detected using avidin or streptavidin conjugated with the enzyme alkaline phosphatase. After hybridization to DNA or RNA immobilized on a membrane, the excess probe is washed off, and the membrane is incubated with the avidin (or streptavidin)–enzyme complex. The biotin–avidin–enzyme complex is visualized by incubation of the membrane with 5-bromo-4-chloro-3-indolyl phosphate, which is enzymatically converted to the colored product indigo.

Another nonradioactive nucleotide analogue commonly used is digoxigenin-11-dUTP, which is incorporated into the DNA probe by the random primer technique. After hybridization of the digoxigenin-labeled probe to immobilized DNA, the probe is detected with an antibody–enzyme conjugate, antidigoxigenin alkaline phosphatase. The location of the antibody–antigen conjugate is visualized as with the biotin alkaline phosphatase complex. These methods are highly sensitive and are capable of detecting target sequences of 1 to 10 pg with incubation periods of 1 h or shorter. The color intensity of the hybridization signal may be increased by extending the incubation with the enzyme up to 24 h.

Biotin may also be chemically incorporated into a nucleic acid sequence by light activation of photobiotin, a photo-activatable analogue of biotin. Irradiation of photobiotin with visible light results in the formation of stable linkages with single- or double-stranded nucleic acids. Forster et al. [47] were able to detect a target sequence as small as 0.5 pg with this technique.

In DNA—protein cross-linking, a small macromolecule, such as a synthetic polymer carrying many primary amino groups, is attached to a protein [46]. The positively charged protein can now bind electrostatically to any polyanion, such as a single-stranded nucleic acid molecule. The nucleic acid—protein complex will hybridize to a target DNA or RNA sequence, and the hybridization can be detected by searching for the protein (e.g., peroxidase or phosphatase) that converts a colorless substrate into a colored product; the site of hybridization may be visualized directly.

B. Nucleic Acid Transfer and Hybridization

Dot- or Slot-Blot Hybridization

In dot- or slot-blot hybridization, nucleic acid solutions are applied directly to nitrocellulose or nylon membranes by vacuum filtration. A multiwell filtration manifold is used that allows sample application either as an array of circles (dot-blots) or slots thus permitting the rapid quantitative screening of multiple samples for target sequences. The sample may be a crude cell lysate or purified nucleic acid (DNA or RNA). After transfer, fixation of DNA fragments can be achieved by baking the membrane for 20 min to 1 h at 80°C or by UV cross-linking (for nylon membranes). Membranes carrying immobilized nucleic acid fragments permit stringent hybridization conditions and reprobing of the blot after removal of the first probe. Immobilization and subsequent detection of fragments of different sizes and conformations may be facilitated by using different transfer media and hybridization conditions [49—56]. Dot- or slot-blots may be probed by both radioisotopic and colorimetric methods. Quantitation of the visualized probe may be accomplished by densitometric scanning.

An example of slot-blot hybridization with a bank of DNA probes is provided in Figure 7. In this example, DNA obtained by in situ lysis and direct extraction was used as target DNA for DNA probes for nitrogenase (*nif*), bacterial photoreaction center (*puf*), fluorescent pseudomonads using a 23S ribosomal RNA gene probe, and both the Nah 7 naphthalene catabolic plasmid (NAH) and the *nahA* gene for naphthalene dioxygenase (A) [20]. Detection of positive sequences in the targets are apparent for each of the probes, ranging from 1 ng for *nahA* to 0.5 μg for *nif*.

Southern Transfer of DNA

The transfer method developed by Southern [57] involves capillary transfer of DNA (or RNA in Northern transfer) fragments of different size from electrophoretic gels onto immobilizing matrices such as nylon or nitrocellulose membranes. Although agarose gels (0.7

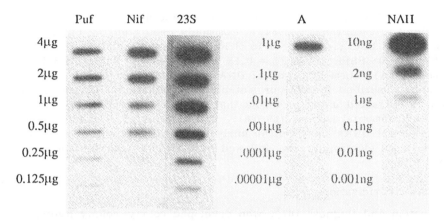

Figure 7 Comparative slot-blot hybridization for the quantitation of specific genes in microbial populations by extraction and analysis of soil and sediment DNA. Lanes 1–4 contain DNA obtained by direct extraction of soil contaminated with polyaromatic hydrocarbons. Lane 5 contains pure *nahA* DNA as a positive control. The soil DNA was probed with ^{32}P-labeled genes for nitrogenase (*nif*), bacterial photo-reaction center (*puf*), fluorescent pseudomonads using a 23S rRNA (23S) and the *nahA* gene for naphthalene dioxygenase (A). The number to the left of each column indicates the quantity of DNA applied to each slot.

to 1.0%) are commonly used for fragment separation, small DNA fragments may be separated better on polyacrylamide gels. Inasmuch as DNA fragments larger than 10 kilobase pairs (kbp) are not easily transferred by capillary action, they are usually fragmented further to ensure quantitative transfer of all bands, irrespective of their molecular mass [58,59]. DNA fragmentation is done by first immersing the gel in 0.25 N HCl for 15 min at room temperature, which causes depurination of the DNA. Subsequent denaturation of the DNA fragments by alkali (1.0 M NaCl, 0.5 M NaOH) hydrolyzes the phosphodiester backbone at the sites of depurination, and denatures double-stranded DNA to the single-stranded form [60]. DNA fragmentation can also be achieved by exposing the gel to UV light (254 nm); optimal exposure time, however, should be empirically determined [59–61]. Several factors, such as the size of eluted fragments, pore size of the transfer medium, and ionic strength of the transfer buffer, can affect retention of nucleic acid fragments on the filter membrane [62,63]. Denaturation of DNA fragments by alkali treatment before transfer is absolutely necessary for the

binding of DNA molecules to filters, as double-stranded DNA does
not bind to membranes [62,64]. Commonly used transfer buffers
include 20× SSC (1× SSC = 0.15 M NaCl, 15 mM Na citrate, pH
7.0) and 20× SSPE (1× SSPE = 0.18 M NaCl, 10 mM NaPO$_4$, pH 7.7,
1 mM EDTA). Immobilization of DNA on the transfer medium is the
same as described for dot- or slot-blot hybridization. Plasmid DNA
can be extracted using protocols that have demonstrated efficiency
for environmental isolates [65–66]. Preparations of plasmid and
chromosomal DNAs may then be analyzed by vertical or submerged
agarose gel electrophoresis, followed by Southern transfer and hybridization.

Colony Hybridization

Grunstein and Hogness [67], who developed colony–colony hybridization, originally used this technique to identify colonies of *E. coli*
that carried recombinant plasmids containing the 18S and 28S rRNA
genes of *Drosphila melanogaster*. The method was later used successfully by Hanahan and Meselson [68] to screen bacterial colonies
present at a high density (approximately 100,000) on culture plates.
Since then, colony hybridization has been used in rapid screening
and identification of a variety of microbial species or specific genes
within a microbial community. These include detection of *Salmonella*
spp. [69] and toxigenic strains of *E. coli* [70–72], *Yersinia* spp.
[73], and *Vibrio cholerae* [74] in foods and water; certain *Rhizobium* strains [75]; toluene, naphthalene, and polychlorinated bi-
phenyl degradative populations in sediments [22,23,76,77]; and
mercury resistance genes in gram-negative bacterial isolates [25].

The method involves the cultivation of bacterial cells on agar
plates followed by their transfer onto membranes of nylon or nitrocellulose. The transfer is accomplished by placing the membrane on
the agar plate surface and, subsequently, removing the filter to
which bacterial colonies have adhered. The membrane-bound colonies are then lysed by placing the disk (with colony side up) in
a 1- to 1.5-ml pool of 0.5 M NaOH for approximately 1 min. After
completion of cell lysis, the membranes are neutralized by placing
them in 1.5 M NaCl, 1 M Tris, pH 8.0. As with dot-blots or
Southern transfers, the membranes are air dried, baked in vacuo
at 80°C for 1 to 2 h, and hybridized with a labeled probe. Sayler
et al. [23] were able to detect one colony containing a toluene-cata-
bolic plasmid (TOL) among 10^6 colonies of *E. coli* containing nonho-
mologous DNA.

Compared with population estimates made from growth on selective substrates (toluene, naphthalene), population estimates obtained
from DNA–DNA colony hybridization more closely agree with estimates
based on actual substrate degradation rates. For instance, in the

case of oil-contaminated sediments, only 74% of the colonies capable of growth on a naphthalene minimal medium were also positive targets for a naphthalene catabolic plasmid (NAH7) probe. Furthermore, when naphthalene or toluene served as the sole carbon source for a microcosm population, the estimates derived from DNA—DNA colony hybridizations using *tol* or *nah* probes corresponded with estimates of substrate mineralization rates and past exposure to environmental contaminants [23].

Although the foregoing examples illustrate the usefulness of colony hybridization as an important gene probe tool, colony hybridization, dot-blotting, and Southern transfer should be used to augment, not replace, classic enrichment selection and differential cultivation methods. In cases in which a bacterium of interest cannot be detected or enumerated by conventional methods, colony hybridization may prove valuable. Two such examples are the detection of debilitated bacteria on a primary nonselective cultivation medium or the detection of organisms carrying defective, poorly expressed or nonselectable genes of interest. Concerning the latter, genetically engineered microorganisms released to the environment are likely candidates for the application of colony hybridization to compliment primary nonselective cultivation of an environmental sample.

C. Hybridization Protocols and Applications

A variety of practical applications of gene probe technology have been applied to environmental analysis [78]. This section describes, with some detail, specific methods and some of their applications.

The three-gene probe techniques described in the foregoing differ mainly in the way the DNA is transferred to the filter membrane. The steps after DNA transfer and immobilization, including hybridization, detection, and probe removal, are essentially the same for all methods. The following describes a procedure commonly used in hybridization on nitrocellulose membranes; when nylon membranes are used, various modifications of this protocol are usually recommended by the manufacturer.

DNA—DNA Hybridization

The DNA—DNA hybridization procedure may be used with both biotinylated or ^{32}P-labeled DNA probes. The amount of prehybridization and hybridization buffers is approxiamtely 100 µl of solution per square centimeter of membrane. In the prehybridization step, the membrane is first placed in a polyethylene heat-sealable bag and wetted with 5× SSC containing 0.1% SDS. The membrane is then incubated in hybridization buffer (without probe) at 42°C for a minimum of 30 min. The hybridization buffer consists of 5× SSC, 5× Den-hardt's solution (1× = 0.02% Ficoll, 0.02% polyvinylpyrrolidone, 0.02% bovine serum albumin), 50% formamide (deionized), 0.1% SDS, 10% dextran sulfate, and 200 μg ml^{-1} denatured herring or salmon sperm [79]. The SSC solution may be replaced by SSPE; indeed, SSPE is a better buffer, especially in the presence of formamide [59]. Other protein solutions, such as heparin [80] or gelatin [81], may also be used instead of Denhardt's. These solutions serve to block sites on the membrane to which the labeled probe can bind, there-by reducing the background. In addition to the protein solutions, detergents, such as SDS, will also block nonspecific binding of the probe to the membrane. Heterologous denatured low-molecular-mass DNA is included in the hybridization buffer, to prevent random nonspecific annealing of the probe to the immobilized DNA [82].

Hybridization and washing stringencies also have important roles in eliminating nonspecific binding of the DNA probe. The optimal hybridization temperature is approximately 20 to 25°C below the melting temperature (T_m) of the hybrid [83]. The melting temper-ature is given by the following formula: $T_m = 81.5 + 16.6 \log[\text{Na}^+] + 0.41$ (% G + C) -0.61 (% formamide) $-(300 + 2000[\text{Na}^+]/\text{length of hybridized duplex in base pairs})$, where [Na$^+$] is the molarity of the sodium concentration and (% G + C) is the percentage of quano-sine and cytosine residues in the DNA. This formula was origi-nally derived from solution hybridization experiments (described later), but it can also be used in mixed-phase hybridization reac-tions. The T_m decreases 0.61°C for each 1% increase in formamide concentration [84]; it also decreases by 1.0°C for every 1% de-crease in homology [85]. Formamide serves two purposes: it pre-serves the filter membrane better than an aqueous solution, and it also decreases the T_m of the hybrid [86]. The addition of dextran sulfate, an anionic polymer, to the hybridization buffer is optional; it increases the rate of hybridization (mixed-phase) by excluding the free probe from the solution, thereby increasing its apparent concentration [58]. Polyethylene glycol may be substituted for dextran sulfate [87]. When a single-copy gene is to be detected, the amount of probe should be 1 to 5×10^6 cpm (or 100 to 200 ng biotinylated probes) per milliliter hybridization buffer; for genes with a high copy number, the probe concentration may be reduced by one order of magnitude. All probes must be denatured by boil-ing in TE (10 mM Tris, pH 7.6, 1 mM EDTA) for 3 to 5 min or by incubating with 0.1 volume 1 N NaOH at 37°C for 5 min before adding to the hybridization solution. Biotinylated probes should not be alkali treated, as biotin is cleaved by alkali.

After completion of hybridization, filter membranes should be washed at 20°C below the T_m of the hybrid, to remove the un-bound probe. To eliminate noncomplementary hybrids, higher temperatures (up to 5°C below the T_m of the hybrid) may also be used [82,88]. A general washing protocol for nitrocellulose membranes using biotinylated or radiolabeled probes is as follows:

1. Twice in 2× SSC or SSPE 0.1% SDS for 2 to 4 min at room temperature.
2. Twice in 0.2× SSC or SSPE 0.1% SDS for 15 to 30 min at 42°C. The second wash may be carried out at 65°C for higher stringency.
3. Final rinse in 0.2× SSC or SSPE.

The signal/noise ratio may be optimized by varying the ionic strength and the SDS concentration of washing solutions.

For blots probed with radiolabeled probes, the membrane is wrapped in a plastic food wrap while still slightly damp and placed in a cassette with X-ray film and an intensifying screen. Depending upon the intensity of the signal (which may be checked by Geiger counter before exposure to the X-ray film), the blot is exposed at -80°C for several hours to a few days.

If the blot needs to be reprobed, the previously bound probe should be removed before hybridization. The membrane can be deprobed by incubation at 60°C for 1 to 2 h in a solution containing 50% formamide, 10 mM Tris-HCl, pH 8.0, and 1 mM EDTA. This solution should be changed once during incubation. After deprobing, the membrane is rinsed with 2× SSC or SSPE at room temperature, and air dried. Alternatively, the membrane can be deprobed by heating to boiling in a solution of 0.2% SSC (SSPE), 0.1% SDS for 15 min. This procedure may be repeated several times until the blot is reasonably free of bound probe. With either protocol, the deprobed membrane should be exposed to X-ray film for several days to ensure complete removal of the radiolabeled probe.

RNA–DNA Hybridization

The RNA probes are generally synthesized by in vitro transcription of DNA. Contrary to DNA probes, which bind to both DNA strands of the target, RNA probes are strand-specific and hybridize only to the (+) or (-) DNA strand. RNA–DNA hybrids are significantly more stable than DNA–DNA hybrids [85]. The T_m for an RNA–DNA duplex [89] is calculated as follows: $T_m = 79.8 + 18.5$ log-$[Na^+] + 58.4$ (% G + C) $+11.8$ (% G + C)^2 -820/no. of base pairs in duplex -0.5 (% formamide).

When RNA is used as the molecular probe in hybridization reactions, higher stringencies are required to eliminate nonspecific annealing to the bound DNA [90–92]. RNase contamination of all materials must be prevented, and protein solutions such as nonfat dry milk, that may contain trace amounts of RNase should be avoided. Prehybridization and hybridization procedures are essentially the same as described for DNA—DNA hybridization, but reactions are carried out at 50 to 60°C. Filter washes are done at 65°C, and background binding of the probe to the membrane may be further eliminated by incubating the blot in 1.0 $\mu g\ ml^{-1}$ RNaseA in 2× SSC for 15 min at room temperature.

Solution Hybridization and DNA Reassociation Kinetic Analysis

In solution hybridization techniques, the target DNA is not immobilized on a solid support, such as a filter membrane, but rather is in a solution to which a radiolabeled probe is added. Hence, the hybridization reaction between the probe and complementary sequences of the target DNA takes place in solution. After completion of hybridization, double- and single-stranded DNAs can be separated by fractionation using hydroxyapatite column chromatography. The radioactivity of each fraction is then determined by scintillation counting, and the percentage of DNA in the sample that is homologous to the probe is calculated [93]. Solution hybridization can be used not only for the detection of specific DNA sequence, but it can theoretically also be used to study the diversity of bacterial communities in soil or sediment through reassociation kinetic analysis [26,77].

In this type of analysis, the purified DNA is sheared, denatured, and allowed to reassociate in solution. The percentage of DNA that is reannealed over time is determined by radioactive labeling of a small fraction of the total DNA, and a graph is produced from these kinetic data [94]. The rate of renaturation is given by the $C_0 t$ value which is defined as the product of the molar concentration of nucleotide residues (C_0) and the time of renaturation (t).

Although DNA reassociation kinetic analysis has been used extensively in the study of genome complexity among eukaryotes, its application into the study of microbial communities is rather recent [26,77]. Reassociation of the two strands follows second-order kinetics, the rate of association of which is dependent upon the number of similar sequences; the greater the similarity between the two strands [5], the faster the reassociation. DNA sequences that occur more frequently in the community will reassociate faster than those that occur at a lower frequency. Conversely, in a community with many species, the greater genetic complexity will result in slow reassociation of homologous sequences. Reassociation analysis of DNA from environmental samples may take weeks as a result of the genetic diversity. This problem can be partially resolved by increasing the amount of DNA used in reassociation experiments. Because the rate of DNA reannealing is directly proportional to the square of the DNA concentration, a twofold increase in DNA concentration would result in a fourfold increase in the reassociation rate. Usually, an environmental sample containing 2 to 5 mg DNA isolated from a community of more than 100 genotypes should be sufficient to conduct a reassociation kinetic analysis within a reasonable time frame (i.e., within approximately 1 week) [77]. If C_0t values are significantly large (over 3 days), probes should be labeled with ^3H rather than ^{32}P, as the high-energy beta-emission of the latter is sufficient to break the phosphodiester bonds of DNA. Alternatively, DNA reassociation can be measured spectrophotometrically by monitoring the hyperchromatic shift of DNA; single-stranded DNA absorbs more strongly at 260 nm than double-stranded DNA.

Most Probable Number DNA Hybridization

In the most probable number (MPN) method, microbial samples are diluted to extinction, and viable cells are allowed to grow in appropriate media. The original density of the target population is then determined by probability theories [94—101]. Fredrickson et al. [24] developed a microliter MPN procedure in conjunction with DNA hybridization for detection and enumeration of specific bacterial genotypes in soil. In this method, a soil mixture and medium optimal for growth of the organisms under study are prepared. Serial dilutions of the mixture are then placed in sterile microliter plates and incubated at the appropriate temperature (optimal for the growth of the organisms) for 3 to 4 days. The supernatant is then removed carefully to avoid disturbance of the sedimented soil particles at the bottom of the tubes and filtered onto a nitrocellulose or nylon membrane using a vacuum manifold, thus creating one spot for each MPN tube. Subsequent steps, including cell lysis, DNA denaturation, hybridization, washing, and detection, are carried out as described for colony hybridization.

Ribosomal RNA Analysis

Ribosomal RNAs (rRNAs) have been used for the phylogenetic analysis of the prokaryotic and eukaryotic kingdoms [18,19,102—125], because they are extremely conserved both in function and structure among all organisms. Nucleotide sequences are also conserved, to varying degrees, across kingdoms, thus providing hybridization targets for gene probing, cloning, and sequencing. Of the three rRNA species present in prokaryotes [5S (~120 bps), 16S (~600 bps), and 23S (~3000 bps)], 5S and 16S rRNAs have been preferentially used for rRNA-based phylogenetic characterization of bacteria present in natural environments. The 5S rRNA is suitable mainly for phylogenetic analysis of populations with limited

complexity. In such cases, a heterogenous population of 5S rRNAs
is extracted from a mixed microbial community and sorted by high-
resolution gel electrophoresis. From the various species-specific
molecules separated on the gel, individual 5S rRNA types can be
purified and sequenced. The sequence data are then used to determine the phylogenetic relations of the contributing microorganisms
by comparison with available 5S rRNA sequences. The lack of sequence variability of the 5S rRNA and its small size limits its use
in the characterization of complex microbial populations.

The 16S and 23S rRNAs are more suitable for phylogenetic studies because of their greater variability and the more information ob-
tainable from these larger molecules. Historically, rRNA analysis
has focused on the 16S subunit because a larger data base exists
for this subunit. Population analysis based on the 16S subunit involves either direct sequencing of the 16S rRNA or cloning of the
DNA that codes for the 16S rRNA, followed by sequencing of the
cloned DNA. The first procedure involves isolating the total RNA
from a mixed population, annealing a primer to a particular 16S
rRNA sequence, followed by dideoxy sequencing utilizing reverse
transcriptase [26].

In the second procedure, total DNA is isolated from a population and "shotgun" cloned into bacteriophage λ to create a recombinant DNA library in which all the genes of the population are represented. The gene library is then screened, using a universal
or "mixed kingdom" probe to obtain only those clones that have the
16S rRNA genes. These rRNA clones are then sequenced, as de-
cribed previously, using specific primers, and the sequences are
compared with other 16S rRNA sequences to determine the phylogenetic identification of the organism. With both of these procedures, population complexity is not a factor because individual
subunits do not have to be electrophoretically separated before sequencing.

In Situ Hybridization

Giovannoni et al. [102] and DeLong et al. [103] have developed an
in situ hybridization assay (i.e., hybridization of fixed whole cells)
for cells from pure or mixed cultures fixed onto glass microscope
slides. Cell suspensions are applied in 5-ml aliquots to glass microscope slides pretreated with gelatin to promote adhesion, air
dried, and subsequently fixed in a glutaraldehyde solution. After
transfer through a series of ethanol solutions, the slides are dried,
prehybridized, as with membranes, and hybridized with ^{35}S-la-
beled oligonucleotide probes. The cellular DNA is visualized by
treating the slides with an autoradiographic emulsion. After development of the emulsion, the slides are additionally stained with

Giemsa to visualize the cells. By using this method and oligonucleotide probes to unique 16S rRNA sequences, Giovannoni et al.
[102] were able to obtain phytogenetic identification of fixed cells.

In addition to isotopically labeled probes, fluorescent probes
have also been used for in situ hybridization. If different fluors
that fluoresce at different wavelengths are covalently bound to different oligonucleotide probes, several probes may be used simultaneously. DeLong et al. [103] synthesized an oligonucleotide that
binds to the universal 16S rRNA sequence labeled with fluorescein
and an oligonucleotide specific for the 16S rRNA of eukaryotes labeled with X-rhodamine. Cells were suspended in buffer, applied to
microscope slides and fixed as described earlier. The fluorescent
probes were lyophilized, resuspended in a hybridization mixture,
and applied to the slides in 30-ml aliquots. After the hybridization period (5 to 16 h), slides were washed in buffer and viewed
immediately. With a microscope equipped with filters specific for
the excitation and emission spectra of each fluor, it was possible
to simultaneously distinguish between eubacterial and eukaryotic
cell mixtures. Potential problems with the use of fluorescent
probes include high autofluorescent background and the low permeability of fixed cells to oligonucleotide probes. If probes of
several hundred nucleotides are used, the cells must be mildly digested with lysozyme or proteinase K [103]. Before applying these
methods to soil or sediment samples, it should be remembered that
the dilution of the environmental sample plus the loss of some cells,
which usually occurs during cell extraction, could lower the sensitivity of the assay. Such assays become even more difficult when
the target sequence has a small copy number. DeLong et al. [103]
however, suggested that this problem may be resolved, to some extent, by using multiple fluorescent probes or oligonucleotide probes
labeled with multiple fluors. A possible protocol for in situ probing
of microbial biofilms associated with soil particles is given in Figure 8.

IV. DETECTION SENSITIVITY USING
NUCLEIC ACID HYBRIDIZATION

The sensitivity of molecular assays in an environmental analysis is
defined as the probability that the procedure will detect positive
targets when these are present in the sample. Each type of hybridization assay under specified conditions has a particular sensitivity that can be influenced by several factors. These include
the length and abundance of the target-sequence probe; the length,
specific activity, and type of nucleic acid label; and the stringencies

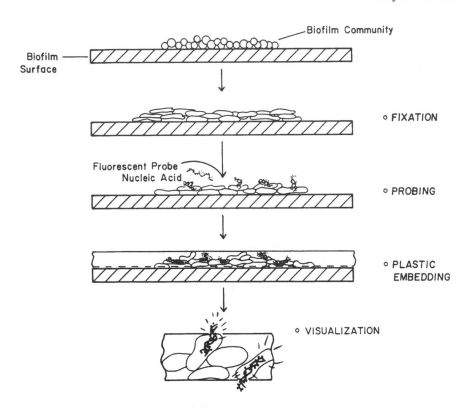

Figure 8 Conceptual protocol for the application of in situ hybridization methods with oligonucleotide probes for the analysis of soil
biofilms.

of the hybridization and washing procedures. Regardless of the
method of hybridization, a reasonable signal strength is necessary
for the investigator to record a positive signal. This creates a
limit on the number of cells in an environmental sample that can
be detected. The cell detection limit, however, can be increased
in various ways, depending upon the hybridization procedure. In
colony hybridization, for example, if the target cells are a low proportion of the total culturable bacterial cells, selective media must
be used to increase the sensitivity of detection [127]. On the
other hand, a decreased sensitivity in colony blots may result if
the transfer of colonies to the membrane or cell lysis was ineffi-
cient. Relative to the latter, a simple alkaline lysis is ordinarily
used for colony blots or colony dot-blots. There are, however,

other possibly more efficient lysis protocols in the literature, such as that used by Zeph and Stotzky [128] for the detection of bacteriophage-transduced bacteria in soil.

In the MPN procedure, after amplification of cells in the MPN tubes, there should be enough cells in each tube to give a positive result, even with one cell per tube. In the MPN–DNA hybridization procedure of soil samples, however, there was some loss in sensitivity as the result of cells that were not recovered, as they were bound on soil particles [24]. Cells used for hybridization were separated from soil particles simply by sedimentation of the particles in the MPN tubes. In sensitivity experiments, Fredrickson et al. [24] reported a detection limit of 10 cells g^{-1} of *Pseudomonas putida* soil, whereas the detection limit for *Rhizobium leguminosarum* was 10^2 to 10^3 cells g^{-1} of soil, presumably as the result of the tighter binding of the *Rhizobium* on soil particles. They also reported a lower recovery of cells from soils high in clay and organic matter than from soils low in clay and organic matter. Bentjen et al. [129] used MPN–DNA hybridization to enumerate Tn5 mutants of *Azospirillum lipoferum* in a soil microcosm; a 35-bp Tn5-specific sequence was used for probe synthesis of an oligonucleotide, and a detection limit of approximately 10^3 cells g^{-1} dry soil was reported.

For hybridization methods that involve direct application of the purified DNA to the membrane, such as dot-blot analysis, there is a limit to the amount of DNA that can bind to the membrane; therefore, there is a limit to the amount of sample which can be used for hybridization. Holben et al. [11] suggested that control experiments be performed to determine a membrane's DNA-binding capacity before DNA from large samples is applied to the membrane. If dilution of a DNA sample is necessary for membrane binding, a loss of sensitivity in the detection of cells in the original sample may result. Here, solution hybridization techniques that do not depend on the DNA-binding capacity of a membrane may prove superior. Also, restriction of the DNA (if it is pure enough) and subsequent transfer of the restricted fragments to membranes by Southern blotting would spread the DNA out, thereby decreasing the amount of DNA bound in each spot.

In dot-blot analysis, where there is no amplification of the target cells (in contrast with colony or MPN–DNA hybridization), a loss of sensitivity can occur as the result of the signal detection limit. The use of probes specific for unique DNA sequences results in detection limits on the order of 10^5 cells, although this number may be lowered by using probes of higher specific activity [11].

A popular approach to increase the sensitivity of DNA hybridization is the polymerase chain reaction (PCR) [130]. The DNA polymerase of *Thermus aquaticus* can synthesize more than 10^5 copies of a specific target sequence (up to 4-kbp long) from primers

annealed to both strands in a reaction mixture containing dNTPs.
DNA amplification is carried out by multiple cycles of heating the
mixture to denature the target DNA and then cooling for polymerase extension of new strands from the primers. Steffan and Atlas
[131] determined that the sensitivity of dot blots for target DNA
sequences could be increased by at least three orders of magnitude
using prior amplification of target sequences by PCR. With use of
an oligonucleotide probe complementary to a region in a species-
specific repeated sequence, they were able to detect (after PCR
amplification) one cell per gram of *Pseudomonas cepacia* sediment
from 100 g of sample in a dot-blot, whereas the detection limit without PCR amplification was 10^4 cells g^{-1} sediment from a 100-g
sample.

V. ENGINEERED MICROORGANISMS: IDENTIFICATION AND ASSESSMENT

The main goal of genetic engineering is to create, through genetic
alteration, organisms with unusual but useful properties. Genetically engineered microorganisms (GEMs) are standard bacterial
strains altered in small or large ways before introduction into the
environment. Proper identification of GEMs and their parental microorganisms and the ways in which they have been altered are
vital to a full understanding of their likely behavior and to assessing the environmental hazards they might pose. The potential adverse ecological impact of GEMs introduced into soil and other natural environments remains largely unknown. Consequently, the
ability to monitor the survival, growth, and genetic transfer by
GEMs in the environment is important.

The applications of molecular techniques in the enumeration and
monitoring of GEMs or recombinant DNA in the environment have
been extensively reviewed [21,132]. Major issues in developing
successful monitoring protocols include (1) the ability to discrim-
inate the recombinant host from recombinant genes in the environment, (2) the masking of GEM populations by the diverse and nu-
merically dominant indigenous soil microbes, (3) developing strategies to assess adverse ecological consequences, and (4) measuring
the transfer dynamics of recombinant genes. Many of these issues
have been recently reviewed relative to GEM introductions to soils
[133].

The need for applications of molecular techniques, such as
gene-probing methods, and their specificity have been documented
for studies on plasmid transfer in soil populations of *Rhizobium*
[134]. The sensitivity of some of the molecular methods has not
yet achieved the practical limits of conventional microbiological

enumeration methods. Devanas and Stotzky [135] state that a practical detection limit of 20 organisms per gram of soil can be achieved with conventional methods (e.g., selective media); whereas molecular techniques, such as nucleic acid extraction, may have practical sensitivity limits of only 10^4 organisms per gram of soil. As a consequence, it may not be possible to rely solely on molecular methods to achieve quantitative risk analysis for GEMs in the environment. However, it is likely that detection thresholds will be continually pushed lower with the development of improved methods, such as the PCR technique.

Two areas of improvement in molecular analysis are needed. These are (1) standardization of protocols to make data comparable across environmental applications and (2) increased levels of sensitivity. These improvements are needed for virtually all environmental applications, regardless of whether they are direct at GEMs or other environmental processes.

VI. CONCLUSION

A major driving force for the development of molecular methods for environmental analysis has been the unknown stability, survival, and mobility of GEMs and their recombinant genes in the environment. A side benefit of this concern has been the demonstration of practical applications of molecular methods for environmental analysis as described in this chapter.

To date, the application of molecular techniques to the field of microbial ecology has been primarily in the structural analysis of microbial communities. However, new methods are on the horizon that will contribute to the analysis of the activity of specific populations and genes in environmentally significant processes. These methods relate, in part, to RNA analysis, as previously indicated, but new molecular "marking" techniques to create reporters for specific physiological activities in situ will augment already existing techniques in environmental analysis.

It appears likely that new molecular techniques will be applicable to the analysis of microbial communities in soils. It is predicted that within the decade, they will be commonplace in their application and as routine as conventional microbial cultivation and enumeration techniques.

ACKNOWLEDGMENTS

We thank M. James for word processing. Parts of this work were supported by USGS 14-08-0001-G1482, U.S. Air Force F49620-89-C-0023,

Gas Research Institute 5087-253-1490, and Electric Power Research
Institute RP-3015-1. We also thank Claudia Werner for assistance.

REFERENCES

1. Kingsbury, D. T. 1969. Estimate of the genome size of various microorganisms. J. Bacteriol. 98:1400–1401.
2. Bak, A. L., C. Christiansen, and A. Stenderup. 1970. Bacterial genome sizes determined by DNA renaturation studies.
   J. Gen. Microbiol. 64:377–380.
3. Faegri, A., V. L. Torsvik, and J. Goksoyr. 1977. Bacterial and fungal activities in soil: separation of bacteria and
   fungi by a rapid fractionated centrifugation technique. Soil
   Biol. Biochem. 9:105–112.
4. Torsvik, V. L., and J. Goksoyr. 1978. Determination of bacterial DNA in soil. Soil Biol. Biochem. 10:7–12.
5. Holm-Hanssen, O., W. H. Sutcliffe, Jr., and J. Sharp. 1968.
   Measurement of deoxyribonucleic acid in the ocean and its ecological significance. Limnol. Oceanogr. 13:507–514.
6. Robertson, F. W., and K. Tait. 1971. Optimal conditions for
   the fluorometric determination of DNA. Anal. Biochem. 41:
   477–481.
7. Lien, T., and G. Knutsen. 1976. Fluorometric determination
   of DNA in *Chlamydomonas*. Anal. Biochem. 74:560–566.
8. Crissman, H. A., and R. A. Tobey. 1974. Cell-cycle analysis in 20 minutes. Science 184:1297–1298.
9. Hill, B. T., and S. Whatley. 1975. A simple, rapid micro-
   assay for DNA. FEBS Lett. 56:20–23.
10. Hill, B. T. 1976. The use of anti-tumor antibiotics for sim-
    ple quantitative assays for DNA. Anal. Biochem. 70:635–638.
11. Holben, W. E., J. K. Jansson, B. K. Chelm, and J. M. Tiedje.
    1988. DNA probe method for the detection of specific micro-
    organisms in the soil bacterial community. Appl. Environ. Mi-
    crobiol. 54:703–711.
12. Ogram, A., G. S. Sayler, and T. Barkay. 1987. Extraction
    and purification of microbial DNA from sediments. J. Micro-
    biol. Methods 7:57–66.
13. Hobbie, J. E., R. J. Daley, and S. Jasper. 1977. Use of
    nucleopore filters for counting bacteria by fluorescence mi-
    croscopy. Appl. Environ. Microbiol. 33:1225–1228.
14. Ogram, A., G. S. Sayler, D. M. Gustin, and R. Lewis. 1988.
    DNA adsorption to soils and sediments. Environ. Sci. Technol.
    22:982–984.
15. Steffan, R. J., J. Goksoyr, A. K. Bej, and R. M. Atlas.
    1988. Recovery of DNA from soils and sediments. Appl. En-
    viron. Microbiol. 54:2908–2915.

16. Bakken, L. R. 1985. Separation and purification of bacteria from soil. Appl. Environ. Microbiol. 49:1482–1487.

17. Balkwill, D. L., D. P. Labeda, and L. E. Casida, Jr. 1975. Simplified procedure for releasing and concentration microorganisms from soil for transmission electron microscopy viewing as thin-section and frozen etched preparations. Can. J. Microbiol. 21:252–262.

18. Stahl, D. A., D. J. Lane, G. J. Olsen, and N. R. Pace. 1984. Analysis of hydrothermal vent-associated symbionts by ribosomal RNA sequences. Science 224:409–411.

19. Stahl, D. A., D. J. Lane, G. J. Olsen, and N. R. Pace. 1985. Characterization of a Yellowstone hot spring microbial community by 5S ribosomal RNA sequences. Appl. Environ. Microbiol. 49:1379–1384.

20. Sayler, G. S., J. Fleming, B. Applegate, C. Werner, and K. Nikbakht. 1989. Microbial community analysis using environmental nucleic acid extracts, p. 658–662. *In* T. Hattoiri, Y. Ishida, Y. Maruyama, R. Morita, and A. Uchida (eds.), Recent advances in microbial ecology. Japan Scientific Societies Press, Tokyo.

21. Jain, R. K., R. S. Burlage, and G. S. Sayler. 1988. Methods for detecting recombinant DNA in the environment. CRC Crit. Rev. Biotechnol. 8:33–84.

22. Pettigrew, C., and G. S. Sayler. 1986. The use of DNA: DNA colony hybridization in the rapid isolation of 4-chlorobiphenyl degradative bacterial phenotypes. J. Microbiol. Methods 5:205–213.

23. Sayler, G. S., M. S. Shields, E. T. Tedford, et al. 1985. Application of DNA:DNA colony hybridization to the detection of catabolic genotypes in environmental samples. Appl. Environ. Microbiol. 49:1295–1303.

24. Fredrickson, J. K., D. F. Bezdicek, F. E. Brockman, and S. W. Li. 1988. Enumeration of Tn5 mutant bacteria in soil by using a most-probable-number-DNA hybridization procedure and antibiotic resistance. Appl. Environ. Microbiol. 54:446–453.

25. Barkay, T., D. L. Fouts, and B. H. Olson. 1985. The preparation of a DNA gene probe for the detection of mercury resistance genes in gram-negative bacterial communities. Appl. Environ. Microbiol. 49:686–692.

26. Sayler, G. S., and G. Stacey. 1986. Methods for evaluation of microorganism properties, p. 35–58. *In* J. Fiksel and V. T. Covello (eds.), Biotechnology risk assessment: issues and methods for environmental introductions. Pergamon Press, New York.

27. Rigby, P. W. J., M. Dieckmann, C. Rhodes, and P. Berg. 1977. Labeling deoxyribonucleic acid to high specific activity

in vitro by nick translation with DNA polymerase I. J. Mol. Biol. 113:237.

28. Lehman, I. R. 1981. DNA polymerase I of *Escherichia coli*, p. 15–37. *In* P. D. Boyer (ed.), The Enzymes, Vol. 14. Academic Press, New York.

29. Kelley, W. S., and K. H. Stump. 1979. A rapid procedure for isolation of large quantities of *Escherichia coli* DNA polymerase I utilizing a pol A transducing phage. J. Biol. Chem. 254:3206–3210.

30. Feinberg, A. P., and B. Vogelstein. 1983. A technique for radiolabelling DNA restriction endonuclease fragments to high specific activity. Anal. Biochem. 132:6–13.

31. Feinberg, A. P., and B. Vogelstein. 1984. Addendum. Anal. Biochem. 137:266–269.

32. Miller, E. C., U. Juhl, and J. A. Miller. 1966. Nucleic acid guanine: reactions with the carcinogen N-acetoxy-2-acetylaminofluorene. Science 153:1125–1127.

33. Tchen, P., R. P. P. Fuchs, E. Sage, and M. Leng. 1984. Chemically modified nucleic acids as immunodetectable probes in hybridization experiments. Proc. Natl. Acad. Sci. USA 81:3466–3470.

34. Landegent, J. E., N. Jansen in de Wal, R. A. Baan, J. H. J. Hoeijmakers, and M. van der Ploeg. 1984. 2-Acetylaminofluorene-modified probes for the indirect hybridocytochemical detection of specific nuclei acid sequences. Exp. Cell Res. 153:61–72.

35. Bayer, E. A., and M. Wilchek. 1980. The use of the avidin-biotin complex as a tool in molecular biology. Methods Biochem. Anal. 26:1–45.

36. Langer, P. R., A. A. Waldrop, and D. C. Ward. 1981. Enzymatic synthesis of biotin labeled polynucleotides: novel nucleic acid affinity probes. Proc. Natl. Acad. Sci. USA 78: 6633–6637.

37. Leary, J. J., D. J. Brigati, and D. C. Ward. 1983. Rapid and sensitive calorimetric method for visualizing biotin-labeled DNA probes hybridized to DNA or RNA immobilized on nitrocellulose: bio-blots. Proc. Natl. Acad. Sci. USA 80:4045–4049.

38. Langer-Safer, P. R., M. Levine, and D. C. Ward. 1982. Immunological method for mapping genes on *Drosophila* polytene chromosomes. Proc. Natl. Acad. Sci. USA 79:4381–4385.

39. Hutchison, N. J., P. R. Langer-Safer, D. C. Ward, and B. A. Hamkalo. 1982. In situ hybridization at the electron microscope level: hybrid detection by autoradiography and colloidal gold. J. Cell Biol. 95:609–618.

40. Manuelidis, L., P. R. Langer-Safer, and D. C. Ward. 1982. Hig-resolution mapping of satellite DNA using biotin-labelled DNA probes. J. Cell Biol. 95:619—625.

41. Singer, R. H., and D. C. Ward. 1982. Acetin gene expression visualized in chicken muscle tissue culture by using in situ hybridization with a biotingalted nucleotide analog. Proc. Natl. Acad. Sci. USA 79:7331—7335.

42. Brigati, D. J., D. Myerson, J. J. Leary, et al. 1983. Detection of viral genomes in cultured cells and paraffin-embedded tissue sections using biotin-labelled hybridization probes. Virology 126:32—50.

43. Manning, J. E., N. D. Hershey, T. R. Broker, M. Pellegrini, H. K. Mitchell, and N. Davidson. 1975. A new method of in situ hybridization Chromosoma. 53:107—117.

44. Pellegrini, M., D. S. Holmes, and J. Manning. 1977. Application of the avidin biotin method of gene enrichment to the isolation of long double-stranded DNA containing specific gene sequences. Nucleic Acids Res. 4:2961—2973.

45. Renz, M. 1983. Polynucleotide—histone H1 complexes as probes for blot hybridization. EMBO J. 2:817—822.

46. Renz, M., and C. Kurz. 1984. A colorimetric method for DNA hybridization. Nucleic Acids Res. 12:3435—3444.

47. Forster, A. C., J. L. McInnes, D. C. Skingle, and R. H. Symons. 1985. Non-radioactive hybridization probes prepared by the chemical labelling of DNA and RNA with a novel reagent, photobiotic. Nucleic Acids Res. 13:745—761.

48. Kafatos, F. C., C. W. Jones, and A. Efstratiadis. 1979. Determination of nucleic acid sequence homologies and relative concentrations by a dot hybridization procedure. Nucleic Acids Res. 7:1541—1552.

49. Thomas, P. S. 1980. Hybridization of denatured RNA and small DNA fragments transferred to nitrocellulose. Proc. Natl. Acad. Sci. USA 77:5201—5205.

50. Bresser, J., J. Doering, and D. Gillespie. 1983. Quick-blot; selective mRNA or DNA immobilization from whole cells. DNA 2:243—254.

51. Brandsma, J., and G. Miller. 1980. Nucleic acid spot hybridization: rapid quantitative screening of lymphoid cell lines for Epstein—Barr viral DNA. Proc. Natl. Acad. Sci. USA 77: 6851—6855.

52. White, B. A., and F. C. Bancroft. 1982. Cytoplasmic dot hybridization. Simple analysis of relative mRNA levels in multiple small cell or tissue samples. J. Biol. Chem. 257:8569—8572.

53. Dooley, T. P., J. Tamm, and B. Polisky. 1985. Isolation and characterization of mutants affecting functional domains of *Col*E1 RNA I. J. Mol. Biol. 180:87—96.

54. Gergen, J. P., R. H. Stern, and P. C. Wensink. 1979. Filter replicas and permanent collections of recombinant DNA plasmids. Nucleic Acids Res. 7:2115–2134.

55. Mushinski, J. F., S. R. Bauer, M. Potter, and R. P. Reddy. 1983. Increased expression of *myc*-related oncogene mRNA characterizes most BALB/c plasmacytomas induced by pristine or Abelson marine leukemia virus. Proc. Natl. Acad. Sci. USA 80:1073–1077.

56. Dobner, P. R., E. S. Kawasaki, L. Yu, and F. C. Bancroft. 1981. Thyroid of glucocorticoid hormone induces pre-growth-hormone mRNA and its probable nuclear precursor in rat pituitary cells. Proc. Natl. Acad. Sci. USA 78:2230–2234.

57. Southern, E. M. 1975. Detection of specific sequences among DNA fragments separated by gel electrophoresis. J. Mol. Biol. 98:503–517.

58. Wahl, G. M., M. Stern, and G. R. Stark. 1979. Efficient transfer of large DNA fragments from agarose gels to diazobenzloxymethal-paper and rapid hybridization by using dextran sulfate. Proc. Natl. Acad. Sci. USA 76:3683–3687.

59. Meinkoth, J., and G. M. Wahl. 1984. Hybridization of nucleic acids immobilized on soilid supports. Anal. Biochem. 138:267–284.

60. Kochetkov, N. K., and E. I. Budovski (eds.). 1972. Organic chemistry of nucleic acids, B:477–532. Plenum Press, London.

61. Lis, J. T., L. Prestidge, and D. S. Hogness. 1978. A novel arrangement of tandemly repeated genes at a major beat shock site in *D. melanogaster*. Cell 14:901–919.

62. Gillespie, D., and S. Spiegelman. 1965. A quantitative assay for DNA–RNA hybrids with DNA immobilized on a membrane. J. Mol. Biol. 12:829–842.

63. Nagamine, Y., A. Sentenac, and P. Fromageot. 1980. Selective blotting of restriction DNA fragments on nitrocellulose membranes at low salt concentrations. Nucleic Acids Res. 8:2453–2460.

64. Nygaard, A. P., and B. D. Hall. 1963. A method for the detection of RNA–DNA complexes. Biochem. Biophys. Res. Commun. 12:98–104.

65. Anderson, D. G., and L. L. McKay. 1983. Simple and rapid method for isolating large plasmid DNA from lactic streptococci. Appl. Environ. Microbiol. 46:549–552.

66. Kado, C. I., and S. T. Liu. 1981. Rapid procedure for detection and isolation of large and small plasmids. J. Bacteriol. 145:1365–1373.

67. Grunstein, M., and D. S. Hogness. 1975. Colony hybridization: a method for the isolation of cloned DNAs that contain a specific gene. Proc. Natl. Acad. Sci. USA 72:3961–3965.

68. Hanahan, D., and M. Meselson. 1980. Plasmid screening at high colony density. Gene 10:63–67.

69. Fitts, R., M. Diamond, C. Hamilton, and M. Neri. 1983. DNA–DNA hybridization assay for detection of *Salmonella* spp. in foods. Appl. Engiron. Microbiol. 46:1146–1151.

70. Hill, W. E., J. M. Madden, B. A. McCardell, et al. 1983. Foodborne anterotoxigenic *Escherichia coli*: detection and enumeration by DNA colony hybridization. Appl. Environ. Microbiol. 45:1324–1330.

71. Moseley, S. L., I. Huq, A. R. M. A. Alim, M. So, M. Samadpour-Motalebi, and S. Falkow. 1980. Detection of enterotoxigenic *Escherichia coli* by DNA colony hybridization. J. Infect. Dis. 142:892–898.

72. Scheverria, P., J. Seriwatana, O. Chityothin, W. Chaicumpa, and C. Tirapat. 1982. Detection of enterotoxigenic *Escherichia coli* in water by filter hybridization with three gene probes. J. Clin. Microbiol. 16:1086–1090.

73. Hill, W. E., W. L. Payne, and C. C. G. Aulisio. 1983. Detection and enumeration of virulent *Yersina enterocolitica* in food by DNA colony hybridization. Appl. Environ. Microbiol. 46:636–641.

74. Kaper, J. B., S. L. Moseley, and S. Falkow. 1981. Molecular characterization of environmental and nontoxigenic strains of *Vibrio cholerae*. Infect. Immun. 32:661–667.

75. Hudgson, A. L. M., and W. P. Roberts. 1983. DNA colony hybridization to identify *Rhizobium* strains. J. Gen. Microbiol. 129:207–212.

76. Sayler, G. S., R. K. Jain, A. Ogram, et al. 1986. Applications for DNA probes in biodegradation research. Proc. 4th Int. Symp. Microb. Ecol. 499–508. Sloverne Society for Microbiology, Ljubljana, Yugoslavia.

77. Ogram, A., and G. S. Sayler. 1988. The use of gene probes in the rapid analysis of natural microbial communities. J. Ind. Microbiol. 3:281–292.

78. Layton, A. C., and G. S. Sayler. 1990. Environmental application of nucleic acid hybridization. Annu. Rev. Microbiol. 44:625–648.

79. Denhardt, D. T. 1966. A membrane-filter technique for the detection of complementary DNA. Biochem. Biophys. Res. Commun. 23:641–646.

80. Singh, L., and K. W. Jones. 1984. The use of heparin as a simple cost-effective means of controlling background in nucleic acid hybridization procedures. Nucleic Acids Res. 12:5627–5638.

81. Renart, J., J. Reiser, and G. R. Stark. 1979. Transfer of proteins from gels to diazobenzyloxymethyl paper and detection with antisera: a method for studying antibody specificity and antigen structure. Proc. Natl. Acad. Sci. USA 76:3116–3120.

82. Jeffreys, A. J., and R. A. Flavell. 1977. A physical map of the DNA regions flanking the rabbit-globin gene. Cell 12:429–439.

83. Beltz, G. A., K. A. Jacobs, T. H. Eickbush, P. T. Cherbas, and F. C. Kafatos. 1983. *In* R. Wu, L. Grossman, and K. Moldave (eds.). Methods Enzymol. 100:266–285.

84. Casey, J., and N. Davidson, 1977. Rates of formation and thermal stabilities of RNA:DNA and DNA:DNA duplexes at high concentrations of formamide. Nucleic Acids Res. 4:1539–1552.

85. Bonner, T. I., D. T. Brenner, B. R. Neufeld, and R. J. Britten. 1973. Reduction in the rate of DNA reassociation by sequence divergence. J. Mol. Biol. 81:123–135.

86. Howley, P. M., M. A. Israel, M. Law, and M. A. Martin. 1979. A rapid method for detecting and mapping homology between heterologous DNAs. Evaluation of polyoma virus genome. J. Biol. Chem. 254:4876–4883.

87. Amasino, R. M. 1986. Acceleration of nucleic acid hybridization rate by polyethylene glycol. Anal. Biochem. 152:304–307.

88. Maniatis, T., E. F. Fritsch, and J. Sambrook. 1982. Molecular cloning: a laboratory manual. Cold Spring Harbor Laboratory, Cold Spring Harbor, New York.

89. Birnstiel, M. L., B. H. Sells, and I. F. Purdom. 1972. Kinetic complexity of RNA molecules. J. Mol. Biol. 63:21–39.

90. Melton, D. A., P. A. Krieg, M. R. Rebagliatti, T. Maniatis, K. Zinn, and M. R. Green. 1984. Efficient in vitro synthesis of biologically active RNA and RNA hybridization probes from plasmids containing a bacteriophage SPG promoter. Nucleic Acids Res. 12:7035–7056.

91. Zinn, K., D. DiMaio, and T. Maniatis. 1983. Identification of two distinct regulatory regions adjacent to the human β-interferon glue. Cell 34:865–879.

92. Alwine, J. C., D. J. Kemp, B. A. Parker, et al. 1979. *In* R. Wu (ed.). Methods Enzymol. 68:220–224.

93. Young, B. D., and M. L. M. Anderson. 1985. *In* E. D. Hanes and S. J. Higgins (eds.), Quantitative analysis of solution hybridization: a practical approach. IRL Press, Oxford.

94. Halvorson, H. O., and N. R. Ziegler. 1933. Application of statistics to problems in bacteriology. I. A means of determining bacterial population by the dilution method. J. Bacteriol. 25:101.

95. Halvorson, H. O., and N. R. Ziegler. 1933. Application of statistics to problems in bacteriology. III. A consideration of the accuracy of dilution data obtained by using several dilutions. J. Bacteriol. 26:559.

96. Moran, P. A. P. 1954. The dilution assay of viruses. J. Hyg. 52:189.

97. Taylor, J. 1962. The estimation of numbers of bacteria by ten-fold dilution series. J. Appl. Bacteriol. 25:54.

98. de Man, J. C. 1975. The probability numbers. Eur. J. Appl. Microbiol. 1:67.

99. de Man, J. C. 1977. MPN tables for more than one test. Eur. J. Appl. Microbiol. 4:307.

100. Russek, E., and R. R. Colwell. 1983. Computation of most probable numbers. Appl. Environ. Microbiol. 45:1646.

101. Colwell, R. R. 1979. Enumeration of specific populations by the most-probable-number (MPN) method, p. 56. *In* J. W. Costerton and R. R. Colwell (eds.), Native aquatic bacteria: enumeration, activity and ecology. American Society for Testing and Materials, Special Technical Publication No. 695, Philadelphia.

102. Giovannoni, S. J., E. F. DeLong, G. J. Olsen, and N. R. Pace. 1988. Phylogenetic group specific oligonucleotide probes for identification of single microbial cells. J. Bacteriol. 170:720–726.

103. DeLong, E. F., G. S. Wickham, and N. R. Pace. 1989. Phylogenetic strains: ribosomal RNA based probes for the identification of single cells. Science 243:1360–1367.

104. Olsen, G. J., D. J. Lane, S. J. Giovannoni, and N. R. Pace. 1986. Microbial ecology and evolution: a ribosomal RNA approach. Annu. Rev. Microbiol. 40:337–365.

105. Fox, G. E., K. R. Pechman, and C. R. Woese. 1977. Comparative cataloging of 16S ribosomal ribonucleic acid: molecular approach to prokaryotic systematics. Int. J. Syst. Bacteriol. 27:44–57.

106. Gupta, R., J. M. Lanter, and C. R. Woese. 1983. Sequence of the 16S ribosomal RNA from *Halobacterium volcanii*, an archaebacterium. Science 221:656–659.

107. Gutell, R. R., B. Weiser, C. R. Woese, and H. F. Noller. 1985. Comparative anatomy of 16S-like ribosomal RNA. Prog. Nucleic Acid Res. MOl. Biol. 32:155–216.

108. Hasegawa, M., Y. Iida, T. Yano, F. Takaiwa, and M. Iwabuchi, 1985. Phylogenetic relationships among eukaryotic kingdoms inferred from ribosomal RNA sequences. J. Mol. Evol. 22:32–38.

109. Jarsch, J., and A. Bock. 1985. Sequence of the 16S ribosomal RNA gene from *Methanococcus vannielli*: evolutionary implications. Syst. Appl. Microbiol. 6:54–59.

110. Klotz, L. C., and R. L. Blanken. 1981. A practical method for calculating evolutionary trees from sequence data. J. Theor. Biol. 91:261–272.

111. Komiya, H., M. Hasegawa, and S. Takemura. 1983. Nucleotide sequence of 5S rRNAs from sponge *Halichondria japonica* and tunicate *Halocynthia roretzi* and their phylogenetic positions. Nucleic Acids Res. 11:1969–1974.

112. Lane, D. J., B. Pace, G. J. Olsen, D. A. Stahl, M. L. Sogin, and N. R. Pace. 1985. Rapid determination of 16S ribosomal RNA sequences for phylogenetic analyses. Proc. Natl. Acad. Sci. USA 82:6955—6959.

113. Lane, D. J., D. A. Stahl, G. J. Olsen, D. J. Heller, and N. R. Pace. 1985. A phylogenetic analysis of the genera *Thiobacillus* and *Thiomicrospira* by 5S ribosomal RNA sequences. J. Bacteriol. 163:75—81.

114. Lane, D. J., D. A. Stahl, G. J. Olsen, and N. R. Pace. 1985. Analysis of hydrothermal vent-associated symbionts by ribosomal RNA sequences, *In* M. L. Jones (ed.), The hydrothermal vents of the East Pacific rise: an overview. Bull. Biol. Soc. Wash. 6:389—400.

115. Leffers, H., and R. A. Garrett. 1984. The nucleotide sequence of the 16S ribosomal RNA gene of the archaebacterium *Halococcus morrhuae*. EMBO J. 3:1613—1619.

116. Mackay, R. M., D. F. Spencer, M. N. Schnare, W. F. Doolittle, and M. W. Gray. 1982. Comparative sequence analysis as an approach to evaluating structure, function, and evolution of 5S and 5.8S ribosomal RNAs. Can. J. Biochem. 60:480—489.

117. Olsen, G. J., N. R. Pace, M. Nuell, B. P. Kaine, R. Gupta, and C. R. Woese. 1985. Sequence of the 16S rRNA gene from the thermoacidophilic archaebacterium *Sulfolobus solfataricus* and its evolutionary implications. J. Mol. Evol. 22:301—307.

118. Sogin, M. L., H. J. Elwood, and J. H. Gunderson. 1986. Evolutionary diversity of eukaryotic small subunit rRNA genes. Proc. Natl. Acad. Sci. USA 83:1383—1387.

119. Stackebrandt, E., W. Ludwig, and G. E. Fox. 1985. 16S ribosomal RNA oligonucleotide cataloging. Methods Microbiol. 18:75—107.

120. Stackebrandt, E., and C. R. Woese. 1981. The evolution of prokaryotes, p. 1—31. *In* M. J. Carlisle, J. R. Collins, and B. E. B. Moseley (eds.), Molecular and cellular aspects of microbial evolution. Cambridge University Press, Cambridge.

121. Tomioka, N., and M. Sugiura. 1983. The complete nucleotide sequence of a 16S ribosomal RNA gene from a blue-green alga, *Anacystis nidulans*. Mol. Gen. Genet. 191:46—50.

122. Woese, C. R., and G. E. Fox. 1977. Phylogenetic structure of the prokarytic domain: the primary kingdoms. Proc. Natl. Acad. Sci. USA 74:5088—5090.

123. Woese, C. R., M. Sogin, D. Stahl, B. J. Lewis, and L. Bonen. 1976. A comparison of the 16S ribosomal RNAs from mesophilic and thermophilic bacilli: some modifications in the Sanger method for RNA sequencing. J. Mol. Evol. 7:197—213.

124. Woese, C. R., E. Stackebrandt, T. J. Macke, and G. E. Fox. 1985. A phylogenetic definition of the major eubacterial taxa. Syst. Appl. Microbiol. 6:143–151.

125. Woese, C. R., and R. S. Wolfe (eds.). 1985. The bacteria, Vol. 8, Archaebacteria, pp. 582. Academic Press, Orlando, Florida.

126. Stahl, D. A., M. L. Sogin, and B. Pace. 1985. Rapid determination of 16S ribosomal RNA sequences for phylogenetic analysis. Proc. Natl. Acad. Sci. USA 82:6955–6959.

127. Datta, A. R., B. A. Wentz, D. Shook, and M. W. Trucksess. 1988. Synthetic oligodeoxyribonucleotide probes for detection of *Listeria monocytogenes*. Appl. Environ. Microbiol. 54:2933–2937.

128. Zeph, L. R., and G. Stotzky. 1989. Use of a biotinylated DNA probe to detect bacteria transduced by bacteriophage p1 in soil. Appl. Environ. Microbiol. 55:661–665.

129. Bentjen, S. A., J. K. Fredrickson, P. van Voris, and S. W. Li. 1989. Impact soil-core microcosms for evaluating the fate and ecological impact of the release of genetically engineered microorganisms. Appl. Environ. Microbiol. 55:198–202.

130. Saiki, R. K., D. H. Glenfund, S. Stoffel, et al. 1988. Primer-directed enzymatic amplification of DNA with a thermostable polymerase. Science 239:487–494.

131. Steffan, R., and R. M. Atlas. 1988. DNA amplification to enhance detection of genetically engineered bacteria in environmental samples. Appl. Environ. Microbiol. 54:2185–2191.

132. Stotzky, G., M. A. Devanas, and L. R. Zeph. 1990. Methods for studying bacterial gene transfer in soil by conjugation and transduction. Adv. Appl. Microbiol. 35:57–169.

133. Stotzky, G. 1989. Gene transfer among bacteria in soil, p. 152–222. *In* S. B. Levy and R. V. Miller (eds.), Gene transfer in the environment. McGraw-Hill Book Co., New York.

134. Schofield, P. R., A. H. Gibson, W. F. Duclman, and J. M. Watson. 1987. Evidence for genetic exchange and recombination of *Rhizobium* symbiotic plasmids in a soil population. Appl. Environ. Microbiol. 53:2942–2947.

135. Devanas, M. A., and G. Stotzky. 1986. Fate in soil of a recombinant plasmid carrying a *Drosophila* gene. Curr. Microbiol. 13:279–283.

136. Adams, R. L. P., J. T. Knowler, and D. P. Leader. 1986. The biochemistry of nucleic acids, p. 12–16. Chapman & Hall, New York.

137. Stevenson, F. J. 1985. Geochemistry of soil humic substances, p. 13–31. *In* G. R. Aiken, D. McKnight, and R. L. Wershaw (eds.), Humic substances in soil, sediment and water. John Wiley & Sons, New York.

138. Keoleian, G. A., and R. L. Curl. 1987. Effects of humic acid on the adsorption of tetrachlorobiphenyl by kaolinite, p. 233. *In* I. H. Suffet and P. MacCarthy (eds.), Aquatic humic substances. American Chemical Society, Washington, D.C.

5

Desert Varnish:
A Biological Perspective

JAMES T. STALEY, JOHN B. ADAMS, and FRED E. PALMER
University of Washington, Seattle, Washington

I. INTRODUCTION

In this chapter we review and discuss the evidence for a biological
origin for desert varnish. Desert varnish is a black to brown or
orange coating found on rocks in arid and semiarid environments
(Fig. 1). The coating ranges in thickness from micrometers to
millimeters and is rich in oxides of manganese and iron and in
clay minerals. Evidence indicates that desert varnish is formed
by an accretional process in which airborne dust serves as the
principal source of the elements in the coating.

This review examines varnishes that occur above ground level
on rock surfaces in arid and semiarid areas. Varnishes also occur
on rocks in alpine regions, glacial moraines, and in streams and
intertidal marine environments. For this reason, a more general
term "rock varnish" has been coined to refer collectively to these
various coatings and polished surfaces [1]. Rock varnishes that
lie outside the scope of this review include stains that penetrate
rock surfaces, siliceous glazes, coverings of algae or lichens, manganese dendrites, manganese nodules, and crack deposits. These
deposits are superficially similar to desert varnish, but typically
are different in mineralogy. There are other varnishes that are
morphologically and mineralogically similar to desert varnish, but
which are formed under different conditions. These include stream
varnishes, drip varnishes, and subglacier varnishes.

Within the past decade, there has been a significant increase
in the amount of research on desert varnish. Varnish has been

Figure 1   Desert varnish on basalt rocks from the Deem Hills in the Sonoran Desert near Phoenix, Arizona.   The darkest areas on the rocks have the heaviest varnish.

described from deserts throughout the world, although a comprehensive global comparison of occurrences and properties of varnish has not been made. Much of the work to date has been directed toward characterization of desert varnish and understanding its origin. For years the origin of varnish has been viewed as a puzzle, but the significance of the problem was not clear. More recently there has been a growing appreciation that desert varnish preserves a valuable record of past environmental conditions. Today, there is increased interest in varnish as a means of determining the exposure ages of natural rock surfaces and ancient manmade objects on time scales of thousands to tens or hundreds of thousands of years and as a way of inferring past climatic conditions.

To understand the significance of the ages of varnishes and to infer past environments it is essential to understand the mechanism by which varnish is produced. Both chemical and biological processes have been proposed for varnish formation, and although considerable progress has been made in narrowing the choices among various hypotheses, there remain important unanswered questions. For example, if wind-deposited soil is the principal source of the varnish constituents, what is the mechanism by which manganese, iron, and clays are selectively concentrated relative to other components in the source materials.

Largely because chemical mechanisms have not provided an explanation for the high relative concentrations of manganese, iron, and clays, several groups have explored the role of microorganisms in varnish formation. There is now abundant evidence that a variety of microorganisms inhabit desert varnish and desert rocks and soils. Many of the types of organisms have been shown to be capable of oxidizing manganese, suggesting to some investigators a mechanism whereby manganese oxides can be deposited from solution. Other investigators have reported the production of varnish in bacterial cultures. However, a full explanation of the proposed causative role of microorganisms remains elusive, and there is still no general agreement that desert varnish is a biological phenomenon.

II. EARLY WORK

Observations of rock varnish in stream environments date back at least to Humboldt [2] and Darwin [3]. American desert varnishes were discussed by Loew [4], Merrill [5], and Surr [6]; however, these and other early accounts are not specific about the chemistry or origin of the coatings. Francis [7] appears to have provided the first credible argument for the biological origin of rock

varnish. He described black iron and manganese oxides on rocks
on land and in streams on cleared areas of former dense subtropical
rain forests in Australia, and he attributed the coatings to the activities of lichens and algae. He noted the similarity of his stream
varnish with that reported by Humboldt [2].

White [8] suggested that in the Mojave and Sonoran Deserts windblown plant materials, especially pollen, might carry the ingredients
of desert varnish to the rocks where they could be released following decomposition. The desert varnish in the Mojave Desert near
Stoddard's Well in California was described in detail by Laudermilk
[9]. He agreed with Francis [7] that lichens may provide a mechanism for the formation of desert varnish. Microscopic observations showed that small (0.2-mm diameter average) dark "lichens"
occurred on the varnish near its margins, on the rock near the
edge of the varnish, and, less abundantly, on the rock distant
from the varnish. They were nearly absent from the surface of
well-developed varnish, except at the margins.

Laudermilk proposed a model of varnish formation based on the
dissolution of minerals in the rock by lichen acids. The minerals
would be absorbed from the rock by lichens and some of the manganese and iron precipitated as hydroxides. Eventually, the lichens
would be "choked" and killed by the iron and manganese oxides.
With the decomposition of the lichens, acids would again be produced, which would aid in the smooth distribution of the minerals
on the surface. Although Laudermilk described the small lichen
colonies as being about 0.2 mm in diameter, he noted that there
were many that were much smaller. He cultivated fungi from the
larger type and apparently assumed that the smaller ones were
younger colonies of the same type. He made no attempt to examine the smaller microbial biota, the bacteria, algae, and fungi.
His observational data is very suggestive of a biological mechanism
for varnish deposition, but causal relationships were not established.

Laudermilk [9] also reported the chemical analysis of a varnish
sample from a rhyolite. He was surprised to find that 90% of the
varnish was SiO_2. As an explanation, he suggested that much fine
dust was imbedded in the varnish. Although not recognized at the
time, this was an indication that clay minerals, as well as iron and
manganese oxides, are an important characteristic of the varnish.

III. TEXTURE AND STRUCTURE

Among those who have not studied desert varnish in the field, there
is a common misconception that varnish occurs as a uniform deposit
that totally obscures the underlying rock. Varnish deposits in some

localities are uniform and continuous; however, most are discontinuous, patchy, and variable in thickness, texture, and structure. The variability of varnish at both macroscopic and microscopic scales is partly the result of variability in the substrate, as the first requirement for varnish formation is that the substrate be stable over the period of varnish formation. Even the most durable rocks show evidence of cracking, spalling, or abrasion; therefore, some non-varnished areas commonly are exposed on even the most ancient and durable varnished surfaces.

Varnish also is texturally variable by its nature. The deposits commonly accumulate in cracks and pits in the substrate and on porous surfaces, as opposed to smooth ones, such as quartz grains. In thin section (Fig. 2) and in scanning electron microscope (SEM)

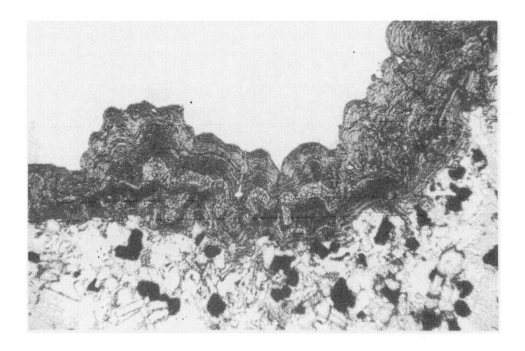

Figure 2 Ultrathin section (ca. 5-μm thick) of desert varnish on basalt from Deem Hills, Arizona. Note the laminations in the varnish layer, which is about 100-μm thick. The contact with the underlying basalt is sharp. Basalt consists of feldspar, pyrexene, and glass (white), with scattered grains of magnetite (black).

images, varnish ranges from thin subparallel layers to highly diverse botryoidal and pinnacle structures. Field and microscopic observations indicate that desert varnish begins as discrete spots that coalesce as they enlarge, rather than as uniform thin films. These observations have suggested a biological origin [1,10,11].

The predominant biological forms observed on desert varnish with the light microscope (Fig. 3) and SEM (Fig. 4) are microcolonial fungi (MCF). They commonly are abundant in all but the driest desert conditions. Hyphae and colonies of MCF, and possibly other fungi, may become varnished and contribute to the varnish surface texture and structure [11]. Honeycomblike structures that appear to be remnants of MCF colonies have been observed under the varnish surface where sections have been created by breaking the varnish [11]. It also appears that MCF and lichens can create pits in the varnish surface (Fig. 5), probably through the production of organic acids [12].

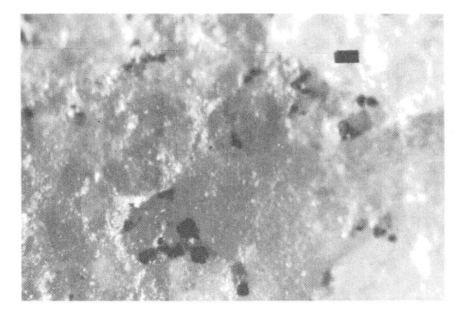

Figure 3 Macrophotograph illustrating the appearance of microcolonial fungi (MCF) on an unvarnished rock surface. The MCF are more difficult to distinguish by light microscopy on the black varnished surface because they are also black, owing to melanin-type pigments. Bar represents 100 μm.

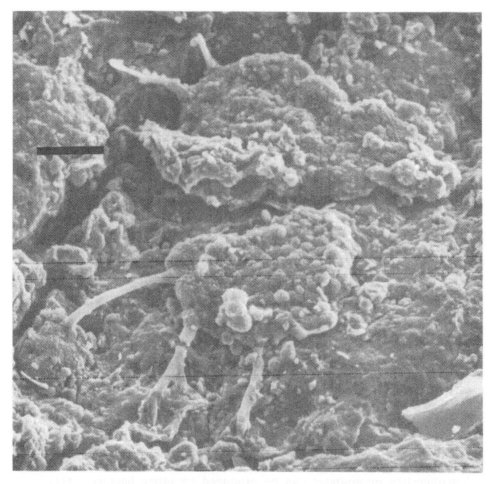

Figure 4   MCF found on a varnished rock from Deem Hills, Arizona. Note the "strawberry runner" hyphae. Bar represents 20 µm.

Bacteria are rarely seen on varnishes, even with the SEM. This may be the result of the small fields of view obtained at the required magnifications, of the difficulty in distinguishing bacteria from other particles, and perhaps because critical-point drying generally has not been used in sample preparations. Dorn and Oberlander [13] suggest that structures shown in an SEM image are *Metallogenium*-like bacteria. These structures appear to be bacteria; however, it should be noted that there is some doubt about

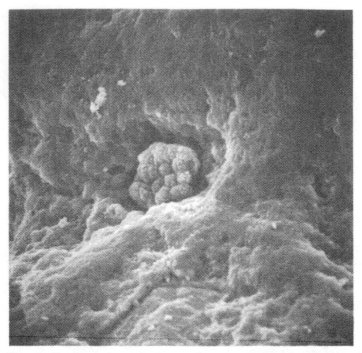

Figure 5 Scanning electron micrograph illustrating an MCF colony
on varnish from an Australian rock. Note that the colony appears
in a recessed area on the surface.

the identification of *Metallogenium* (see later discussion), and *Metallogenium*-like precipitates can be produced by other bacteria [14].

IV. CHEMISTRY AND MINERALOGY

Before the development of the electron microprobe and energy-dis-
persive X-ray analysis (EDAX) combined with SEM, Engel and Sharp
[15] analyzed varnish from the Mojave Desert and from Death Valley
with wet chemical and spectrographic methods. The varnish was
scraped from the substrate and, therefore, the analyses required
correction for contamination by the host rock. Extensive analyses
of varnish awaited the development of the electron microprobe and
SEM-EDAX. During the past decade several studies using these
methods have provided detailed chemical information on the internal

structure of varnish and on the relationship of varnish to the adjacent substrate (Perry, MS thesis, University of Washington, Seattle, 1979) [10,16—21].

Engel and Sharp [15] showed that the main elements in order of abundance were O, H, Si, Al, Fe, Mn, Mg. Ca, Na, P, Ti, and K. The concentrations of Mn and Fe were notably high, relative to the substrates and to airborne dust. These chemical results, which have been confirmed in many subsequent studies, are consistent with data that indicate that the mineralogy of varnish consists largely of mixed-layer illite—montmorillonite clay that is cemented by birnessite and hematite [18—20].

There is considerable variability in the proportions of Mn and Fe in varnishes, even within a single sample, and in varnishes from different localities. The familiar dark brown to black or purplish varnish may have over 20% MnO_2 and variable amounts of Fe_2O_3. Varnish with a low concentration of Mn is typically orange to brick red, as the result of the presence of hematite. The orange varnishes usually are not richer in iron, but, instead, have less of the black MnO_2 that masks the color of the iron. In some samples variations in the proportions of Mn and Fe produce black and orange layers that are a few micrometers thick [10,21].

Potter and Rossman [20] found that the average oxidation state of Mn in three varnish samples from the Mojave Desert ranged from +3.76 to +3.93. They also concluded that "Fe must be predominantly in the +3 oxidation state based on mineralogy and absence of spectroscopic features due to Fe^{2+} near 1000 nm." There appears to be an absence of chemical data on the oxidation state of iron in varnish. Engel and Sharp [15] measured total iron in their varnish samples as Fe_2O_3, although they determined FeO and Fe_2O_3 for larger samples of the host rocks. On the basis of the evidence presented by Potter and Rossman [20], virtually all of the manganese in varnish occurs in the +4 state and essentially all of the iron occurs in the +3 state.

Engel and Sharp [15] also determined that the significant trace elements in varnish were Cu, Co, Ni, Pb, Ba, Cr, Yb, B, Y, Sr, and V. Additional samples from Death Valley and from Nevada were analyzed spectrographically by Lakin et al. [22] who found that the amounts of Co, Ba, La, Mo, Ni, Pb, and Y were correlated with the concentration of Mn. Knauss and Ku found that Hg, Zn, As, Cu, Ba, Co, V, Mo, and Ce were the relatively abundant trace elements in a varnish sample from Utah [23]. Bard (PhD thesis, University of California, Berkeley) and co-workers analyzed varnishes from the Great Basin using neutron activation and X-ray fluorescence [24, 25]. Bauman noted the similarity between the elemental abundances of deep-sea ferromanganese nodules and varnish [26]. Dorn and Oberlander [1] concluded that the high concentrations of Cu, Ni, and Zn in older varnishes from the western United States were consistent with cation scavenging by manganese oxides and hydroxides.

There is now general agreement that varnish is chemically unrelated to the substrate and that varnish commonly occurs as micrometer-scale layers that vary in composition, notably in the Mn/Fe ratio [1,10,21]. The chemical data are not consistent with models in which typical varnish is derived from its substrate by leaching or from adjacent soil by capillary action. Attention instead has been turned to airborne dust as the primary source for the chemical constituents of varnish.

Engel and Sharp [15] calculated enrichment ratios for varnish relative to the host rocks, noting that MnO was enriched by a factor of 66.2 to 292 and that Fe_2O_3 was enriched by a factor of 1.7 to 6.1. Although they favored the conclusion that the rocks were the source of the varnish, they also considered soil and airborne dust as sources. Relative to the airborne dust, MnO was enriched by a factor of over 117, and Fe_2O_3 was enriched by a factor of 3.6; relative to local soils MnO was enriched by 29 to 43, and Fe_2O_3 by 2.0 to 2.1.

Dorn and Oberlander [1] pointed out that the Fe/Mn ratios for airborne material and for soils in the desert areas of the western United States are similar, typically ranging from 40:1 to 70:1, suggesting that these materials have a common origin. The chemical data for airborne dust are in general agreement with these values (C. D. Elvidge, MS thesis, Arizona State University, Tempe) [15, 27]. In contrast, Fe/Mn ratios for black varnish are typically in the range of 1:1 to 3:1, reflecting the relative enrichment of Mn.

In their study of the chemistry and mineralogy of varnish, Potter and Rossman [19] concluded that "the oxides appear to be transported by water, and it is most likely that the clays are transported by wind or water." Others [1,10,11] have cited morphological evidence from thin sections and SEM data that varnish has the form of a precipitated deposit, rather than that of only an accumulation of detrital grains. The chemical and morphological observations place constraints on the possible mechanism of formation of varnish.

To produce varnish from airborne dust at the surface of the substrate at least three main steps seem to be required: (1) Mn, Fe, and other elements, such as Co, Ni, Cu, must be taken into solution; (2) the same elements must be selectively precipitated onto the substrate itself or onto materials, such as clay minerals, that become attached to the substrate; and (3) the airborne dust from which the elements have been extracted must be removed from the substrate, permitting a new supply of the source material. The process of solutional extraction and precipitation, furthermore, must be able to concentrate Mn by a factor of 100 or more, Fe by a factor of 3 or 4, and certain other elements by factors of 2 to 20 or more.

Virtually all chemical studies of the origin of varnish have focused on the problem of how Mn is concentrated. The range of E_h and pH conditions under which Mn is soluble in water is well known, therefore, attention has been given to possible natural E_h—pH fluctuations, especially in desert regions that might cause solution and precipitation of Mn. Elvidge and associates have proposed a physicochemical model for the formation of varnish that entails small fluctuations in E_h and pH within the airborne dust on a substrate (C. D. Elvidge, MS thesis, University of Arizona, Tempe) [29,30]. They suggested that under neutral or acid conditions Mn^{2+} will be dissolved from the dust, whereas a change to more alkaline conditions will cause Mn^{2+} to be oxidized to the Mn^{4+} state, thereby precipitating MnO_2 onto clays and onto the substrate. Dorn and Oberlander [1], however, argued that Fe-rich orange varnishes are associated with arid alkaline environments, whereas the Mn-rich black ones are characteristic of moderately arid near-neutral environments.

The high concentration of Mn in varnish relative to possible source materials has led several workers to propose that biological agents are involved. Dorn and Oberlander [31] were the first to present a detailed discussion of a biological model for varnish that entails oxidation of Mn by bacteria and concomitant precipitation of Mn-oxide to cement clay particles to the substrate. The model presumes that Mn is available in aqueous solution through the interaction of moisture and airborne dust on the substrate and that Mn-depleted dust is removed from the substrate by the action of wind or water.

Curiously, there has been little study of the mechanism responsible for the concentration of Fe in varnish, although Fe is the most abundant element after Si and Al, and Fe commonly is concentrated by a factor of 3 to 4 relative to dust or soil source materials. Iron in the 3+ state is essentially insoluble under the E_h—pH conditions expected at rock surfaces [28]; therefore, to be dissolved from the airborne dust the Fe must be in the 2+ state or in an organic molecule such as a siderophore that complexes Fe^{3+}. Most Fe in desert soils and in airborne dust is in the 3+ state [27], suggesting that in the formation of varnish, the extraction of any Fe^{2+} would require very large quantities of source material or that organic complexes are important in mobilizing Fe.

The chemistry of varnish, although not fully understood in terms of origin, has been used to date the time of formation of the varnish itself and, thereby, to constrain the exposure age of the

substrate. The uranium series of radioisotopes was used to date
varnish from the Colorado Plateau in Utah at greater than 300,000
years BP [23]. This method, however, assumes that the varnish
is a closed system and that there is no external source of thorium
and protactinium. The dates obtained by this method are considerably older than those obtained by other methods from other deserts. It is now considered doubtful that varnish is a closed system, given the evidence for selective leaching of such elements as
K and Ca [31—33] and, therefore, it is unlikely that reliable dating
with uranium series radioisotopes is possible.

Older varnish samples that have been dated independently were
observed to be selectively depleted in K and Ca relative to Ti,
presumably by slow leaching of cations over time [32,34]. By
analyzing cation ratios for several varnish samples on substrates
of known age, and for adjacent soils, an empirical dating curve
was produced for each locality. This dating method entails scraping
samples of varnish for chemical analysis and requires "calibration"
to local soils, as the $(K + Ca)/Ti$ values varies with the local and
regional geology. The method is empirical in that the complete
leaching profile of elements in varnish has not been established,
and little is known about rates of leaching and how they vary with
changing environmental conditions. Nevertheless, the method has
been used to date a variety of rock surfaces in deserts [25,32,36,
37].

Dorn et al. [36] pioneered the use of ^{14}C as a means of dating
varnish, taking advantage of accelerator mass spectrometer (AMS)
technology to analyze small amounts of carbon. Organic carbon is
trapped in varnish in the form of detrital material, including pollen
grains, and as microorganisms. Carbon derived from carbonate
minerals in the varnish (especially from caliche dust) was removed
by treating samples with HCl.

Carbon dioxide from the atmosphere is a potential source of contamination for dating varnish by the ^{14}C method. If varnish is
open to cation leaching, it can be expected that it is also open to
atmospheric carbon. Dorn et al. [38] found that ^{14}C dates for
clay-bearing samples were consistently younger than for the same
samples after the silicates were removed by treatment with HF.
Carbon in the remaining material, the manganese and iron oxides,
gave older dates that were considered to be more correct, based
on independent field evidence. These results suggest that varnish
is at least a partially open chemical system, with the manganese
and iron oxides being the least subject to chemical modification
with time. This view is also consistent with the results of Dorn
and co-workers [39—41] who found variations in the ^{13}C content
in manganese and iron oxides in varnish layers, which they related to changes in paleoenvironmental conditions.

Because ^{14}C dating is a well-developed technology, its application to desert varnish has been generally accepted. However, the difficulty of sample preparation and limited access to AMS facilities has hampered widespread use of the technique. The details of the analytical method and of the applications are beyond the scope of this chapter, and these can be found in Dorn [35] and Dorn et al. [38].

V. EVIDENCE FOR BIOLOGICAL ORIGIN

If desert varnish is produced by a biological mechanism, then organisms must be directly or indirectly responsible for its typical elemental, mineralogical, and textural characteristics. As with the chemical studies discussed previously, most biological research has been directed at a biological agent responsible for manganese concentration. Furthermore, most research has centered on the hypothesis of the immobilization of manganese by a simple one-step biological oxidation and precipitation.

If microorganisms are the causative agents of desert varnish, they would be expected to be found in all varnishes that are actively being formed. Not only should microorganisms be present, but it should be possible to detect their metabolic activities on active varnish surfaces and to cultivate them. Furthermore, it should be possible to use cultures in the laboratory to produce reactions that are characteristic of those of varnish. Indeed, it might even be possible to produce a varnish in the laboratory under conditions that simulate those of the natural environment by using microorganisms. The foregoing is a modified statement of Koch's postulates, which are the criteria used to prove that a microroganism is the causative agent of a disease. Microbial ecologists have also used these postulates to demonstrate that a process has a biological origin. As will be discussed, investigators have followed some or all of these steps to address the question of a biological origin of desert varnish.

Although early studies, such as those of Laudermilk [9], provided suggestive evidence for the involvement of microorganisms in desert varnish formation, little attention was directed to this issue until the last half of this century. Varnish contains small amounts of organic material (<1% [39]) as well as living and metabolically active microorganisms.

Taylor-George et al. [11] measured microbial activities in desert varnish on basalt outcrops in the Sonoran Desert near Phoenix, Arizona. Activity measurements showed no detectable carbon dioxide fixation (measured with ^{14}C-bicarbonate) on varnish samples collected from the southern exposures at this site. This suggested that there was little or negligible potential for autotrophic activities by phototrophic or chemolithotrophic bacteria. The rocks studied were devoid of any evidence of lichens, although some lichens were found on varnished rocks from northern exposures at this same site. These observations discounted the necessity for lichens or other phototrophic organisms in desert varnish formation, assuming, of course, that this was an area of active varnish development, an observation that was supported by radiocarbon dating results (unpublished data). In contrast, active respiration of radiolabeled sodium acetate was measured on these same varnish surfaces indicating the presence of heterotrophic organisms that were metabolically active. The heterotrophic bacterial concentrations ranged from 1100 to 3200/cm^2 of varnish scrapings, actinomycetes from 400 to 630/ cm^2, and slow-growing fungi, now called microcolonial fungi (MCF), occurred at concentrations of 600 to 1500/cm^2 of varnish.

A similar study of rocks in the Negev Desert of Israel indicated that from 250 to 2900 heterotrophic bacteria were present per square centimeter of scraped varnish from limestone rocks characteristic of that region [42]. Counts from scrapings were almost always considerably higher than surface swab counts, suggesting that these organisms were indigenous to the varnish and not merely airborne flora that had settled at the site. Other microbial groups were not detected in high concentrations from these Negev samples.

Considerable microbiological work has centered on the possible role that manganese-oxidizing microorganisms might have in desert varnish formation. It has been argued that, because the final deposit contains abundant Mn^{4+} oxide, bacteria and other microorganisms capable of oxidizing Mn^{2+} may be responsible for the deposition of Mn^{4+} in the varnish. Consequently, particular attention has focused on determining the presence and concentrations of manganese-oxidizing bacteria in desert varnish. Thus, when Krumbein [43] found manganese- and iron-precipitating microorganisms, including lichens, algae, fungi, and bacteria, on varnished rock surfaces in the Negev Desert, he concluded that these organisms were important in the formation of varnish. Similarly, in the study by Taylor-George et al. [11] some 9 to 33% of the bacteria tested were able to oxidize manganese. Even more striking results were recorded for varnish samples from the Negev Desert in which 83 to 100% of the bacteria isolated were able to oxidize manganese in the laboratory [42]. Although these findings implicate manganese-oxidizing bacteria in varnish formation, without other evidence, it cannot be concluded that they are responsible for the process.

The identification of the manganese-oxidizing bacteria from varnish also has been undertaken. Palmer et al. [14] found that most of the isolates from the Sonoran and Mojave Deserts were gram-positive bacteria. Representatives of the genera, *Micrococcus, Planococcus, Arthrobacter,* and *Bacillus,* were reported from that study (one isolate, first thought to be a member of the budding bacteria, was subsequently identified as a member of the genus, *Geodermatophilus*). A *Bacillus* sp. also was reported earlier in varnish samples from Sonoran Desert rocks examined by Perry and Adams [10]. Some gram-negative rods that were manganese oxidizers and that did not appear to be members of the genus *Arthrobacter,* also were found in the Sonoran Desert varnish [14], but these were not identified. A study of manganese oxidizing bacteria from the Negev Desert resulted in the identification of some of the same genera including *Micrococcus, Arthrobacter, Bacillus,* and *Geodermatophilus* [14].

The bacteria that have been identified as manganese oxidizers from varnishes are common heterotrophic bacteria. They are not known to derive energy from the process of manganese oxidation and they are not known to be able to carry out carbon dioxide fixation. Instead, they must rely on organic materials for their growth. These results are consistent with not finding primary productivity occurring on varnish rock samples from the Sonoran Desert, using radiolabeled carbon dioxide as a carbon source [11]. Thus, microorganisms in varnish presumably derive their organic nutrients from the same windblown dust that is the source of the iron, manganese, and clays.

The bacteria that have been isolated from varnish are principally gram-positive organisms. In contrast, bacteria isolated from aquatic environments, and more traditionally viewed as iron- and manganese-depositing bacteria, are predominantly gram-negative and include such genera as *Leptothrix* and other sheathed bacteria and the exotic unicellular prosthecate bacteria, such as *Pedomicrobium* and *Hyphomonas* [44]. In addition, "*Metallogenium*" has been described primarily from aquatic habitats, but some have questioned whether or not it is a living organism [45,46].

Microcolonial fungi also have been implicated in varnish formation. These appear typically as microscopic (ca. 100 µm diameter), black, colonies on desert rock surfaces. They occur on unvarnished as well as on varnished rocks [47]. They may well be the "small lichens" that Laudermilk observed in his early work on desert varnish of the Mojave Desert [9]. It is thought that MCF also must rely on organic carbon sources brought to the rock surface in windblown dust [47]. In any event, rock is not usually regarded as a substratum for fungal growth, and this, coupled with the difficulty of cultivating these microorganisms, has deterred mycologists from extensively investigating this group. However, studies that have been performed have allocated this group to a genus called *Lichenothelia* [48,49]. Because it is not yet known whether all

MCF are members of this genus, the vernacular term, microcolonial fungus, is used to denote this group in this chapter.

The MCF are able to survive desiccation and exposure to high temperatures. Laboratory investigations of rock chips colonized by MCF from Oregon and Arizona indicate that these microorganisms can readily survive 70 and 80°C, respectively, for periods of 21 days [50]. Indeed, some MCF from Arizona survived 100°C for 2 days. Survival at high temperatures was generally favored at low relative humidities (0% vs 100%). Interestingly, although these MCF can survive high temperatures, they grow best at room temperature, and most do not grow above 35°C. Moreover, those MCF studied require water activities of at least 0.95 for growth. Therefore, the MCF that have been studied are able to survive the harsh conditions of high temperature and low water activity encountered in the summer in desert environments, but their growth is restricted primarily to periods when moisture is available and the temperatures are lower.

Evidence for the involvement of MCF in desert varnish formation is observational. Taylor-George et al. [11] found them to be physically incorporated within developing varnish layers when observed by SEM. This implies that the MCF may be providing favorable conditions for the cementing of the varnish onto the rock or underlying varnish substratum. In an investigation of MCF from Australian varnish samples, MCF were found on some areas on rocks in which the manganese deposition occurred only within the colony of the individual MCF (Fig. 6) [51]. However, it was not possible to isolate MCF from these rocks, and it is not known whether they were able to oxidize manganese or grow in association with bacteria that are. Alternatively, it could be hypothesized that these fungi prefer to grow in areas where manganese oxides accumulate by chemical or physical means.

Although the discovery of manganese-oxidizing bacteria and manganese-depositing MCF is suggestive of their precipitation in desert varnish formation, it does not prove that these organisms are responsible for manganese deposition and concentration. Two separate studies have undertaken a "proof of process" by using the approach of Koch's postulates. Manganese-oxidizing or precipitating microorganisms have been isolated from varnish, and these pure cultures produced varnishes under laboratory conditions. In one study, Krumbein and Jens [52] isolated 17 strains of manganese- and iron-precipitating fungi and 28 strains of bacteria, including members of the genus *Arthrobacter*, from Negev Desert varnish samples. In addition, some cultures of cyanobacteria that were able to precipitate manganese were also isolated. To demonstrate varnish formation in the laboratory, rock particles

Figure 6 An SEM of an MCF colony from an Australian rock. EDAX analyses indicate that manganese is found only in the interior of the colony, not in any of the designated outlying areas. Bar represents 10 μm.

and acid-cleaned quartz sand were embedded in fungal growth medium, such that the substrates were exposed above the agar medium. The medium was inoculated with isolates of the bacteria, fungi, and cyanobacteria. Iron and manganese oxide films were observed to form on the surfaces of the rock and sand particles. The mineralogy of the surfaces was not investigated.

Dorn and Oberlander [1,31] followed a similar approach. They isolated manganese-oxidizing bacteria from desert varnish from the Mojave Desert near Barstow and in Death Valley. Their cultures included a budding *Pedomicrobium*-like bacterium and a "*Metallogenium*"-like bacterium, neither of which has been reported from Sonoran Desert sources. They conducted their experiments in liquid cultures containing sterile chips of gneiss. They reported that varnish was formed on the rock chips in 6 months. The artificial varnish contained about 80% clay from the bentonite that was added to the medium. To our knowledge the mineralogy of the manganese and iron oxides and hydroxides was not characterized.

The production of artificial varnishes in the laboratory is a logical approach to the resolution of whether desert varnish is formed by a biological process. However, the experiments cited could not reproduce the natural conditions under which desert varnish forms. Because microorganisms grow slowly under desert conditions and because varnish may be deposited very slowly, it may not be possible to conduct an entirely realistic experiment in the laboratory.

There are the additional problems that manganese deposition constitutes only one of the steps involved in the deposition of iron-, manganese-, and clay-rich coatings, and that the oxidation of manganese alone may not be the sole process responsible for its deposition. Even if manganese deposition is caused by microorganisms, this does not explain the process for iron deposition, which may not involve an oxidation step or even microorganisms. In addition, there remains the question of how microorganisms interact with clays to form varnish.

VI. SUMMARY AND CONCLUSIONS

There has been an increase in interest in using varnish to decipher the paleoenvironmental record in deserts. Geomorphologists are now assessing varnish as one way to decipher the exposure ages of landscapes. Stable carbon isotopes in varnish may provide a way to reconstruct climatic and vegetation histories. Deserts are especially sensitive to climatic changes, and varnish deposits may help to infer past climatic conditions on the time scales of tens to hundreds of thousands of years. Varnished artifacts may provide a unique means of assigning ages to early human cultures.

To realize the full potential of desert varnish as a paleoenvironmental indicator, it is necessary to understand more fully the mechanisms of formation and alteration of varnish and the responses of the mechanisms to environmental conditions. Several key indicators depend critically on the accumulation or exchange of chemical components in varnish over long periods.

At the most basic level, it is necessary to know whether desert varnish is fundamentally a biological deposit and, therefore, whether its origin is an ecological problem, rather than a geological one. Although recent biological research implicates microorganisms as the causative agents of desert varnish, many basic problems still remain. For example, it is not yet possible to obtain from a culture collection a bacterium or a group of bacteria that can be used to produce an artificial varnish in the laboratory in fulfillment of Koch's postulates. Microbiologists need to demonstrate unequivocally that the microorganisms they culture can be used to produce varnish under conditions that closely simulate those of the natural environment. Furthermore, the varnish produced should be mineralogically similar to natural varnish.

The actual mechanism of varnish deposition needs to be studied at the molecular level. Are enzymes involved? How are the manganese and iron solubilized and reprecipitated to produce the final oxides of the deposit? Is clay important in the process; is it involved through the physical adsorption of oxides or in other ways?

It is also important to continue to explore whether varnish formation or some aspect of its formation is mediated by abiological mechanisms. Not all researchers accept the causative role of microorganisms in varnish formation. For example, Smith and Whalley [53,54] studied iron- and manganese-rich "desert varnish" on North African rocks and concluded that there was no evidence for microbial involvement in the coatings. This study, however, prompted a discussion with Dorn [55] about whether the Moroccan silica glaze studied by Smith and Whalley was a true (i.e., clay-rich) desert varnish. Clearly, there is a need to standardize the terminology and, perhaps, a need for coordinated research on the same samples.

However, a compelling case for an abiotic origin for the Mn- and Fe-rich clay varnishes remains to be made. Dorn [55] and Smith and Whalley [53,54] appear to agree that the absence of organisms does not prove an abiotic origin, and the presence of organisms does not prove a biotic origin. Proposed abiotic mechanisms for producing varnish are still focused on "feasible" changes in E_h–pH conditions for precipitation of manganese. Such proposed changes need to be observed in the field or reproduced in laboratory experiments, along with the conditions necessary to concentrate iron and clays.

Finally, it is possible that there may be a complex interaction between the principal chemical constituents, the biological components, and the physical environment. This, combined with the slow rate at which varnish formation occurs, may make the next level of study of desert varnish particularly challenging.

ACKNOWLEDGMENTS

Paul Bierman provided helpful comments on the manuscript. This work was supported, in part, by NASA grants NAGW-1319 and NAGW-85.

REFERENCES

1. Dorn, R. I., and T. M. Oberlander. 1982. Rock varnish. Prog. Geogr. 6:317–367.

2. Humboldt, A. 1852. Personal narrative, Vol. 2, p. 243–246. Henry G. Bohn, London.

3. Darwin, C. 1871. Natural history and geology, 1st ed., p. 2–13. Appleton and Co., New York.

4. Loew, O. 1876. U.S. geographic and geological surveys west of the 100th meridian, p. 179. Annu. Rep. Chief of Engineers, Appendix JJ.

5. Merrill, G. P. 1898. U.S. Geo. Sur. Bull. 150:389–391.

6. Surr, G. 1909. Granites. Mining Sci. Press 99:712–714.

7. Francis, W. D. 1921. The origin of black coatings or iron and manganese oxides on rocks. Proc. R. Soc. Queensl. 32: 110–116.

8. White, C. H. 1924. Desert varnish. Am. J. Sci. 9:413–420.

9. Laudermilk, J. D. 1931. On the origin of desert varnish. Am. J. Sci. 21:51–66.

10. Perry, R. S., and J. B. Adams. 1978, Desert varnish: evidence for cyclic deposition of manganese. Nature 276:489–491.

11. Taylor-George, S., F. Palmer, J. T. Staley, D. J. Borns, B. Curtiss, and J. B. Adams. 1983. Fungi and bacteria involved in desert varnish formation. Microbiol. Ecol. 9:227–245.

12. Dorn, R. I. 1986. Rock varnish as an indicator of aeolian environmental change, p. 291–307. *In* W. G. Nickling (ed.), Aeolian geomorphology. Allen & Unwin, London.

13. Dorn, R. I., and T. M. Oberlander. 1981. Microbial origin of desert varnish. Science 213:1245–1247.

14. Palmer, F. E., J. T. Staley, R. G. E. Murray, T. Counsell, and J. B. Adams. 1985. Identification of manganese-oxidizing bacteria from desert varnish. Geomicrobiol. J. 4:343–360.

15. Engel, C. G., and R. S. Sharp. 1958. Chemical data on desert varnish. Geol. Soc. Am. Bull. 69:487–518.

16. Hooke, R. L., H. Yang, and P. W. Weiblen. 1969. Desert varnish: an electron microprobe study. J. Geol. 77:275–288.

17. Allen, C. C. 1978. Desert varnish of the Sonoran Desert: optical and electron probe microanalysis. J. Geol. 86:743–752.

18. Potter, R. M., and G. R. Rossman. 1977. Desert varnish: the importance of clay minerals. Science 196:1446–1448.

19. Potter, R. M., and G. R. Rossman. 1979. The tetravalent manganese oxides: identification, hydration, and structural relationships by infrared spectroscopy. Am. Mineral. 64:1199–1218.

20. Potter, R. M., and G. R. Rossman. 1979. The manganese- and iron-oxide mineralogy of desert varnish. Chem. Geol. 25:79–94.

21. Dorn, R. I. 1984. Cause and implications of rock varnish microchemical laminations. Nature 310:767–770.

22. Lakin, H. W., C. B. Hunt, D. F. Davidson, and U. Oda. 1963. Variation in minor-element content of desert varnish. USGS Professional Paper 475-B:B28—B31.

23. Knauss, K. G., and T. Ku. 1980. Desert varnish: potential for age dating via uranium-series isotopes. J. Geol. 88: 95—100.

24. Bard, J. C., F. Asaro, and R. F. Heizer. 1976. Perspectives on dating of Great Basin petroglyphs by neutron activation analysis of the patinated surfaces. Lawrence Berkeley Laboratory Report No. 4475.

25. Bard, J. C., F. Asaro, and R. F. Heizer. 1978. Perspectives on the dating of prehistoric Great Basin petroglyphs by neutron activation analysis. Archaeometry 20:85—88.

26. Bauman, A. J. 1976. Desert varnish and marine ferromanganese oxide nodules: congeneric phenomena. Nature 259: 387—388.

27. Pewe, T. L., E. A. Pewe, R. H. Pewe, A. Journaux, and R. M. Slatt. 1981. Desert dust: characteristics and rates of deposition in central Arizona, p. 169—190. *In* Desert dust: origin, characteristics, and effects on man. Geol. Soc. Am. Spec. Paper 186.

28. Nielands, J. B. 1984. Siderophores of bacteria and fungi. Microbiol. Sci. 1:9—14.

29. Elvidge, C. D., and C. B. Moore. 1979. A model for desert varnish formation. Geol. Soc. Am. Abstr. Prog. 11:271.

30. Elvidge, C. D., and C. J. Collet. 1981. Desert varnish in Arizona: distribution and spectral characteristics. ASP-ACSM fall meeting, pp. 215—222.

31. Dorn, R. I., and T. M. Oberlander. 1981. Rock varnish origin, characteristics, and usage. Z. Geomorphol. 25:420—436.

32. Dorn, R. I. 1983. Cation-ratio dating: a new rock-varnish age-determination technique. Quat. Res. 20:49—73.

33. Dragovich, D. 1984. Desert varnish as an age indicator for Aboriginal rock engravings: a review of problems and prospects. Archaeol. Oceania 19:48—56.

34. Dorn, R. I. 1989. Quaternary alkalinity fluctuations recorded in rock varnish microlaminations on western U.S.A. volcanics. Paleogeogr. Paleoalimatol. Paleoegol. (in press).

35. Dorn, R. I. 1989. A critical evaluation of cation-ratio dating of rock varnish, and evaluation of its application to the Yucca Mountain Respository by the Department of Energy and its subcontractors. *In* J. Bell (ed.), Yucca Mountain site investigation, Nevada Bureau of Mines and Geology, Reno, Nevada.

36. Dorn, R. I., D. B. Bamforth, T. A. Cahill, et al. 1986.
     Cation-ratio and accelerator radiocarbon dating of rock var-
     nish on Mojave artifacts and landforms. Science 231:830–836.

37. Harrington, C. D., and J. W. Whitney. 1987. Scanning elec-
     tron microscope method for rock varnish dating. Geology 15:
     967–970.

38. Dorn, R. I., A. J. T. Jull, D. J. Donahue, T. W. Linick,
     and L. J. Toolin. 1989. Accelerator mass spectrometry radio-
     carbon dating of rock varnish. Geol. Soc. Am. Bull. 101:
     1363–1372.

39. Dorn, R. I., and M. J. DeNiro. 1985. Stable carbon isotope
     ratios of rock varnish organic matter: a new paleoenviron-
     mental indicator. Science 227:1472–1474.

40. Dorn, R. I., M. J. DeNiro, and J. O. Ajie. 1987. Isotopic
     evidence for climatic influence on alluvial-fan development in
     Death Valley, California. Geology 15:108–110.

41. Dorn, R. I. 1988. A rock varnish interpretation of alluvial-
     fan development in Death Valley, California. Natl. Geogr.
     Res. 4:56–73.

42. Hungate, B., A. Danin, N. Pellerin, et al. 1987. Charac-
     terization of manganese-oxidizing (Mn2-Mn4) bacteria from
     Negev Desert rock varnish: implications in desert varnish
     formation. Can. J. Microbiol. 33:939–943.

43. Krumbein, W. E. 1969. Uber den Einfluss der Mikroflora auf
     die Exogene Dynamik (Verwitterung und Krustenbildung).
     Geol. Rundsch. 58:333–363.

44. Ghiorse, W. C. 1984. Biology of iron- and manganese-de-
     positing bacteria. Annu. Rev. Microbiol. 38:515–550.

45. Klaveness, D. 1977. Morphology, distribution, and signifi-
     cance of manganese-accumulating microorganism *Metallogenium*
     in lakes. Hydrobiologia 56:25–33.

46. Gregory, E., R. S. Perry, and J. T. Staley. 1980. Charac-
     terization, distribution and significance of *Metallogenium* in
     Lake Washington. Microbiol. Ecol. 6:125–140.

47. Staley, J. T., F. Palmer, and J. B. Adams. 1982. Micro-
     colonial fungi: common inhabitants on desert rocks? Science
     215:1093–1095.

48. Hawksworth, D. L. 1981. *Lichenothelia*, a new genus for
     the *Microthelia aterrima* group. Lichenologist 13:141–153.

49. Henssen, A. 1987. *Lichenothelia*, a group of microfungi on
     rocks. Progress and problems in lichenology in the eighties.
     Bibl. Lichenol. 25:257–293.

50. Palmer, F. E., D. R. Emery, J. Stemmler, and J. T. Staley.
     1987. Survival and growth of microcolonial rock fungi as af-
     fected by temperature and humidity. New Phytol. 107:155–
     162.

51. Staley, J. T., M. J. Jackson, F. E. Palmer, D. J. Borns, B. Curtiss, and S. Taylor-George. 1983. Desert varnish coatings and microcolonial fungi on rocks of the Gibson and Great Victoria Deserts. BMR J. Geol. Geophys. 8:83—87.

52. Krumbein, W. E., and K̆. Jens. 1981. Biogenic rock varnishes of the Negev Desert (Israel): an ecological study of iron and manganese transformation by cyanobacteria and fungi. Oecologia 50:25—38.

53. Smith, B. J., and W. B. Whalley. 1988. A note on the characteristics and possible origins of desert varnishes from southeast Morocco. Earth Surf. Proc. Landforms 13:251—258.

54. Smith, B. J., and W. B. Whalley. 1989. A note on the characteristics and possible origins of desert varnishes from southeast Morocco. A reply to comments by R. I. Dorn. Earth Surf. Proc. Landforms 14:171—172.

55. Dorn, R. I. 1989. A comment on "A note on the characteristics and possible origins of desert varnishes from southeast Morocco" by Drs. Smith and Whalley. Earth Surf. Proc. Landforms 14:167—169.

6

Extraction of Enzymes
from Soils

M. ALI TABATABAI and MINHONG FU* *Iowa State University, Ames, Iowa*

I. INTRODUCTION

Ninety years have passed since the first report was presented by Woods [1] on oxidizing enzymes, especially peroxidase, in soils at the annual meeting of the American Association for the Advancement of Science in Columbus, Ohio. Although the progress in soil enzymology was extremely slow until 1950, exponential progress has been made in this field within the past four decades. The history of abiontic soil enzyme research has been elegantly prepared by Skujins [2]. Reviews of recent advances and state of knowledge in this field are presented in a book edited by Burns [3] and chapters by Kiss et al. [4] and Skujins [5]. To our knowledge, no specific review article has been prepared on the work done on extraction of soil enzymes. This review considers mainly the procedures used for extraction of enzyme-containing substances from soils. Details of these procedures will be discussed and the steps used, when available, to purify the extracted enzymes will be presented. The problems encountered in the extraction and purification of soil enzymes will be discusssed.

II. STATE OF ENZYMES IN SOILS

The enzyme activity of soils results from the activity of accumulated enzymes and from those in proliferating microorgansms. As defined by Kiss et al. [4], accumulated enzymes in soils are regarded

---

Current affiliation: Cornell University, Ithaca, New York.

as enzymes present and active in a soil in which no microbial proliferation occurs. Sources of enzymes in soils are primarily the microbial biomass, although they can also originate from plant and animal residues. Enzyme activities in soils are derived from free enzymes, such as exoenzymes released from living cells, endoenzymes released from disintegrating cells, and enzymes bound to cell constituents (enzymes present in disintegrating cells, in cell fragments, and in viable but nonproliferating cells). Proliferating microorganisms produce enzymes that are released to the soil, or that remain within the multiplying cells. Schematic diagrams showing the components of the enzyme activity in soils are presented by Skujins [5] and Kiss et al. [4].

The term "state of enzymes in soils" was used by Skujins [5,6] to describe the phenomena whereby enzymes are stabilized in soils. Describing the state of an enzyme in soil is an attempt to describe the location and microenvironment in which it functions and how the enzyme is bound or stabilized within that environment [4].

Free enzymes in soils are adsorbed on organic and mineral constituents or complexed with humic substances, or both. The amount of free enzyme in the soil solution is minute compared with that in the adsorbed state. Microbial cells or cell fragments also may exist in an adsorbed state or in suspension.

The stability of enzymes within the soil matrix has resulted in several theories to explain the protective influence of soil constituents on enzyme activity. Enzymes can be immobilized on organic and inorganic constituents of soils. It is well known that microorganisms, the most significant sources of soil enzymes, are adsorbed by soil organic and inorganic constituents [7]. Other studies have shown that proteins are adsorbed on clay minerals. An early work by Ensminger and Gieseking [8] showed that proteins adsorbed on montmorillonite were stabilized against microbial attack. These results gave support to the concept that the most important factor in enzyme stabilization is the clay fraction of soils. A study by Haig (A. D. Haig, Ph.D. dissertation, University of California, Davis, 1955) supported this hypothesis. He found that acetylesterase activity in a fine sandy loam soil was associated with primarily the clay fraction. Several studies within the past three decades have shown that clay minerals adsorb enzyme protein, but the exact mechanism(s) involved in binding proteins on clay minerals has not been determined [9—12]. However, Albert and Harter [13] reported that adsorption of lysozyme and ovalbumin by Na—clay minerals resulted in an increase in Na^+ concentration of the clay—protein suspensions, and they interpreted this as evidence of a cation-exchange adsorption mechanism.

Stabilization of enzymes in the soil environment by organic matter, rather than inorganic components, has also been suggested. Much of the information dealing with this hypothesis has been gathered from studies involving synthetic polymer—enzyme complexes [reviewed in 14]. It has been suggested that proteins bind to humus by hydrogen, ionic, or covalent bonds. The extent that enzymes are bound by each of these mechanisms is difficult to determine. Studies by Simonart et al. [15] suggested that hydrogen bonding is a minor factor in enzyme stabilization in soils. By using phenol as a hydrogen bond-breaking solvent, they were able to desorb only a small amount of proteinaceous material.

Ionic bonding seems to have a role in enzyme stabilization, and such a mechanism for the binding of enzymes to soil organic matter was proposed by Butler and Ladd [16]. They proposed that the organic matter binds enzymes by amino-carboxyl salt linkages. Such linkages, however, should be easily broken by many of the extraction reagents (e.g., urea, pyrophosphate) used to remove active enzyme materials from soil. However, the small yields of active enzyme materials that have been extracted from soil indicate that mechanisms that propose ionic bonding may not be entirely valid [17—19]. However, Burns et al. [20] extracted approximately 20% of the original native soil urease activity by using urea (urea being hydrolyzed subsequently by the extracted urease). The clay-free precipitate obtained contained urease activity that was not destroyed by the addition of the proteolytic enzyme, pronase. Burns et al. concluded that native soil urease is located in organic colloidal particles that contain pores sufficiently large for water, urea, ammonia, and carbon dioxide to pass freely, but small enough to exclude pronase [20].

A clear hypothesis of enzyme immobilization by covalent attachment has not been described. Ladd and Butler [14] suggested that linkage of soil quinones by nucleophilic substitution to sulfhydryl and to terminal and ε-amino groups of enzyme proteins may result in active organoenzyme derivatives, provided that these groups do not form a part of the active site of the enzyme.

A hypothesis that has received little attention is that the enzymes in soils are glycoproteins. Malathion esterase extracted from soil by Satyanarayana and Getzin [21] was thought to be a glycoprotein, given the following evidence: (1) amino acids constituted only 65% of the purified enzyme, and (2) a carbohydrate-splitting enzyme, hyaluronidase, enhanced the catalytic activity of the esterase, presumably by loosening the carbohydrate shield and allowing the protein core to gain easier access to the substrate. The evidence gained by incubating the esterase with hyaluronidase suggested that the carbohydrate-protein linkage occurs through N-acetylhexoseamine-tyrosine bonds. Mayaudon et al. [22] reached a similar conclusion, when they observed that diphenol oxidase activity was not affected by pronase alone, but the activity was destroyed when the enzyme was incubated in the presence of both lysozyme and pronase.

III. STABILITY OF ENZYMES IN SOILS

Numerous studies have demonstrated that free enzymes are readily decomposed or are inactivated very rapidly when added to soils. Because of the importance of urease in the hydrolysis of the fertilizer urea, and because of its availability in pure form, most studies on the stability of enzymes in soils have been conducted with this enzyme. The first evidence that native soil enzymes are more stable than those added to soil was obtained by Conrad in 1940 [23] using urease. He hypothesized that organic soil constituents protect native urease against microbial degradation and other processes that result in the decomposition or inactivation of enzymes. This hypothesis has been supported by numerous studies that have shown that enzyme activities in soil are significantly correlated with the organic matter content of surface soils and soil profiles [24–27]. Other studies have shown that the urease activity associated with the clay-free organic fraction extracted from soils is not destroyed by proteolytic enzymes and that urease added to soils is degraded with time [28,29]. The instability of the added enzyme to soils is expected, because soils contain proteases, which degrade the added free enzymes. Studies on the effects of soils on acid phosphatase and inorganic pyrophosphatase of corn roots have shown that a major portion of the phosphatases in plant roots released into the soil environment is rapidly degraded by soil proteases or that their activities are inhibited by soil constituents, or both [30]. Other studies have indicated that phosphatases in plant roots may form complexes with clay, suggesting that such complexation may be a mechanism by which some portion of enzyme activity in soil becomes stabilized [31]. The native enzyme activity is a product of many processes that lead to partial incorporation of locally produced enzymes in the soil environment into a three-dimensional network of clay–organic matter–enzyme complexes [32].

The dominant mechanisms of immobilization of enzymes have been summarized by Weetall [33] and are diagramatically reproduced in Figure 1. Essentially all the mechanisms suggested to explain the immobilization and protection of enzymes by soil organic and inorganic colloids are included. These mechanisms include adsorption, microencapsulation, cross-linking, copolymerization, entrapment, ion exchange, adsorption and cross-linking, and covalent attachment.

Many mechanisms have been proposed to explain the stability of enzymes when in association with soil humic colloids [14,34,35]. These include ion exchange, H-bonding, lipophilic reactions, and covalent bonding to humates during synthesis [34]. Although it is possible that more than one, and even all these processes, are

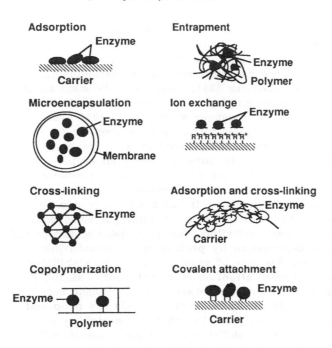

Figure 1 Schematic representation of methods for immobilizing enzymes. From Ref. 33.

involved in binding enzymes to humic substances in soils, attempts to release enzymes from enzyme—humic complexes have not been entirely successful. It is likely that enzymes are held to humic substances through yet unknown mechanism(s) that defy current methods used for the extraction of enzymes from other biological materials. The numerous attempts that have been made to extract enzymes from soils have shown that pH shifts, treatment with buffers, precipitation with ammonium sulfate, agitation, and ion exchange are only partially, and most often unpredictably, successful in extracting enzyme activity from soils. As Burns [36] has pointed out, it is not surprising, given the remarkable stability of immobilized enzymes in soils, that enzyme complexes with humic materials are extremely resistant to extraction and subsequent purification by conventional techniques. Although it is possible to extract humic substances that contain limited enzyme activites from soils [5,36,37], extensive studies in this laboratory have shown that it is not possible to separate active enzyme proteins from the associated carbohydrates in extracts obtained from soils (M. A. Tabatabai and D. E. Stott, unpublished data, 1984).

The mechanisms by which enzymes are stabilized in soils may involve one or more of the many methods available for immobilizing enymzes on water-insoluble matrices [33]. However, the main mechanism appears to be adsorption on clay or on complex organic heterocondensates and polymers in a manner analogous to that proposed for the stabilization of organic nitrogenous compounds on soil organic matter [37]. Unlike nitrogenous substances, however, stabilization of exocellular enzymes must involve more than simple protection from biological attack. Enzyme proteins must be protected by a mechanism that maintains the catalytic properties, although not necessarily the kinetic properties. It has been amply demonstrated that reaction of enzyme with clay or organic polymers usually results in decreases in enzyme activities, but may stabilize the bound enzyme protein against degradation by proteinases or denaturing agents [38—41]. The possible role of organic colloids in stabilizing soil enzymes has received more attention in recent years, as the result, in part, of a perceived analogy between enzyme—humic complexes and enzymes insolubilized by support materials, as proposed by Weetall [33]. Such reactions may (1) enhance the rigidity of the enzyme molecules, thereby stabilizing them against denaturation by heating or drying; (2) render the enzymes inaccessible to attack by free proteinases; and (3) if the bound enzyme is itself a proteinase, prevent destruction by autodigestion [37].

IV. MODEL ENZYME COMPLEXES

Because it has been difficult to understand the nature of enzyme—humic complexes in soils, an alternative, synthetic approach has been used to understand the relationship between immobilized enzymes and humic substances in soils. Early work by Ladd and Butler [42] showed that soil humic acids and trypsin react ionically to form an insoluble precipitate with initially decreased enzyme activity, as compared with untreated trypsin, but with enhanced stability against autodigestion. Comparisons by Ladd and Butler [43] of some properties of soil humic acids and synthetic phenolic polymers that incorporated amino derivatives showed that products from the oxidative polymerization of quinones or phenols in the presence of amino compounds exhibit many chemical and physical properties that were similar to those of humic acids extracted from soils. This finding resulted in the preparation of enzymatically active complexes of proteases and humic analogues in which the enzymes exist in an active state [44]. The enzyme—humic acid analogues were prepared by adding 100 mg of solid p-benzoquinone to 40 mg of either pronase or trypsin enzymes dissolved in 25 ml of 0.1 M phosphate buffer, pH 8.0. The mixture was shaken for 16 h at 4°C and centrifuged at 10,000 g for 20 min to separate soluble products. The supernatants were fractionated by ultrafiltration by using an XM-50 membrane, with a nominal molecular weight retention limit of 50,000. Under these conditions, uncombined trypsin or pronase should pass through the membrane. The soluble materials of higher than 50,000 molecular weight could be termed, by convention, humic acids, because they were of high molecular weight, dark brown, and could be precipitated by mineral acids.

Application of this procedure by Rowell et al. [44] showed that each proteinase yielded a soluble and an insoluble enzymatically active product that differed from each other and the respective free enzyme in their nitrogen contents and kinetic properties. In complexes produced by this procedure, binding of the enzymes to the aromatic polyanionic moieties may involve covalent as well as hydrogen bonds, and none of the complexes apparently contained appreciable amounts of ionically bonded enzyme. The addition of $CaCl_2$ to solutions or suspensions of all types of enzyme preparations failed to dissociate significant amounts of enzyme activity or color from the benzoquinone polymer. The model proteinase–humic acid analogues had specific activities less than those of the free enzymes, irrespective of whether or not proteins or low-molecular-weight compounds were used as the assay substrate, but they were more stable after incubation at 25°C or at elevated temperatures. The pronase complexes were, however, less stable to lyophilization than free pronase.

Studies by Maignan [35] on the preparation of a range of insoluble humic acid–invertase–Ca^{2+} complexes showed that varying the concentration and order of addition of reactants and the time of incubation of the organic components with Ca^{2+} resulted in preparations that differed in their initial specific activities and in their stabilities after repeated incubation with substrates. When complexes were formed by adding invertase to suspensions of insoluble Ca–humate, the invertase was bound by adsorption on the surface of the support. Mixing the enzyme with the soluble neutralized humic acid before the addition of Ca^{2+} resulted in invertase being partly adsorbed on the surface, possibly by covalent bond formation, and partly entrapped within micelles.

More recent studies by Sarkar and Burns [45,46] on the synthesis and properties of β-D-glucosidase–phenolic copolymers as analogues of soil–enzyme–humic complexes showed that β-D-glucosidase can be fixed on phenolic copolymers containing L-tyrosine, pyrogallol, and resorcinol. The copolymers were similar to naturally occurring soil enzyme–humic complexes in many ways: E_4/E_6

ratios (the ratio of absorbances at 465 and 665 nm); C, H, N, and
S content; and IR spectra. The enzyme activity of the copolymers
showed varying degrees of resistance to proteolysis, organic solvents, and storage at high temperatures. As expected, all immobilized enzymes showed increased K_m values and decreased V_{max}
values in comparison with soluble β-D-glucosidase; the β-D-gluco-
sidase—resorcinol copolymer was the most active. They reported
that β-D-glucosidase activity was completely resistant to protease
when the copolymer was fixed on bentonite clay, but V_{max} values
were reduced further. When added to soil, soluble β-D-glucosidase
was rapidly inactivated, whereas β-D-glucosidase—resorcinol/pyro-
gallol and β-D-glucosidase-L-tyrosine copolymers were comparatively
stable. From these experiments, Sarker and Burns [46] concluded that the copolymerization of enzymes during humic matter
formation is a major factor in the stabilization of soil enzymes and
that adsorption and entrapment are comparatively insignificant.

V. ENZYMES EXTRACTED FROM SOILS

Many enzymes have been extracted from soils (Table 1). These include several oxidoreductases, many hydrolases, and a lyase. It is
likely that this list is not complete, because the literature in this
field is scattered in many journals, some of which are not international in scope. Nevertheless, the list indicates the type of enzymes present in soils, the methods and procedures employed in
their extraction, and the potential problems encountered in their
separation and purification. It is important to note that no enzyme proteins, as such, have been extracted from soils, but rather,
organic substances have been extracted from soils that contain enzyme activities. The classification of these enzymes, the reactions
that they catalyze, and the substrates used in their assays are summarized in Table 1.

VI. EXTRACTION PROCEDURES

The first report on extraction of an enzyme activity from soil is
that by Antoniani et al. [47]. They precipitated protein with a
cathepsinlike activity from soil with ammonium sulfate and sodium
tungstate. Briggs and Segal [48] were the first to report, in 1963,
the isolation of urease activity-containing protein from soil. They
isolated 12 mg of protein with urease activity that was equivalent
to 75 Sumner units mg^{-1} from 25 kg of soil. In 1963, Martin-Smith
[49] isolated from soils two enzymes that decomposed uric acid. A
few years later, Getzin and Rosefield [55] reported the extraction

Table 1 Enzymes Extracted From Soils[a]

Class/EC number[b]	Recommended name	Reaction	Substrate
Oxidoreductases			
1.7.3.3	Urate oxidase	Urate + O_2 → unidentified products	Uric acid
1.10.3.2	Laccase[c]	See reaction of 1.14.18.1	Hydroquinone, p-phenylenediamine
1.11.1.6	Catalase	$2H_2O_2$ → O_2 + $2H_2O$	H_2O_2
1.11.1.7	Peroxidase	Donor + H_2O_2 → oxidized donor + $2H_2O$	H_2O_2, pyrogallol, chloroanilines, o-dianisidine
1.14.18.1	Monophenol monooxygenase (polyphenol oxidase, o-diphenol oxidase, p-diphenol oxidase)	Tyrosine + dihydroxyphenylalanine + O_2 → dihydroxyphenylalanine + dioxophenylalanine + H_2O	d-Catechol, p-quinol, p-cresol, 3,4-dihydroxyphenylalanine, p-phenylenediamine, pyrogallol, hydroquinone
Hydrolases			
3.1.1.1	Carboxylesterase	A carboxylic ester + H_2O → an alcohol + a carboxylic acid anion	Malathion, hydroxymethylcoumarin butyrate
3.1.1.2	Arylesterase	A phenyl acetate + H_2O → a phenol + acetate	Phenyl acetate, phenyl butyrate, naphthyl acetate

(continued)

Table 1 Continued

Class/EC number[b]	Recommended name	Reaction	Substrate
[Hydrolases]			
3.1.1.3	Triacylglycerol lipase	Triacylglycerol + H_2O → diacylglyceral + a fatty acid anion	4-Methylumbelliferone nonanoate
3.1.3.1	Alkaline phosphatase	An orthophosphoric monoester + H_2O → an alcohol + orthophosphate	p-Nitrophenyl phosphate
3.1.3.2	Acid phosphatase	An orthophosphoric monoester + H_2O → an alcohol + orthophosphate	p-Nitrophenyl phosphate
3.1.4.1	Phosphodiesterase	An orthophosphoric diester + H_2O → an orthophosphoric monoester + alcohol or phenol or nucleoside	bis-p-Nitrophenyl phosphate
3.1.6.1	Arylsulfatase	A phenol sulfate + H_2O → a phenol + sulfate	p-Nitrophenyl sulfate
3.2.1.4	Cellulase	Endohydrolysis of 1,4-β-glucosidic linkages in cellulose, lichenin and cereal β-glucose	Cellulose, carboxymethyl-cellulose
3.2.1.8	Endo-1,4-β-xylanase	Endohydrolysis of 1,4-β-xylosidic linkages in xylans	Xylan

3.2.1.21	β-Glucosidase	Hydrolysis of terminal, non-reducing β-D-glucose residues with release of β-D-glucose	p-Nitrophenyl-β-D-glucoside, cellobiose, p-nitrophenyl-β-D-glucopyranoside
3.2.1.26	β-Fructofuranosidase (Invertase)	Hydrolysis of terminal non-reducing β-D-fructofuranoside residues in β-fructofuranosides	Sucrose
3.4.21–24	Proteinases	Hydrolysis of proteins to peptides and amino acids	Casein, gelatin, albumin
3.5.1.5	Urease	Urea + H_2O → CO_2 + $2NH_3$	Urea
Lyases 4.1.1.28	Aromatic-L-amino-acid decarboxylase	L-Tryptophan → tryptamine + CO_2	dl-3,4-Di-hydroxyphenyl-alanine, dl-tyrosine, dl-tryptophan, dl-phenyl-alanine, tryptophan

aFor references, see Table 2.

bEC number, Enzyme Commission number authorized by the International Union of Biochemistry.

cNow classified as monophenol monooxygenase (1.14.18.1).

of an enzyme that decomposed malathion. Since then, several enzymes have been extracted from soils. Isolation procedures, including the reagents involved, and the reaction properties of the
isolated enzyme are important in providing the information needed
for understanding the reaction properties of the soil enzyme, bonding of the enzyme to clay and soil organic matter and, therefore,
their state and stability in soils. Such information is essential for
attempting to explore the potential usefulness of soil enzymes in
nutrient cycling and enhancement of degradation of waste materials
added to soils.

Several procedures for the extraction of enzymes from soils have
been proposed (Table 2). The solvent systems include salt solutions,
such as 0.1 M phosphate at pH 6 or 7, which extract small amounts
of organic matter. Other procedures employ a strong solubilizing
agent (e.g., 0.1 M NaOH) or treatments involving ultrasonic vibration, which removes enzymes as well as a relatively large amount of
soil organic matter [21,28,48,53]. Salt solutions are mild extractants and probably remove soil enzymes that are loosely bound to
organomineral complexes and free enzymes not associated with soil
colloids [4,5]. Because the activity of the enzymes extracted with
salt solutions is low, the yield of enzymes is often not reported.

Table 2   Extraction Procedures for Soil Enzymes

Enzyme	Extraction procedure	Purifi-cation[a]	Ref.
Urate oxidase	10 g soil, 15 ml 0.1 M phosphate buffer, pH 7.0, 10 drops tol-uene, 16 h without shaking at 24°C, filtered.	–	49
Laccase[b]	75 g soil, 100 ml of 0.3 M K_2HPO_4, 50 ml 0.3 M, pH 7 EDTA, shaken 1 h, stood 16 h at at 0°C, cen-trifuged, filtered (NAFS extract).[c]	+	50,53,71
	To 2 g soil sample, add 2 ml buf-fer, either acetate, aconitate-NaOH, citrate, citrate-phosphate, glycine-HCl, phosphate, phthal-ate—NaOH, succinate—NaOH, Tris—HCl or Tris—malate—NaOH,	–	75

(continued)

Table 2 (Continued)

Enzyme	Extraction procedure	Purification[a]	Ref.
[Laccase][b]	shake 30 min and centrifuge for 30 min. Repeat the procedure 3 times.		
Catalase	A mixture of soil and 0.01 M pyrophosphate in a ratio of 1:2 (w/v), adjust pH to 7.0–7.3, 18 h reciprocating shaking, centrifuged (4°C, 30 min, 21,000 g) and filtered. Residue extracted 4 more times.	–	51
Peroxidase	50 g soil, 50 ml of 0.05 M phosphate buffer, pH 6.0, mixed 5 min, filtered. Further filtration (0.2 μm) for sterilization.	–	52
	250 g soil suspended in 200 ml of 50 mM phosphate buffer (pH 6.0), shaken at room temperature for 5 min, centrifuged for 5 min at 4°C, filtered.	+	76
Polyphenol oxidase[b]	75 g soil, 100 ml of 0.3 M K_2HPO_4, 50 ml 0.3 M, pH 7 EDTA, shaken 1 h, stood 16 h at 0°C, centrifuged, filtered. (NAFS extract)	+	53
	100–200 g soil, 0.05 M citrate buffer, pH 6 (1 ml buffer/g soil), shaken for 30 min at room temperature. Centrifuged, filtered to remove coarse and floating debris, recentrifuged.	–	54
Carboxylesterase	120 g soil, 600 ml 0.2 M NaOH, shaken for 30 min, centrifuged, filtered, pH adjusted to 7.9 with 3 M HCl.	+	55

(continued)

Table 2   (Continued)

Enzyme	Extraction procedure	Purification[a]	Ref.
Arylesterase	0.25 M, pH 7, phosphate buffer, 2 M urea, 4 M NaCl, 0.013 M EDTA, 0.06 M mercaptoethanol; biostatic agents: $CHCl_3$ and $CH_3C_6H_5$. After 4.5 h, suspension filtered, residue extracted once with 0.25 M, pH 7 phosphate buffer for 3 h, twice with 0.5 M phosphate buffer for 3–4 h.	+	56
Lipase	10 g soil, 20 ml 0.1 M phosphate buffer, pH 7.5, shaken 30 min at 30°C, filtered.	–	57
Phosphatase	0.14 M sodium pyrophosphate, pH 7.1, soil/sol'n = 1:10, 37°C for 2–4 h in shaking water bath. Centrifuged, bacteriological filtration.	–	58
	0.2 M phosphate, pH 8, 0.2 M EDTA, 1 h (NAFS extract).	+	59,65
Phosphodi- esterase	0.2 M phosphate, pH 8, 0.2 M EDTA 1 h (NAFS extract).	–	59
	0.1 M phosphate buffer, pH 7, with the use of 0.3 M KCl and 0.01 M EDTA (2.5 ml extractant per gram soil), shaken at 30°C for 30 min, filtered and dialyzed vs. water at 5°C for 24 h.	+	17
Arylsulfatase	0.2 M phosphate, pH 8, 0.2 M EDTA 1 h (NAFS extract).	+	59
Cellulase	0.2 M phosphate, pH 8, 0.2 M EDTA 1h (NAFS extract).	+	59

(continued)

Table 2   (Continued)

Enzyme	Extraction procedure	Purifi-cation[a]	Ref.
Xylanase	0.2 M phosphate, pH 8, 0.2 M EDTA 1 h (NAFS extract).	–	59
β-Glucosidase	0.2 M phosphate, pH 8, 0.2 M EDTA 1 h (NAFS extract).	+	59
	0.1 M phosphate buffer, pH 7, 0.3 M KCl and 10 mM EDTA, 40°C, 1 h shaking, filtered, concentrated.	+	18
Invertase	0.2 M phosphate, pH 8, 0.2 M EDTA 1 h (NAFS extract).	+	59
Proteinases	0.2 M phosphate, pH 8, 0.2 M EDTA 1 h (NAFS extract).	+	59
	0.14 M sodium pyrophosphate, pH 7.1, soil/sol'n = 1:10, 37°C for 2–4 h in a shaking water bath. Centrifuged, bacteriological filtration.	–	58
	0.1 M Tris-borate buffer, pH 8.1 (5 ml buffer/g soil), shaken at 40°C for 60 min, centrifuged for 30 min at 2000 *g*.	–	60
Urease	25 g soil, 250 ml water sonicate for 20 min in water bath, slowly add 0.95 M citrate, 0.05 M sodium dihydrogen phosphate, 0.05 M glycine and 2 M NaCl mixture w/stirring, adjusted pH to 6.3, treated with 1 ml toluene, agitated 2 h at 10°C, centrifuged for 30 min, filtered.	+	28

(continued)

Table 2 (Continued)

Enzyme	Extraction procedure	Purification[a]	Ref.
[Urease]	0.25 M, pH 7 phosphate buffer, 2 M urea, 4 M NaCl, 0.013 M EDTA, 0.06 M mercaptoethanol, 4.5 h 4°C. (Soil suspended in above solution.) $CHCl_3$ and toluene added as biostatic agents, filtered, residue extracted once w/0.25 M pH 7 phosphate for 3 h, twice w/0.05 M phosphate for 3—4 h.	+	20
	Soil suspended in equal volume of 0.02 M tetra-Na pyrophosphate, shaken for 30 min, centrifuged, filtered (0.8-μm membrane).	−	61
	0.1 M sodium pyrophosphate, pH 7.1, 37°C, soil/extractant = 1:10, 1 ml toluene, incubated in water bath, 18 h, oscillated, then centrifuged for 30 min at 4°C, filtered twice.	−	58,62
Aromatic-*L*-amino acid decarboxylase	75 g soil, 100 ml 0.3 M K_2HPO_4, 50 ml 0.3 M pH 7 EDTA, shaken 1 h, let stand 16 h at 0°C, centrifuged, filtered.	+	53
	75 g soil, shaken for 2 h with 150 ml 0.2 M Na_2CO_3, let stand 16 h at 0°C, centrifuged for 20 min, filtered solution brought to pH 7 w/1 M HCl.	+	63

[a]+, purified; −, not purified.
[b]The international Union of Biochemistry has classified these enzymes monophenol monooxygenase (1.14.18.1).
[c]NAFS extract = neutral, sterile, and lyophilized extract of fresh soil.

An extracting solution that was made 0.1 or 0.2 M relative to sodium pyrophosphate is also used for extraction of soil enzymes. The enzyme activities that have been extracted from soils with this reagent and have been assayed include casein, catalase, urease, phosphatase, and protease [51,65,68]. A solution containing Na_2CO_3 was used by Chalvignac and Mayaudon [63] for the extraction of enzymatically active humic acid from a forest mull. The dialyzed humic acid acted on *L*-tryptophan—carboxyl-^{14}C, but not on *D*-tryptophan.

Ladd [60] used a Tris-borate buffer for the extraction of enzymatically active components from dry soils. The extracts contained brown materials, which were precipitated with 0.1 M $CaCl_2$. The Tris-borate buffer extracts contained from 50 to 75% of the protease activity of the unextracted soil. Distilled water extracts were less active than buffer extracts, and extracts from stored, remoistened soils were less active than those from dry soils. The proteases were optimally active near 55°C and pH 7.5, were stable on storage at 25°C for at least 10 days (but not at 50°C for 1 to 2 days), were relatively stable to gamma-irradiation, and were unaffected by incubation with microbial proteases.

McClaugherty and Linkins [69] reported that enzymes (endocellulase, exocellulase, exochitinase, laccase, and peroxidase) were extracted from soils and litter by mixing 40 to 50 g fresh weight of sample in 300 ml of cold, 50 mM, acetate buffer, pH 5, in a blender. The resulting suspensions were centrifuged for 10 min at 4000 *g* at 4°C, and the supernatant was filtered through a glass fiber filter. The pellet was resuspended in 60 ml of cold 1 M NaCl fortified, 50 mM acetate buffer at pH 5, allowed to stand 15 min, and then centrifuged and filtered as before. The two supernatants were combined, resulting in a 200 mM NaCl solution. The two-stage extraction yielded higher amounts of both extractable and bound enzyme activities than any single-stage extraction that they tested.

Present knowledge indicates that most of the accumulated enzymes in soils are strongly associated with soil organic matter and can be extracted in high yield only under conditions in which humic materials are also brought into solution. However, the use of alkali reagents or ultrasonic vibration can cause pronounced lysis of soil organisms. Therefore, these methods may yield enzymes recently present in living organisms as well as the extracellular enzymes associated with soil colloids.

VII. PURIFICATION METHODS

Since 1970, several groups have successfully extracted organic substances from soils that have active enzyme activities. The extraction procedures involve shaking the soil suspension with a buffer for a period ranging from 5 min to 24 h at ambient, or below, temperatures, but more usually at 37 to 40°C (Table 3). The extraction conditions are selected to favor extraction of abiotic enzyme proteins and to avoid extensive damage to live cells. The extraction procedures may extract several enzymes, but investigators have assayed for one or a few specific ones.

Getzin and Rosefield [55] were the first to report the partial purification and properties of a soil enzyme that degrades the insecticide, malathion. An unusually stable esterase that catalyzes the hydrolysis of malathion to its monoacid was extracted from radiation-sterilized and nonirradiated soil with 0.2 M NaOH and partially purified by $MnCl_2$ treatment, $(NH_4)_2SO_4$ precipitation, dialaysis, and ion-exchange chromatography on Bio-Rex 70 cation-exchange resin. A 240-g sample of Chehalis clay loam yielded 20 mg of final preparation, which contained 9.8 mg of protein and hydrolyzed 31 μmol of malathion in 4 h at pH 7.0 and 37°C. The remainder of the preparation was primarily carbohydrates that, upon acid hydrolysis, yielded a mixture of pentoses and hexoses. Assuming that the enzyme activity of the soil extract was equal to 100%, the recovery values by the $MnCl_2$ treatment, $(NH_4)_2SO_4$ precipitation, and chromatography on Bio-Rex 70 cation-exchange resin were 82, 63, and 32%, respectively. The corresponding purification values were 1.0-, 9.3-, 12.3-, and 21.6-fold. Lineweaver–Burk plots obtained for two concentrations of the enzyme gave identical K_m values of 2.12×10^{-4} M. The enzyme was optimally active around pH 7.0 and very stable, being denatured only at temperatures above 70°C and by 24-h exposures at pH <2.0 and >10.0. Lyophilization partially destroyed the activity, but no loss of activity occurred during extended storage of enzyme solutions at 4° or -10°C. When this partially purified enzyme was applied to soil, activity was detected for the 8-week duration of the experiment. From the experimental evidence of the persistence and adsorption characteristics of the partially purified enzyme in soil, they concluded that this enzyme exists in soils as a stable, cell-free enzyme.

Further studies on the properties of this stable, cell-free esterase [21] showed that it can be extracted from soil with 0.2 M alkali and purified 560-fold by $MnCl_2$ treatment, protamine sulfate treatment, and QAE Sephadex A-50 chromatography. Phosphonate and phenyl thiophosphate anticholinesterase insecticides were potent competitive inhibitors of soil esterase activity. Inhibition was also observed with mono- and dithiols, but not with diisopropyl fluorophosphate or sulfhydrol compounds. The esterase was not susceptible to enzymatic proteolysis or easily inactivated by metal ions.

Table 3   Soil Enzymes Purification Procedures[a]

Enzyme	Purification procedure	Ref.
Laccase	NAFS extract[b], G-100 Sephadex column DEAE DE52 cellulose.	50,64
	NAFS extract, treated with salmine and SP Sephadex C25 to remove humic matter (EFS extract).[c]	
	EFS extract, G-100 Sephadex and DEAE DE52 cellulose.	
Peroxidase	DEAE-cellulose chromatography, dialysis, gel filtration chromatography (Sephadex G-75) SDS polyacrylamide gel electrophoresis.	76
Monophenol monooxygenase	$(NH_4)_2SO_4$ precipitation, 0°C, pH adjusted to 7 by 1 M KH_2PO_4, 2 h, 0°C, centrifuged. To precipitate, add 0.2 M K_2HPO_4, dialyze vs. 0.02 m. phosphate buffer, centrifuged, filtered (NAFS extract).	53
	NAFS, G-100 Sephadex column DEAE DE 52 cellulose chromatography.	64
	NAFS, treated with salmine and SP Sephadex C25 to remove humic matter (EFS extract).	
	EFS, G-100 Sephadex and DEAE DE52 cellulose.	
Carboxylesterase	Partially purified by $MnCl_2$ treatment, $(NH_4)_2SO_4$ precipitation, protamine sulfate treatment, dialysis, ion-exchange chromatography, and electrophoresis.	21,55

(continued)

Table 3   (Continued)

Enzyme	Purification procedure	Ref.
Arylesterase	Extracts dialyzed 48 h vs. water with 5 mM mercaptoethanol. Concentrated by dialyses vs. polyethylene glycol, pH adjusted to 7, filtered through 0.45 µm, stored at -18°C.	56
Phosphatase	Same as cellulase below.	59
	Exhaustive ultrafiltration with 0.1 M pyrophosphate, pH 7.1, and dialysis of the extracts.	65
Phosphodiesterase	Solid $(NH_4)_2SO_4$ added to soil extract, after standing for 1 day, the precipitate collected by centrifuging at 1500 g for 10 min and 5% protamine sulfate solution was added followed by dialysis for 18 h. The precipitate was removed by centrifugation.	17
Cellulase	Extract treated with $(NH_4)_2SO_4$, precipitate dialyzed vs. phosphate buffer (0.1 mM, pH 7) 36 h, lyophilized and stored at -20°C (NAFS extract).	53
	Further purification: solubilized in 10 mM phosphate buffer, pH 7, treated with salmine sulfate. Supernatant treated with cation exchanger, centrifuged, dialyzed vs. 0.1 mM pH 7 phosphate buffer, lyophilized (EFS extract).	59
(Invertase, phosphatase, protease, β-glucosidase)	Fractionate these enzyme activities by gel and anion-exchange chromatography	
	NAFS, DEAE (DE52) cellulose chromatography EFS, Sephadex G-100 or G-75.	

(continued)

Table 3 (Continued)

Enzyme	Purification procedure	Ref.
β-Glucosidase	Extract concentrated, dialyzed against distilled water for 24 h, at 5°C, $(NH_4)_2SO_4$ added, 1 day, centrifuged, precipitate dissolved in water. Then 2 ml of 5% protamine sulfate added, dialysis vs. water, 18 h, 5°C, precipitate removed by centrifuging, supernatant analyzed.	18
Protease	Exhaustive ultrafiltration with 0.1 M pyrophosphate, pH 7.1, and dialysis of the extracts.	65
Urease	Filtered, dialyzed vs. tap water at 15–16°C, 3 days, vs. distilled water at 4°C, 1 day, precipitate concentrated by centrifuging for 60 min. The enzyme active residues are suspended in 1 mM Na phosphate, pH 7.	20,28
	Modification of above. Dialyzed vs. a mixture of 0.04 M sodium phosphate and 0.01 M EDTA sodium-salt, pH 7 (prevents precipitation).	66
	Exhaustively ultrafiltrating the soil extract vs. 0.1 M pyrophosphate at pH 7.1, separating the retained material into fractions of MW > and < 10^5 followed by gel chromatography.	19,65,68
Aromatic-L-amino acid decarboxylase	$(NH_4)_2SO_4$ precipitation at 0°C, pH adjusted to 7 by 1 M KH_2PO_4, 2 h, 0°C, centrifuged. To precipitate, 0.2 M K_2HPO_4 added, dialyzed vs. 0.02 M phosphate buffer, centrifuged, filtered (NAFS extract).	53,63

[a]For extraction procedures, see Table 2.
[b]NAFS extract = neutral, sterile, and lyophilized extract of fresh soil.
[c]EFS extract = NAFS extract treated with salmine sulfate and Sephadex C25 to remove humic matter.

The characteristic absorption peak at 280 nm for proteins was not present unless the enzyme was first hydrolyzed in 6 M HCl or digested with testicular hyaluronidase. Hyaluronidase digestion increased the esterase activity almost twofold, but it also decreased the stability of the enzyme. On the basis of its chemical composition and response to hyaluronidase treatment, the enzyme was classified as a glycoenzyme. The proposed carbohydrate—protein comlex may account for the unusual stability and persistence of the enzyme in soil as an extracellular entity.

Several groups have worked on the extraction of urease from soils. Burns et al. [20] reported the extraction of an urease-active organocomplex from soils. The extracted colloidal organic matter complex had urease activity and based on X-ray diffraction analysis, was free of clays. The extraction procedure involved suspending a soil sample at 4°C in a solution containing phosphate buffer (0.25 M, pH 7.0), urea (2 M), NaCl (4 M), ethylenediaminetetraacetate (EDTA, 0.013 M), and mercaptoethanol (0.06 M.) A small amount of chloroform and toluene was added as a biostatic agent. After 4.5 h, the suspension was filtered through a porous candle to remove clays, and the residue was extracted once (3 h) with 0.25 M phosphate buffer (pH 7) and twice (3 h and 4 h) with 0.05 M phosphate buffer. The four filtrates, which showed increasing brownness in order of extraction, were dialyzed against tap water for 4 days. Precipitate formation occurred in the last three filtrates. The greatest urease activity, which was increased by the addition of cysteine or by further dialysis, was found in the precipitate from the fourth extract. Only a small amount of urease activity was extractable from this precipitate with 0.8 M phosphate—citrate buffer (pH 6.5). Additional colored material, with little loss in urease activity, was removed by extraction of the sediment with 1 M phosphate buffer, pH 6.5. Overall, this extraction procedure removed 20% of the soil urease activity. Burns et al. [20] estimated, by extrapolating the plot of activity versus numbers of viable ureolytic microbes to zero, that the extracted urease activity represented about one-third the total extracellular soil urease. The extracted urease was apparently not free protein, but a protein—carbohydrate complex, because urease activity in the clay-free precipitate, as in the soil, was not destroyed by the addition of the proteolytic enzyme, pronase. The pronase itself was fairly stable in the presence of soil and destroyed added jackbean urease. Because studies by Estermann et al. [70] showed that pronase markedly reduced urease activity of a complex produced from bentonite and jackbean urease, but not of a complex produced from bentonite, jackbean urease, and lignin, Burns et al. [20] concluded that extracellular urease is present in soils in the latter form.

A method used by Nannipieri et al. [62] for the extraction of urease from a podzol soil involved the use of 0.1 M pyrophosphate. Sodium pyrophosphate (0.1 M), at pH 7.1 and 37°C, extracted a significant fraction of urease from a podzol. The ratio between the extraction solution and the amount of soil used was 10:1, and 1 ml toluene was added for each 10 g of soil as a biostatic agent. After incubation in a water-bath at 100 oscillations per minute at different temperatures and for variable times, the mixture was centrifuged at 18,000 *g* for 30 min at 4°C. The supernatant was filtered twice; first through a Pyrex G2 filter (pore size 10 to 20 μm), and then through a bacteriological filter (Sartorius SM 11307 type, pore size 0.22 μm). Maximum extraction values were reportedly obtained after 18 h. The yields of soil organic matter and urease activity during the extraction showed different patterns; the extraction of nonspecific organic matter preceded and may have facilitated the subsequent extraction of an active urease—organo complex. The urease extracted by pyrophosphate was equivalent to about 30 to 40% of the total soil urease activity. The number of ureolytic microorganisms was unaffected by pyrophosphate, and the extracted urease was assumed to be extracellular.

A similar extraction procedure was reported by Lloyd [61], who used 0.2 M sodium pyrophosphate that extracted urease from an alpine humus soil but not from four other soils. The alpine humus soil was high in humified organic matter (17%). Lloyd [61] suggested that urease may form an enzyme—organic matter complex that protects the enzyme from breakdown by proteases. However, studies by McLaren et al. [66] showed that sedimentation of extracted urease activity was greater than that of the total organic matter, suggesting that urease was not uniformly associated with the organic matter.

Numerous attempts have been made to purify humus—urease complexes [19]. Some techniques commonly used for enzyme purification were found to be unsuitable for the purification of urease extracted by pyrophosphate from soils. The techniques evaluated were ammonium sulfate precipitation, chromatography on polyvinylpyrrolidone, Sephadex gel chromatography, fractionation on chromatographic columns conditioned at pH 4.0 with acetate buffer (CM-cellulose) or at pH 8.0 with Tris-HCl buffer (DEAE-cellulose), and ultrafiltration of soil extract. The most effective purification of soil urease was achieved by exhaustive ultrafiltration of the soil extract against 0.1 M pyrophosphate at pH 7.1, separating the retained material into fractions of molecular weight higher (A_I) and lower (A_{II}) than 10^5, followed by gel chromatography. As also observed in our laboratory, Ceccanti et al. [19] reported that DEAE-cellulose and DEAE-Sephadex irreversibly adsorbed both organic matter and urease activity, but that urease activity increased after both ultrafiltration and gel chromatography. Ultrafiltration increased the total activity of the extracted urease by about 8%. The specific activity of fraction A_I increased by fourfold and that of fraction A_{II} increased by more than threefold. The fractions obtained by gel chromatography accounted, in toto, for only 13.5% of the organic carbon in the soil extract; the total urease activity increased by 45.6%; and the specific activity increased by 6.9 to 18 times that of the soil extract [19].

Mayaudon and his associates were among the early workers to apply modern biochemical techniques to extraction and purification of soil enzymes [22,59,71,72]. Salmine and SP-Sephadex C-25 were used to separate diphenol oxidases associated with humic materials in neutral extracts of fresh soils (NAFS extract) [71]. Electrophoresis on polyacrylamide gels showed that this preparation was heterogeneous. Elementary analysis showed that the enzyme protein contained 43.13% C, 5.09% N, 7.21% H, and 44.58% O; and chromatographic analyses indicated that the enzyme contained 53% amino acids, 36% sugars and amino sugars, and 10% ammonium and inorganic materials. The purified soil enzyme had a maximum absorption at 270 to 280 nm and degraded several substrates, including d-catechin, p-cresol, catechol, p-phenylenediamine, p-quinol, as measured by rate of oxygen uptake. Further studies on purification of the NAFS extracts showed that, in addition to diphenol oxidase activity, it contained proteolytic activity [22]. In the presence of lysozyme and pronase, the activity of diphenol oxidase was destroyed. Pronase alone had only a small effect on this activity. They suggested that polysaccharides associated with extracts of fresh soil and the NAFS extract had a role in stabilizing the "humoenzymes" system in soils.

Mayaudon and Sarkar [72] reported that a lyophilized neutral sterile extract from the fermentation layer of beech litter (NALF extract) exhibited the chemical characteristics of humic acids, and possessed diphenol oxidase activities. The NALF extract was polydispersed by G-100 Sephadex column chromatography. The first peak had a $K_D \sim 0.05$ (fraction I), the intermediate band (fractions II and III) and the second peak had a $K_D \sim 1.02$ and 1.38 (fractions IV and V). Diphenol oxidase activity was localized in fractions I, II, and III. Electrophoretic studies showed that fractions I, II, and III were heterogeneous. Chromatography on DEAE-cellulose of fraction I permitted the separation of 30% of the laccase activity in a form that was free from humic material.

Invertase, cellulase, phosphatases, protease, and β-glucosidase have been extracted from a permanent pasture soil with 0.2 M phosphate buffer (pH 8) in the presence of 0.2 M EDTA [59]. Attempts were made to fractionate these enzyme activities by gel and anion-exchange chromatography. Batistic et al. [59] estimated the specific activities in all fractions and determined some characteristics of the purified enzymes (optimum pH, temperature, and substrate concentration; K_m and V_{max}). Their results indicated that the extracted enzyme activities occurred partly in the soil as a carbohydrate—enzyme complex and partly as a humo—carbohydrate—enzyme complex.

Phosphodiesterase together with brown-colored compounds have been extracted from a forest soil by using 0.1 M phosphate buffer (pH 7) containing KCl and EDTA [17]. Distilled water extracted only small amounts of enzyme activity. A curvilinear relationship, such as the Langmuir type, has been reported between solution volume and phosphodiesterase activity of the extract. Hayano [17] concluded that the phosphodiesterase extracted was extracellular and was adsorbed on the surface of soil particles by ionic bonding. When he removed the brown compounds in the extract by precipitation with protamine sulfate, the phosphodiesterase activity of the extract was lost when held at 80°C for 20 min and was optimal at pH 5.2 to 6.0. Applying the same extraction and purification methods, Hayano and Katami [18] reported the extraction of β-glucosidase. With the relative activity in parentheses, the extract hydrolyzed p-nitrophenyl β-glucoside (100), phenyl β-glucoside (19), salicin (8), amygdalin (33), cellobiose (52), and gentiobiose (55). The extract, however, had no effect on methyl-β-glucoside and phloridzin (phlorizin). The substrate specificity and optimum pH of the enzymatic activity of the soil extract were similar to those of β-glucosidases derived from various fungi.

Decomplexation of phosphatase from extracted soil humic substances with electron-donating reagents is possible [73]. The isolation procedure involves two steps. First, ammonium humates are formed by intermittent shaking of soil in 1.0 M ammonium acetate NH_4OAc (pH 7.0) for 20 h and extracted by repeated rinsing with distilled water, after the NH_4OAc has been removed by filtration. The filtrate is dialyzed and freeze-dried. The phosphatase-active part is then isolated from the humic substances by decomplexation with 17 mM tetramethyl-p-phenylenediamine (TMePDA). Studies by Gosewinkel and Broadbent [73] showed that the dialyzed filtrate accounted for 3.2% of the soil phosphatase activity. Freeze-drying destroyed 95.4% of the phosphatase activity of the filtrate. A total of 1.77 g of freeze-dried filtrate was obtained by this method from each kilogram of soil, with a phosphatase activity equivalent to 9.85 μmol g^{-1} h^{-1} of p-nitrophenol. The percentage of phosphatase activity recovered in a clear supernatant in the step with TMePDA accounted for 24%, and the amount of total solids precipitated was equivalent to 97.3%. The freeze-dried filtrate was free of clay, as determined by X-ray diffraction analysis.

As Ceccanti et al. [19] concluded, procedures to purify a soil enzyme must allow for experimental criteria adopted for the extraction and for the possible presence in the preparation of an excess of humic and inorganic compounds. If most of the extracted enzyme is strongly associated with soil colloids, its behavior during purification would probably follow that of the extracted organometallic complexes. In fact, it is reasonable to assume that the extraction procedure can give organometallic complexes with the trapped enzyme. Evidence has been presented showing that much of the Fe and Al, important parts of the ash component, is not removed by strong cation-exchange resins from alkali-soluble organic matter preparations [74].

VIII. CONCLUSIONS

Activities of a number of enzymes have been detected in soils. During the past 20 years, many of those activities have been extracted by a variety of reagents, ranging from water to salt solutions or buffers to strong organic matter-solubilizing reagents, such as NaOH or sodium pyrophosphate. The extracted activities are usually associated with carbohydrate—enzyme complexes and are often difficult to purify. Modern biochemical techniques have been used in the purification of the extracted enzymes, but little progress has been made in obtaining pure enzyme proteins from soils. Several of the enzymes extracted from soils could be present in soils as glycoproteins. Methods are needed to separate the small amounts of enzyme proteins from the dominant carbohydrates in any extract obtained from soils. Although several investigators have demonstrated that clay-free extracts could be obtained from soils, the major problem appears to be the strong affinity of the carbohydrate—enzyme complexes for chromatographic columns that makes the separation difficult. Nevertheless, Sephadex (G-100, G-75) gels and ion-exchange (DEAE-cellulose) chromatography have been applied, with some success, in the purification of hydrolases extracted from soils. It appears that various carbohydrates in soils adsorb the enzyme proteins and are responsible for their stabilization against denaturation or proteolysis. The presence of carbohydrate complexes with enzyme proteins probably affects the kinetics of the enzyme action by decreasing the affinity of the enzyme for its substrate. The reactions of free enzymes in soils will not fully be understood until methods are developed to obtain carbohydrate-free, enzyme-active proteins from soils.

REFERENCES

1. Woods, A. F. 1899. As cited by J. Skujins. 1978. History of abiontic soil enzyme research, p. 1–49. *In* R. G. Burns (ed.), Soil enzymes. Academic Press, New York.
2. Skujins, J. 1978. History of abiontic soil enzyme research, p. 1–49. *In* R. G. Burns (ed.), Soil enzymes. Academic Press, New York.
3. Burns, R. G. 1978. Soil enzymes. Academic Press, New York.
4. Kiss, S., M. Dragan-Bularda, and D. Radulescu. 1975. Biological significance of enzymes accumulated in soil. Adv. Agron. 27:25–87.
5. Skujins, J. 1976. Extracellular enzymes in soil. CRC Crit. Rev. Microbiol. 4:383–421.
6. Skujins, J. J. 1967. Enzymes in soil, p. 371–414. *In* A. D. McLaren and G. H. Peterson (eds.), Soil biochemistry, Vol. 1. Marcel Dekker, New York.
7. Hattori, T. 1976. The physical environment in soil microbiology: an attempt to extend principles of microbiology to soil microorganisms. CRC Crit. Rev. Microbiol. 4:423–461.
8. Ensminger, L. E., and J. E. Gieseking. 1942. Resistance of clay-adsorbed proteins to proteolytic hydrolysis. Soil Sci. 53:205–209.
9. McLaren, A. D. 1954. The adsorption and reaction of enzymes and proteins on kaolinite. II. The action of chymotrypsin on lysozyme. Soil Sci. Soc. Am. Proc. 18:170–174.
10. Armstrong, D. E., and G. Chesters. 1964. Properties of protein–bentonite complexes as influenced by equilibrium conditions. Soil Sci. 98:39–52.
11. Hughes, J. D., and G. H. Simpson. 1978. Arylsulfatase-clay interactions. II. The effect of kaolinite and montmorillonite on arylsulfatase activity. Aust. J. Soil Res. 16:35–40.
12. Hamzehi, E., and W. Pflug. 1981. Sorption and binding mechanisms of polysaccharide cleaving soil enzymes by clay minerals. Z. Pflanzenernaehr. Bodenkd. 144:505–513.
13. Albert, J. T., and R. D. Harter. 1973. Adsorption of lysozyme and ovalbumin by clay: effect of clay-suspension pH and clay mineral type. Soil Sci. 115:130–136.
14. Ladd, J. N., and J. H. A. Butler. 1975. Humus-enzyme systems and synthetic, organic polymer–enzyme analogs, p. 143–194. *In* E. P. Paul and A. D. McLaren (eds.), Soil biochemistry, Vol. 4. Marcel Dekker, New York.
15. Simonart, P., L. Batistic, and J. Mayaudon. 1967. Isolation of protein from humic acid extracted from soil. Plant Soil 27:153–161.

16. Butler, J. H. A., and J. N. Ladd. 1969. The effect of methylation of humic acids on their influence on proteolytic enzyme activity. Aust. J. Soil Res. 7:263–268.

17. Hayano, K. 1977. Extraction and properties of phosphodiesterase from a forest soil. Soil Biol. Biochem. 9:221–223.

18. Hayano, K., and A. Katami. 1977. Extraction of β-glucosidase activity from pea field soil. Soil Biol. Biochem. 9:349–351.

19. Ceccanti, B., P. Nannipieri, S. Cervelli, and P. Sequi. 1978. Fractionation of humus–urease complexes. Soil Biol. Biochem. 10:39–45.

20. Burns, R. G., M. H. El-Sayed, and A. D. McLaren. 1972. Extraction of an urease-active organo-complex from soil. Soil Biol. Chem. 4:107–108.

21. Satyanarayana, T., and L. W. Getzin. 1973. Properties of a stable cell-free esterase from soil. Biochemistry 12:1566–1572.

22. Mayaudon, J., L. Batistic, and J. M. Sarkar. 1975. Proprietes des activites proteolytiques extraides des sols frais. Soil Biol. Biochem. 7:281–286.

23. Conrad, J. P. 1940. Hydrolysis of urea in soils by thermolabile catalysis. Soil Sci. 49:253–263.

24. Myers, M. G., and J. W. McGarity. 1968. The urease activity in profiles of five great soil groups from northern New South Wales. Plant Soil 28:25–32.

25. Tabatabai, M. A. 1977. Effects of trace elements on urease activity in soils. Soil Biol. Biochem. 9:9–13.

26. Juma, N. G., and M. A. Tabatabai. 1978. Distribution of phosphomonoesterases in soils. Soil Sci. 126:101–108.

27. Frankenberger, W. T., Jr., and M. A. Tabatabai. 1981. Amidase activity in soils: III. Stability and distribution. Soil Sci. Soc. Am. J. 45:333–338.

28. Burns, R. G., A. H. Pukite, and A. D. McLaren. 1972. Concerning the location and persistence of soil urease. Soil Sci. Soc. Am. Proc. 36:308–311.

29. Zantua, M. I., and J. M. Bremner. 1977. Stability of urease in soils. Soil Biol. Biochem. 9:135–140.

30. Dick, W. A., N. G. Juma, and M. A. Tabatabai. 1983. Effects of soils on acid phosphatase and inorganic pyrophosphatase of corn roots. Soil Sci. 136:19–25.

31. Dick, W. A., and M. A. Tabatabai. 1987. Kinetics and activities of phosphatase–clay complexes. Soil Sci. 143:5–15.

32. McLaren, A. D. 1975. Soil as a system of humus and clay immobilized enzymes. Chem. Scr. 8:97–99.

33. Weetall, H. H. 1975. Immobilized enzymes and their application in the food and beverage industry. Process Biochem. 10:3–24.

34. Burns, R. G. 1978. Enzyme activity in soil: some theoretical and practical considerations, p. 295–339. *In* R. G. Burns (ed.), Soil enzymes. Academic Press, New York.

35. Maignan, C. 1982. Activite des complexes acides humiques-invertase: influence du mode de preparation. Soil Biol. Biochem. 14:439–445.

36. Burns, R. G. 1986. Interaction of enzymes with soil mineral and organic colloids, p. 429–451. *In* P. M. Huang and M. Schnitzer (eds.), Interactions of soil minerals with natural organics and microbes. SSSA Spec. Publ. No. 17. American Society of Agronomy, Madison, Wisconsin.

37. Ladd, J. N. 1985. Soil enzymes, p. 175–221. *In* D. Vaughan and R. E. Malcolm (eds.), Soil organic matter and biological activity. Martinus Nijhoff, Boston.

38. Aomine, S., and V. Kobayashi. 1964. Effects of allophane on the enzymatic activity of a protease. Soil Plant Food (Tokyo) 10:28–32.

39. Kobayashi, V., and S. Aomine. 1967. Mechanisms of the inhibitory effect of allophane and montmorillonite on some enzymes. Soil Sci. Plant Nutr. 13:189–194.

40. Ladd, J. N., and J. H. A. Butler. 1969. Inhibition and stimulation of proteolytic enzyme activities by soil humic acids. Aust. J. Soil Res. 7:253–261.

41. Gould, W. D., F. D. Cook, and J. A. Bulat. 1978. Inhibition of urease activity by heterocyclic sulfur compounds. Soil Sci. Soc. Am. J. 42:66–72.

42. Ladd, J. N., and J. H. A. Butler. 1970. The effect of inorganic cations on the inhibition and stimulation of protease activity by soil humic acids. Soil Biol. Biochem. 2:33–40.

43. Ladd, J. N., and J. H. A. Butler. 1966. Comparison of some properties of soil humic acids and synthetic phenolic polymers incorporating amino derivatives. Aust. J. Soil Res. 4:41–54.

44. Rowell, M. J., J. N. Ladd, and E. A. Paul. 1973. Enzymatically active complexes of proteases and humic acid analogues. Soil Biol. Biochem. 5:699–703.

45. Sarkar, J. M., and R. G. Burns. 1983. Immobilization of β-D-glucosidase and β-D-glucosidase–polyphenolic complexes. Biotechnol. Lett. 5:619–624.

46. Sarkar, J. M., and R. G. Burns. 1984. Synthesis and properties of β-D-glucosidase–phenolic copolymers as analogues of soil humic–enzyme complexes. Soil Biol. Biochem. 16:619–625.

47. Antoniani, C., T. Montanari, and A. Camoriano. 1954. Investigations in soil enzymology. I. Cathepsin-like activity. A preliminary note. Ann. Fac. Agrar. Univ. Milano 3:99–101.

48. Briggs, M. H., and L. Segal. 1963. Preparation and properties of a free soil enzyme. Life Sci. 2:69–72.

49. Martin-Smith, M. 1963. Uricolytic enzymes in soil. Nature 197:361–362.

50. Mayaudon, J., and J. M. Sarkar. 1975. Laccases de *Polyporus Versicolor* dans le sol et la litiere. Soil Biol. Biochem. 7:31–34.

51. Perez-Mateos, M., S. Gonzalez-Carcelo, and M. D. Busto Nunez. 1988. Extraction of catalase from soil. Soil Sci. Soc. Am. J. 52:408–411.

52. Bartha, R., and L. Bordeleau. 1969. Cell-free peroxidase in soil. Soil Biol. Biochem. 1:139–143.

53. Mayaudon, J., M. El Halfawi, and C. Bellinck. 1973. Decarboxylation des acids amines aromatiques-1-^{14}C par les extraits de sol. Soil Biol. Biochem. 5:355–367.

54. Suflita, J. M., and J.-M. Bollag. 1980. Oxidative coupling activity in soil extracts. Soil Biol. Biochem. 12:177–183.

55. Getzin, L. W., and I. Rosefield. 1971. Partial purification and properties of a soil enzyme that degrades the insecticide malathion. Biochim. Biophys. Acta 235:442–453.

56. Cacco, G., and A. Maggioni. 1976. Multiple forms of acetyl-naphthyl-esterase activity in soil organic matter. Soil Biol. Biochem. 8:321–325.

57. Pancholy, S. K., and J. Q. Lynd. 1972. Quantitative fluorescence analysis of soil lipase activity. Soil Biol. Biochem. 4:257–259.

58. Nannipieri, P., B. Ceccanti, S. Cervelli, and E. Matarese. 1980. Extraction of phosphatase, urease, protease, organic carbon, and nitrogen from soil. Soil Sci. Soc. Am. J. 44:1011–1016.

59. Batistic, L., J. M. Sarkar, and J. Mayaudon. 1980. Extraction, purification and properties of soil hydrolases. Soil Biol. Biochem. 12:59–63.

60. Ladd, J. N. 1972. Properties of proteolytic enzymes extracted from soil. Soil Biol. Biochem. 4:227–237.

61. Lloyd, A. B. 1975. Extraction of urease from soil. Soil Biol. Biochem. 7:357–358.

62. Nannipieri, P., B. Ceccanti, S. Cervelli, and P. Sequi. 1974. Use of 0.1 M pyrophosphate to extract urease from a podzol. Soil Biol. Biochem. 6:359–362.

63. Chalvignac, M. A., and J. Mayaudon. 1971. Extraction and study of soil enzymes metabolising tryptophan. Plant Soil 34:25–31.

64. Mayaudon, J., and J. M. Sarkar. 1974. Chromatographie et purification des diphenol oxydases du sol. Soil Biol. Biochem. 6:275–285.

65. Nannipieri, P., B. Ceccanti, C. Conti, and D. Bianchi. 1982. Hydrolases extracted from soil: their properties and activities. Soil Biol. Biochem. 14:257–263.
66. McLaren, A. D., A. H. Pukite, and I. Barshad. 1975. Isolation of humus with enzymatic activity from soil. Soil Sci. 119:178–180.
67. Nannipieri, P., B. Ceccanti, D. Bianchi, and M. Bonmati. 1985. Fractionation of hydrolase-humus complexes by gel chromatography. Biol. Fert. Soils 1:25–29.
68. Nannipieri, P., B. Ceccanti, S. Cervelli, and C. Conti. 1982. Hydrolases extracted from soil: kinetic parameters of several enzymes catalyzing the same reaction. Soil Biol. Biochem. 14: 429–432.
69. McClaugherty, C. A., and A. E. Linkins. 1990. Temperature responses of enzymes in two forest soils. Soil Biol. Biochem. 22:29–33.
70. Estermann, E. F., G. H. Peterson, and A. D. McLaren. 1959. Digestion of clay–protein, lignin–protein and silica–protein complexes by enzymes and bacteria. Soil Sci. Soc. Am. Proc. 23:31–36.
71. Mayaudon, J., M. El Halfawi, and M. A. Chalvignac. 1973. Proprietes des diphenol oxydases extraites des sols. Soil Biol. Biochem. 5:369–383.
72. Mayaudon, J., and J. M. Sarkar. 1974. Etude des diphenol oxydases extraites d'une litiere de foret. Soil Biol. Biochem. 6:269–274.
73. Gosewinkel, U., and F. E. Broadbent. 1986. Decomplexation of phosphatase from extracted soil humic substances with electron donating reagents. Soil Sci. 141:261–267.
74. Sequi, P., G. Guidi, and G. Petruzzelli. 1975. Influence of metals on solubility of soil organic matter. Geoderma 13:153–161.
75. Leonowicz, A., and J.-M. Bollag. 1987. Laccases in soil and the feasibility of their extractions. Soil Biol. Biochem. 19:237–242.
76. Bollag, J.-M., C.-M. Chen, J. M. Sarker, and M. J. Loll. 1987. Extraction and purification of a peroxidase from soil. Soil Biol. Biochem. 19:61–67.

7

Biochemical Analysis of Biomass, Community Structure, Nutritional Status, and Metabolic Activity of Microbial Communities in Soil

ANDERS TUNLID   *Lund University, Lund, Sweden*

DAVID C. WHITE   *University of Tennessee and Oak Ridge National Laboratory, Knoxville, Tennessee*

I. PROBLEMS IN ASSAYING MICROORGANISMS IN SOILS

The determination of microbial biomass and the activity of microorganisms in soil presents a complex analytical problem for assays. Numerous studies have demonstrated that classic methods, which require the isolation and subsequent culture of microorganisms, are not adequate for enumerating microorganisms in soil. Viable counts underestimate the microbial community when compared with direct count techniques [1–3] or with estimates of muramic acid in the prokaryotic cell wall [4]. This discrepancy has been attributed to the selective growth of microbes on artificial media, the formation of a single colony from bacterial aggregates, and the difficulty of quantitative removal of organisms from soil particles [4,5]. Direct counting methods are also subject to technical difficulties when applied to soil systems [5]. For example, cells may be hidden in soil particles or overlapping organisms, particularly in aggregates of organisms attached to particles, and the conversion from counts to biomass, by estimating volumes of the microorganisms, can result in large errors.

Furthermore, classic methods provide a limited insight into the metabolic function and activity of the microorganisms in soil. Specific fluorescent dyes (e.g., fluorescein diacetate) [6,7] have been used to determine the metabolically active cells of microorganisms in soils. Active cells can also be determined by a combination of autoradiography and microscopy [8,9]. However, this methodology requires enzymatic activity in the presence of substrates and is subject to the limitations associated with the density of organisms and the thickness of the biofilm in the field of view. Moreover, when substrates are introduced to measure metabolic activity, the process of introduction can induce artificially high levels of activity, with a possible disturbance artifact.

The problems associated with the use of viable counts and microscopic examination to estimate biomass of microorganisms in soil have stimulated research on the development of new methods. One approach is to estimate the biomass of soil microorganisms by measuring the concentration of specific biochemical components of the microbial cells [5,10–12]. Components that are generally present in all cells are utilized as a measure of biomass, the components that are restricted to subsets of the microbial community are utilized to define the community structure. The validity of the concept of "signatures" for subsets of the microbial community, which is based on the limited distribution of specific components, has been shown for several groups of microorganisms [10–12]. Biochemical methods have also been utilized to indicate the nutritional status of microorganisms in natural environments, and the metabolic activity of the microbial community can be estimated by measuring the rate of isotope incorporation from labeled precursors [10–12].

The biochemical methods do not have the problems associated with the classic methods because they do not depend on growth, with the inherent problem of microbial selection, nor do they require removing cells from surfaces. Biochemical methods examine the community as a whole with the structure of the consortia left intact. In contrast with the chloroform fumigation methods, the biochemical methods give information on biomass and community structure, as well as on the metabolic activity of the microflora.

II. BIOCHEMICAL METHODS FOR BIOMASS MEASUREMENTS

When determining biomass by measuring the concentration of a cellular component, several requirements must be fulfilled [5,10–13]: (1) the measured component must only occur in living microorganisms and not exist in dead cells nor in nonliving parts of the soil organic matter; (2) it must be possible to quantitatively extract and analyze the component in soil samples with appropriate sensitivities; and (3) the component should exist in fairly uniform concentrations in the cell. Several cellular components have been used for biomass measurements in soils. These components include adenosine triphosphate (ATP), microbial membrane components, and constituents of microbial cell walls (Table 1).

Some skepticism exists among several microbiologists concerning estimation of biomass of soil microorganisms by measuring specific cellular constituents. They assume that no cellular components fulfill the foregoing criteria [5,14]. Despite these admonitions, many papers have been published surveying the developments and applications of biochemical methods for assaying the biomass of soil microorganisms. The results of these studies are summarized below.

A. ATP

The rapid metabolism of ATP in living cells, as well as its quantification by the very sensitive luciferin—luciferase assay, have led to the widespread use of ATP analysis as a biochemical tool of biomass determination in soil [5]. The uses of ATP and other nucleotide analyses in microbial ecology have been extensively reviewed by Karl [15], and we will only briefly discuss the application of this method to soil systems.

Table 1   Biochemical Components Used as Signatures for Estimating Biomass of Microorganisms in Soil

Component[a]	Organisms	Ref.
ATP	All cells	16,17,20—25,28, 32—35
Membrane components		
Phospholipids	All cells	45,51,52,67,68,129, 130,135,145—148
LPS	Gram-negative bacteria	51,52,62,67,68
Ergosterol	Fungi	79,80
Cell wall components		
Muramic acid	Bacteria	4,52,91
D-alanine	Bacteria	92
DAP	Gram-negative bacteria	79,80,90,92—95
Glucosamine	Fungi	89,90,106—108,113—117

[a]LPS, lipopolysaccharides; DAP, diaminopimelic acid.

Extracellular ATP has been reported to rapidly decompose in soils [16,17]. However, several studies of aquatic systems have demonstrated the presence of freely dissolved ATP [15], and growth studies of *Escherichia coli* have shown that, during the death phase, viable counts decreased more rapidly than the concentration of ATP. This observation suggests that a sizable portion of the total ATP content originates from dead cells [18,19].

A major problem of ATP measurements in soil is the variability of extraction efficiency [5,20]. Several methods have been used, including cold H_2SO_4 [17,21—25], trichloroacetic acid with paraquat and phosphate [17,20], but none of these is entirely efficient for extracting ATP from all soil types. Losses during the extractions can be accounted for by using a recovery standard of authentic ATP [15].

The most common procedure for measuring ATP in environmental samples utilizes a luciferin—luciferase system, which has a sensitivity of about 10^{-14} mol of ATP [26]. It has been calculated that this sensitivity corresponds to the ATP content of approximately 10^3 bacterial cells by assuming an ATP content of 4.0 nmol mg^{-1} dry weight [15] and an average dry weight of soil bacteria of 6.4 $\times 10^{11}$ cells g^{-1} dry weight [27]. The light emission from the enzymatic reaction can be measured with a spectrophotometer, a liquid scintillation counter, or a special photometer designed for ATP measurements. The reaction can be inhibited by various ions and other components in the extracted sample [28]. Procedures have been developed for eliminating ionic interference by sorption of the ATP to charcoal [29] or ion-exchange columns [21], followed by elution of the ATP for assay. However, hydrolysis of ATP can occur as a result of adsorption to charcoal [30]. Moreover, the enzyme must be purified to prevent reactivity with nonadenosine nucleotides [15]. These and other problems with the luciferin—luciferase method for measuring ATP content in environmental samples led Davis and White [31] to develop a high-performance liquid chromatography (HPLC) procedure for the isolation of ATP, which also made it possible to analyze several other adenine-containing components, including adenine, adenosine, AMP, cyclic-AMP, ADP, and NAD.

The concentrations of ATP in microbes vary, depending on the species, growth rate, and media composition. Karl [15] compiled data of ATP content in various exponentially grown microorganisms and found that the range of ATP was 0.5 to 18 nmol mg^{-1} dry weight. Furthermore, the ATP content varies with the concentrations of several nutrients. For example, growth under phosphate deficiency decreased the cellular content of ATP by more than 90%, compared with cultures grown with adequate phosphate [15,21,32]. The ATP content of microorganisms in soil increases after the addition of glucose to soil [32,33].

To establish a factor for the conversion of ATP content to biomass, the ATP concentrations in various microorganisms grown in vitro were determined. The mean values obtained from these measurements were then used as conversion factors for the soil microbiota [5]. Using such conversion factors, biomass measurements estimated from ATP contents in various soils showed relatively good agreement with other biomass measurements that are based on the chloroform fumigation—incubation method and direct counts [5,33—35]. Several stipulations have to be made for this to be a valid comparison; for example, all ATP has to be in the biomass, and the values of the carbon content of the biomass, as well as the conversion factors used in the fumigation and direct count methods, have to be defined.

B. Phospholipids

Measurements of the content of phospholipids (PL) have been used to estimate the biomass of the microorganisms, especially in aquatic sediments, but also in soil systems [10,11] (see Table 1). Phospholipids are found in the membranes of all living cells, but not in the storage products of microorganisms [36]. Phospholipids are actively metabolized during the growth of bacterial monocultures [37], and they have a relatively rapid turnover in dead bacteria added to aquatic sediments [38]. Similar studies have not been performed in soil systems, although the results from a study on the degradation of labeled phosphatidylcholine in soils suggested a rapid turnover of microbial phospholipids in soils [39].

Our laboratories have developed a suite of methods to analyze phospholipids and other lipid biomarkers of microbial biomass, as well as the community composition and nutritional activity of microorganisms in environmental samples [10,11,40]. These methods are based on an efficient one-phase chloroform:methanol:buffer extraction system, modified from Bligh and Dyer [41,42] (Fig. 1). The one-phase solvent system is then divided into two phases, with the addition of one portion of chloroform and one portion of buffer. The lipids are recovered in the organic phase. The aqueous phase can be used for the analysis of ATP, and the residue from the lipid extraction can be used for analysis of microbial cell wall signatures. The extracted lipids are separated on silicic acid columns into three fractions that contain neutral lipids, glycolipids, and phospholipids [42]. With this procedure, the phospholipids of *E. coli* cells, when added to sediments, were quantitatively recovered [38,43]. No such studies have been performed in soils.

The extraction procedure of Bligh and Dyer [41,42] has been compared with other methods for efficiency in extracting lipid-soluble phosphorus from soils [44]. Somewhat higher amounts of organic phosphorus were recovered with an acid pretreatment, followed by a hexane—acetone extraction or by a series of organic

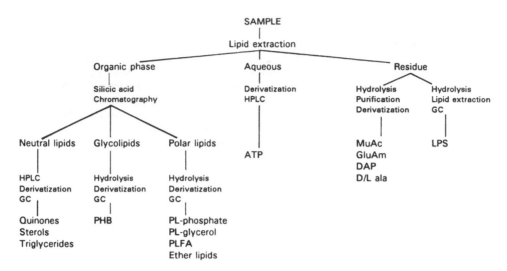

Figure 1 Scheme for the analysis of lipid fractions from soil samples.

solvents different from those used with the modified Bligh and Dyer method. However, more experiments need to be performed to establish the most reliable and accurate method for extracting phospholipids from microorganisms in soil.

Phospholipid in the polar lipid fraction from the silicic acid column is readily measured, after perchloric acid digestion, by colorimetric analysis to a sensitivity of 10^{-9} mol [38]. This enables the detection of about 10^9 bacteria that are the size of *E. coli*. The successive hydrolysis with hydrochloric acid and hydrofluoric acid (HF) and the use of gas chromatography (GC) to assay glycerol increased the sensitivity for glycerol to 10^{-11} mol [45]. The sensitivity of the phospholipid analysis can be significantly improved by analyzing the esterified fatty acids by capillary GC [46] (see Sect. III.A).

The content of phospholipids in microorganisms varies among different taxa [36,47,48]. Kates [36] showed that the phospholipid content of various bacteria varied between 4 and 91 mg g^{-1} dry weight. Furthermore, studies of monocultures have shown that the concentration and composition of phospholipids are affected by growth conditions, such as the temperature, pH, and the nutrient composition of the medium [49]. However, other experiments have shown that when bacteria are grown under conditions in a natural habitat, the cells contain a relatively constant proportion of their biomass as phospholipids [43,50].

Mean values of phospholipid content in various monocultures have been used to convert phospholipid measures to microbial numbers and biomass [e.g., 45,51,52]. The equivalence of microbial biomass measurements using phospholipid analysis, microscopic counts, ATP measurements, and several other chemical methods have been demonstrated in subsurface aquifer sediments (Table 2) [52]. These environments contained a sparse microflora of relatively uniform coccoid bacteria, which made the community ideal for comparing chemical measures and microscopic estimates of biomass.

C. Lipopolysaccharides

The outer cell membrane of gram-negative bacteria contain unique lipopolysaccharide (LPS) polymers consisting of a lipid (lipid A), a core polysaccharide, and an *O*-specific side chain [53]. Analysis of LPS has been used to estimate the biomass of gram-negative bacteria in soils and sediments (see Table 1) [54,58]. Studies have shown that LPS of dead bacteria are rapidly lost from sediments [54]. The decomposition rate of LPS in soils has not been examined.

The most common method used to analyze LPS in environmental samples is based on the limulus amebocyte lysate (LAL) test [59, 60]. In this test, an aqueous extract from the blood cells of the horseshoe crab (*Limulus polyphemus*) reacts specifically with LPS to form a turbid solution, with the amount of gelation being proportional to the LPS concentration. The LAL test is rapid and sensitive, with a detection limit of about 0.4 pg of LPS [60]. There are, however, several problems associated with use of this method for the quantitative analysis of LPS in samples as complex as soil. The LPS has to be quantitatively extracted from the sample, and two extraction procedures have been used: hot phenol—water [62] and trichloroacetic acid [54]. Only about 10% of the LPS in heat-killed *E. coli* cells was recovered from soils when phenol extraction was used [62]. Furthermore, the specificity of the LAL test is controversial, as there are several reports of substances, other than LPS, causing gelation [e.g., 61,64]. It has also been shown that LPS from various bacterial strains can differ in their ability to gel the lysate [66].

The problems associated with the LAL assay can be circumvented by the analysis of specific signature components in LPS. β-Hydroxymyristic acid (HMA) in lipid A and ketodeoxyoctanoate (KDO) in the core region of LPS are two such signatures. Analysis of HMA has been used to determine the LPS content in several soil studies [51, 67,68]. β-Hydroxymyristic acid can be analyzed by acid hydrolysis of the residue from the Bligh and Dyer extraction (see Fig. 1), followed by reextraction with fresh solvent, purification by thin-layer chromatography (TLC), and then separation by capillary GC [58]. This procedure yielded four to ten times more hydroxy acids than the amounts recovered in trichloroacetic acid or hot-phenol—water extracted LPS in samples of marine sediments [58]. The sensitivities of this method are approximately 10^{-12} mol using a flame ionization detector (FID), and 10^{-13} mol using electron-capture detection (ECD), which corresponds to 10^7 or 10^6 *E. coli* cells [58]. The sensitivies of the assays can be significantly improved by the use of special derivatives and mass spectrometric detection [69—71]. Ketodeoxyoctanoate can be analyzed by GC [54,72], but this method has been applied only in studies of gram-negative bacteria in estuarine sediments [54].

Some variation in the content and composition of LPS in bacterial monocultures has been demonstrated, depending on the taxa, temperature, and composition of the growth medium [53,73,74]. The lipid A part appears to be subject to less variation than the sugar parts of LPS. When conversion factors calculated from studies with bacterial monocultures were used, estimates of microbial biomass based on the content of lipid A hydroxy fatty acids showed good agreement with estimates based on direct counts and contents of ATP, phospholipid, and muramic acid in subsurface soils (see Table 2).

D. Ergosterol

Ergosterol (ergosta-5,7,22-trien-3β-ol) is the predominant sterol in most fungi [75]. Analysis of ergosterol has provided a rapid and sensitive method for the quantification of fungal invasions of plant material [76,78], and the method has been introduced by West and Grant for estimating fungal biomass in soil [79,80]. Although the turnover rate of ergosterol in fungi in soils has not been determined, it can be assumed that ergosterol, as a membrane component, is degraded at a rate similar to that of fungal cytoplasm in soil, a rate that is considerably faster than that of microbial cell walls in the same environment [81].

Ergosterol can be recovered from soils by extraction with methanol, saponification, and reextraction with hexane [79]. The ergosterol is then analyzed by HPLC using a UV detector, taking

Table 2 Comparisons Among Microscopic Cell Count, ATP, Cell Wall Components, and Membrane Lipid Biomarkers in the Determination of Subsurface Soil Biomass

Components	Concentration (nmol g^{-1} dry sediment)	Conversion factor (μmol g^{-1} dry wt. cell)	Direct count equivalence[a] (10^6 cells g^{-1} dry wt. sediment)	Dry wt. cells (μg cells g^{-1} dry wt. sediment)
Direct count			7.4 (3.5)	1.3 (1.1)[b]
Membrane				
Phospholipid fatty acids	0.35 (0.08)	100	7.0 (0.2)	3.5 (0.1)
Lipid phosphate	0.11 (0.03)	50	4.4 (0.1)	2.2 (0.05)
Glycerol phosphate	0.22 (0.09)	50	8.8 (0.3)	4.4 (0.15)
LPS-OH-FAME[f]	0.09 (0.04)	15	12.0 (3.0)	6.0 (1.5)
Cell wall				
Muramic acid	0.22 (0.1)	58.5	7.5 (5.0)	3.7 (2.5)
ATP	1.39 (0.42)[c]	1.7[d]/10^7	8.2 (1.1)	1.4 (0.8)[e]

[a]Calculated with 2.0 × 10^{13} cells g^{-1} dry wt.
[b]Assuming a cell volume of 1.07 μm^3 and 1.72 × 10^{-13} g/cell.
[c]ng ATP g^{-1} dry wt. sediment.
[d]ATP given in μg/10^7 cells.
[e]Assuming 10^7 cells weigh 1.72 ng.
[f]Lipopolysaccharide hydroxy fatty acids.
Source: Data from Ref. 52.

advantage of the characteristic UV absorption of ergosterol at 282 nm, which differs significantly from the maximum adsorption of other plant and animal sterols [76]. The sensitivity of this method is approximately 10 ng ergosterol per gram of dry soil, calculated by using values of the ergosterol content in fungi of approximately 2 μg fungal dry weight per gram of soil dry weight. The physical losses in the extraction procedure for ergosterol in soil, estimated by adding [^{14}C]ergosterol, were less than 8% [79]. Ergosterol can also be extracted with the Bligh and Dyer mixture, purified with silicic acid chromatography, recovered from the neutral lipid fraction, and analyzed by capillary GC [83] (see Fig. 1).

The ergosterol content of fungal mycelium varies, depending on species and growth conditions [48,75,76]. For example, Seitz et al. [76] reported a range in ergosterol content between 2.3 and 5.9 mg g^{-1} dry weight from the analysis of three fungal species grown for various times and under different conditions. West et al. [80] demonstrated a high linear correlation between ergosterol content and fungal surface area in stored, air-dried, and substrate-amended grassland and arable soils. Furthermore, the ergosterol/biomass C ratios (estimated by microscopy) for these soils resembled ratios determined in vitro in pure cultures of fungi [80,82,84]. These data indicate that the ergosterol detected in these soils was associated with living mycelia.

E. Muramic Acid, D-Alanine, and Diaminopimelic Acid

The peptidoglycan (PG) of the bacterial cell wall contains several unique components that are not found in other organisms. These unique components have been used as signatures for bacterial biomass in soils (see Table 1). The PG molecule consists of glycan chains of alternating units of N-acetylglucosamine and N-acetylmuramic acid (MuAc). The individual chains are interconnected by short peptide bridges containing specific amino acids, including D-alanine and m-diaminopimelic acid (DAP) [85]. MuAc and D-alanine are present in the PG from all bacteria, whereas DAP is present in all gram-negative bacteria, but only in some gram-positive bacteria [85].

At least parts of bacterial cell walls are resistant to degradation in soils [80,81,86,87]. The half-life of microbial material synthesized from labeled precursors was calculated to be approximately 6 months in soil [86], and the cell walls are more resistant to degradation than is the cytoplasma [81]. The decomposition rate of bacterial cell walls in soil is probably the result of their stabilization by complexing with humid acid polymers [88,90].

Components of the bacterial cell wall are recovered from soil after hydrolysis with strong HCl (4,79,91–94) (see Fig. 1). They are then purified by cation-exchange chromatography [4,96,97]. Muramic acid has been quantified in these soil hydrolysates by analysis of the lactate released from the molecule by subsequent alkaline hydrolysis [4]. The average recovery of MuAc from soils analyzed by this method was 79% [4]. A more sensitive method is to form a derivative and then analyze MuAc by GC [91,96–99] or by HPLC [100]. The sensitivity of these methods is about 10^{-11} mol of MuAc, which corresponds to about 10^8 bacteria with a size comparable with that of *E. coli*. The detection limit of the GC analysis can be substantially improved by forming special derivatives and using electron capture devices (ECD) or mass spectrometers (MS) detection systems [96].

Grant and West [79] developed a simplified procedure for the analysis of DAP in soil samples. They combined the purification and analysis steps by using paper chromatography. The recovery of DAP added to soil using this method was, on average, 80%.

The analysis of *D*-alanine requires special methods to separate the optical isomers of alanine. Two such GC methods have been developed. In one technique, *D*- and *L*-alanine were separated on a GC column coated with an optically active stationary phase [96]. In the other method, the enantiomers were separated on an optically inactive column by forming diastereoisomeric derivatives of *D*- and *L*-alanine [101].

The contents of PG components in bacterial monocultures do not vary greatly with growth rate or medium composition [102]. However, the concentration of MuAc in gram-positive and gram-negative bacteria and cyanobacteria are different. Millar and Casida [4] analyzed the MuAc content of several bacterial isolates from soil and reported an average of 19.4 mg MuAc per gram of biomass C for gram-positive bacteria and 7.2 mg MuAc per gram of biomass C for gram-negative bacteria. Similar values were obtained by Moriarty for some marine and terrestrial bacteria [103].

Millar and Casida [4], from their values for the MuAc content in bacteria, estimated that the MuAc levels in soil were about 100- to 1000-fold higher than could be accounted for by the number of organisms found on plate counts of the examined soils. This discrepancy, in part, might be because only a fraction of the total bacterial population is recovered in plate counts. However, the higher than explainable levels of MuAc could also have resulted from dead bacterial cells and organic material [4]. West et al. [80] reported that the DAP/biomass C ratios in soils (0.32 to 2.55) were more than 100 times higher than those obtained with bacteria grown individually in pure culture, and they suggested a large proportion of the DAP in their soils was also present in nonliving organic matter [87,93]. Manipulations of the soil biomass by storage, air-drying, and glucose-amendment showed, however, that changes in the DAP content and in the populations of bacteria determined by plating were closely related [80]. Studies in subsurface soils have demonstrated an equivalence in biomass estimations between measurements based on MuAc content, other chemical measures, and microscopic counts (see Table 2).

F. Chitin and Glucosamine

Chitin, a polymer of *N*-acetylglucosamine, is a major component of fungal cell walls [104]. Glucosamine is not unique to chitin, but occurs in the bacterial cell wall and in the exoskeleton of invertebrates [105]. Analysis of glucosamine has been used to measure mycelial biomass in wood and leaf litter [106–108], in plants infected by pathogens or mycorrhizal fungi [109–117], and to estimate the biomass of fungi in soil [79,80]. As with bacterial cell wall components, there are data that indicate the presence of glucosamine in nonliving organic material, and experiments have demonstrated a relatively slow degradation of fungal cell wall material in soils [90,98,100,122].

Glucosamine has been recovered from soil samples by strong acid (HCl) hydrolysis, followed by purification by cation-exchange chromatography [108–108]. The hexosamine has been quantified by colorimetric assays [106–108,119], or it can be a derivative formed and analyzed by GC with a sensitivity of about 3×10^{-11} mol [91–98]. The GC methods also analyze MuAc in the sample. Alternatively, glucosamine has been determined in soil hydrolysates by paper chromatography, which is a rapid and simple method that omits the cation-exchange step [79].

The chitin and glucosamine contents of fungi vary over a wide range, depending on species, growth conditions, and age. Aaronson [120] reported that the concentration of chitin ranged between 10 and 250 mg g^{-1} dry weight in various fungal species. The range of glucosamine in three species of salt marsh fungi measured at different ages was 8.5 to 92.8 mg g^{-1} dry weight [121]. In general, the chitin content increased with the age of the mycelium, an observation that has been made in other studies [122,123]. The chitin content in *Ciriolus versicolor*, which was used in Swift's earlier work [106], was 2.4 mg g^{-1} dry weight when grown in wood, but 12.4 mg g^{-1} dry weight when grown in vivo.

Another problem that occurs when using glucosamine analysis to estimate fungal biomass is that the contribution of glucosamine from invertebrates and bacteria must be eliminated or accounted for. Glucosamine from prokaryotes can be accounted for by measuring the amount of MuAc in the sample, and assuming an MuAc/glucosamine molar ratio of 1:1 [85]. However, the main problem of the glucosamine method is, as for the MuAc and DAP analyses, the presence of glucosamine in nonliving organic material. West et al. [80] found that the glucosamine/biomass C ratio was about 100 times above the values reported from the analysis of fungal cultures.

III. ANALYSIS OF COMMUNITY COMPOSITION

Analysis of lipid composition is an important tool for taxonomic and phylogenetic classification of microorganisms, particularly of prokaryotes [48,124,125]. Many lipid compounds are relatively easily extracted and analyzed in environmental samples. Hence, lipids are the most often used signature components for determining the community composition of microorganisms in ecological studies [126–128].

A. Phospholipid Fatty Acids

The ester-linked fatty acids in the phospholipids are currently the most sensitive and the most useful chemical measures of microbial community structure [128]. Analyses of monocultures and consortia of microorganisms isolated from the environment have shown that subsets of a microbial community can be identified by specific "signature" phospholipid fatty acids (PLFA; Table 3). For example, bacteria characteristically contain odd-chain, methyl-branched (e.g., iso- and anteiso-branched), and cyclopropane fatty acids [36,49, 126–130]. Fungi, on the other hand, typically synthesize saturated even-chained and polyenoic fatty acids [48,129]. Many actinomycetes contain methyl-branched tuberculostearic acid (10 Me 18:0) [131]. Signature PLFA have also been identified for methane-oxidizing bacteria [132], sulfate-reducing bacteria [133,134], the soil bacterium, *Flavobacterium balustinum* [135], and for *Franciscella tularensis* [136]. By utilizing fatty acid patterns of bacterial monocultures, Myron Sasser of the University of Delaware, in cooperation with Hewlett Packard, has been able to distinguish over 8000 strains of bacteria [137].

Techniques to Analyze Phospholipid Fatty Acids

Phospholipids are extracted and purified from soils using the Bligh and Dyer extraction procedure and silicic acid chromatography, as described in Section II.C (see Fig. 1). The ester-linked fatty acids are then transesterified to methyl esters by mild alkaline methanolysis, and the fatty acid methyl esters (FAME) are analyzed by capillary GC using FID [128,135,138]. Environmental samples, including soils, usually contain a very complex mixture of PLFA; for example, 30 to 50 different fatty acids can be identified in soils and sediments.

Table 3   Examples of Signature Phospholipid Fatty Acids[a]
for Microorganisms

Eubacteria
  Common signatures: i15:0, a15:0, 15:0, 16:1w5, i17:0, 17:0,
                       18:1w7

  Sulfate reducer: 10Me 16:0, Br17:1, 17:1w6

  Methane-oxidizing bacteria, type 1:   16:1w8*c*, 16:1w8*t*, 16:1w5*c*
                              type 2:   18:1w8*c*, 18:1w8*t*, 18:1w6*c*

  Flavobacterium balustinum: i17:1w7, Br 2OH-15:0

  Francisella tularensis: 24:1w5*c*, 22:1w13*c*, 20:1w11*c*

  Actinomycetes: 10Me18:0

  Fungi: 18:2w6, 18:3w6

---

[a]Fatty acids are designated as total number of carbon atoms: number of double bonds, with the position closest to the aliphatic (w)
end indicated and using *c* for *cis* and *t* for *trans*.   The prefixes
"i," "a," and "Br" refer to *iso*, *anteiso*, and (ms) methyl-branch-
ing in unconfirmed positions.   Cyclopropyl fatty acids are indicated with the prefix "cy."
Source: From Refs. 126—136.

A long (50-m × 0.2-mm id) fused silica column, coated with a cross-
linked nonpolar stationary phase has a satisfactory separation efficiency and is stable and reproducible for long-term analysis of environmental samples [139].   However, for special applications, such
as determining *cis*- and *trans*-isomers in complex mixtures of fatty
acids, columns with more polar stationary phases are needed [140].
The GC retention times give valuable information on the structure
of the FAME.   However, MS is needed for verification of the chemical structures.   Methyl esters of fatty acids have electron impact
(EI) mass spectra which, in most cases, give information on the
presence of methyl branches, hydroxy groups, unsaturations, and
cyclopropyl groups [141,142].   Special techniques are needed to
determine the configurations and localizations of double bonds and
the positions of cyclopropyl rings [143,144].

The sensitivity of these analyses is at the picomolar level, which
corresponds to the content of PLFA in 5×10^6 bacterial cells, such as
E. coli.   Some applications require a substantially higher sensitivity

of PLFA analysis than the GC–FID procedure. Introduction of new mass spectrometric methods have made such analyses possible [40]. One method, called selective ion monitoring (SIM), uses the MS instrument as a highly sensitive and specific detector by measuring only preselected ions that are specific for the analyte. By forming special derivatives and using chemical ionization techniques with negative ion detection, SIM has sensitivities to a few femtomoles (10^{-15}) of fatty acids [69].

The SIM techniques have been used for analyzing PLFA profiles of microorganisms in deep subsurface sediments. When analyzing these samples, it became obvious that special care and techniques have to be used to avoid introducing fatty acid contaminants during the handling and preparation of the samples [146]. The content of PLFA in these samples, was about 6 pmol g^{-1} dry weight, Which corresponds to 10^5 to 10^6 bacterial cells for each gram of dry weight. The application of SIM techniques for the analyses of bacterial signature components have been described in more detail [40].

Applications to Soil Systems

In soil systems, analysis of PLFA has been used to study the dynamics of bacteria associated with the roots of rape plants [146] and to characterize bacteria that suppress damping-off caused by *Rhizoctonia* [135]. The method has also been utilized to examine the biomass and structure of microbial communities in subsurface soils, including soils contaminated with organic pollutants [67,68, 129,130,147,148]. For example, analysis of PLFA demonstrated that degradation of trichloroethylene (TCE) was correlated with the presence of type II methane-oxidizing bacteria [51].

Potential problems with defining microbial community structure by analysis of PLFA could result from a shift in fatty acid composition of some monocultures with changes in temperature and media composition [58]. However, there have been no studies that show that such shifts in PLFA composition substantially changed the interpretation of the community structure during natural growth conditions.

Despite that the analysis of PLFA cannot provide an exact determination of such species or physiological type of microorganism in a given environment, the analyses provide a quantitative description of the overall microbiota in the particular environment sampled. By use of statistical analysis, it is possible to obtain an estimate by PLFA analysis of the differences among various samples and treatments [128].

B. Lipopolysaccharides, Ether Lipids,
    Sphingolipids and Quinones

Analysis of lipid components other than PLFA can provide further insight into the composition of the microbial communities. The composition of hydroxy acids in LPS varies among different groups of gram-negative bacteria [53], and this composition has been used for classifying clinical bacteria and for indicating community composition in marine sediments [57]. This analysis has not been tested in soils.

Archaebacteria are characterized by their unique biphytanyl and di-biphytanyl glycerol ether lipids which are not found in other organisms [152]. These lipids have been used as biomarkers for archaebacteria in sediments, hot spring mats, and fermenters [153–155]. Phytanyl glycerol ether lipids can be analyzed by HPLC after appropriate derivatization [156,157], or by supercritical fluid chromatography [158].

The occurrence of plasmalogens (mono-alk-1-enyl monoacyl glycerophosphatides) in microbes is restricted to specific groups of anaerobic bacteria [150,151]. Plasmalogens can be assayed by their resistance to mild alkaline methanolysis and extreme sensitivity to mild acid [38].

Analysis of respiratory quinones has been used as a sensitive biomarker of aerobic versus anaerobic metabolism in environmental samples [159]. The redox potential of the respiratory quinones suggests that the terminal electron acceptor of those bacteria containing ubiquinones (benzoquinones) should be of higher potential when compared with those of bacteria containing naphthoquinones. Bacteria capable of forming both types of respiratory quinones form ubiquinones when grown aerobically and naphthoquinones when grown anaerobically [160]. Aerobes contain benzoquinones, and some, but not all anaerobes, contain naphthoquinones [161,162]. Hedrick and White [160] analyzed the respiratory quinones from the neutral fraction of the silicic acid-purified lipid extracts (see Fig. 1) with HPLC using electrochemical detection. Manipulation of sediments between aerobic and anaerobic conditions shifted the naphthoquinones/benzoquinones ratio from 0.03 for the aerated consortia to 3.0 for the fermenters, as expected from studies with monocultures.

IV. NUTRITIONAL STATUS

Microoranisms in soil are subjected to the stress of fluxes in nutrients, which may take the form of either the partial or near-complete absence of nutrients [163,164]. Such fluctuations may lead to a transition from balanced to either unbalanced or complete cessation of growth. Unbalanced growth in microorganisms is commonly associated with the accumulation of energy-reserve polymers [165]. Numerous studies have shown that the synthesis of such polymers is induced when the nitrogen, oxygen, phosphorus, potassium, or sulfur supply (depending on the organism) becomes limiting in the presence of an excess of the carbon source [165]. Chemical methods have been utilized to indicate unbalanced growth conditions by measuring the ratio of lipid storage polymers to cellular biomass [166–169].

Microeukaryotes, such as protozoa and fungi, use triglycerides as storage polymers, and the nutritional status of these organisms has been monitored by measuring the triacylglycerols/phospholipid glycerols ratio [166].

Some bacteria form the endogenous lipid storage polymer, poly-β-hydroxybutyrate (PHB), under unbalanced growth conditions [165]. Detailed analyses of PHB extracted from bacterial isolates and environmental samples have shown that this polymer can be a mixture of polymers containing a number of various short-chain fatty acids [167]. This "mixed" polymer of PHB was called poly-β-hydroxyalkanoate (PHA) [167].

The PHB-PHA polymer has been extracted from environmental samples with boiling chloroform and sodium hypochlorite [167–170]. Findlay and White [171], however, demonstrated that PHB can be quantitatively extracted by using the modified Bligh and Dyer lipid extraction method (see Fig. 1). The extracted PHB is further purified by silicic acid chromatography, hydrolyzed, derivatized, and analyzed by GC. The detection limit of this assay using a FID detector is 100×10^{-15} mol of PHB. The sensitivity of this analysis can be significantly improved by using special derivatives and MS detection [172].

PHB analyses, in combination with lipid biomass measurements, have been used in soil systems, for example, to examine the nutritional status of bacteria associated with plant roots [146], to compare the nutritional status of bacteria in uncontaminated and contaminated subsurface soils [67,68], and to monitor bacteria degrading halogenated hydrocarbons in methane-enriched soil columns [172].

Starvation of some bacteria induces the formation of minicells. An examination of PLFA profiles from starved marine bacteria has shown that there is a marked increase in the proportion of PLFA with double bonds in the trans configuration during the formation of such minicells induced by nutrient deprivation [174]. It has been suggested that the nutritional status of some bacteria can be monitored by measuring the ratio of *cis/trans*-PLFA [174].

V. METABOLIC ACTIVITY

The analyses of the biochemical signatures described in the foregoing involve the isolation of the biological components from microbial communities. Inasmuch as each of the components are isolated,
the incorporation of labeled isotopes from precursors can be utilized to provide rates of synthesis or turnover.

Measurements of the rates of DNA synthesis with [^{3}H]thymidine
provides an estimate of the rates of heterotrophic bacterial growth.
When short incubation times are utilized, isotope dilution experiments can be used to estimate precursor concentrations, and DNA
can be purified [175] as long as the thymidine is not catabolized
significantly. This method was first applied to soil systems by
Thomas et al. [176]. The basic assumptions underlying this method
were recently examined in greater detail [177]. Soil DNA was extracted and separated from other macromolecules, such as proteins,
which contained a significant amount of labeling, by using an acid—
base hydrolysis method. The recovery of added [^{14}C]DNA was 58%
in humus soil and 75% in sandy loam soil. Isotope dilution experiments were utilized to estimate the pool of exogenous thymidine;
the uptake of added [^{3}H]thymidine was linear with time for 60 min
[177]. The thymidine method has also been utilized to estimate the
growth rates of bacteria in the rhizosphere [178] and in subsurface soils [179—181]. In estuarine sediments, thymidine was shown
to be rapidly catabolized, with [^{14}C]carbon dioxide appearing 10
min after exposure to [^{14}C]thymidine [182]. This finding suggests that thymidine catabolism could complicate the accuracy of
using thymidine incorporation to estimate bacterial growth in soils.

The incorporation rates of $H_3{}^{32}PO_4$ and [^{14}C]acetate into PL have
been utilized to measure the activity of the total microbiota [183].
The contribution of the microeukaryotic portion to PL synthesis has
been estimated by measuring synthesis in the presence of cyclo-
heximide [184]. Incorporation of [^{14}C]acetate has been utilized to
examine the synthetic activity of subsurface soils and sediments
[180,181]. The technique enables measurements of microbial activities that span more than five orders of magnitude.

Labeling experiments with [^{14}C]acetate have also been used to
measure the formation rate of PHB and PLFA [185]. This ratio was
shown to be a sensitive measure of the nutritional status of the bacterial habitat, and the technique made it possible to measure the
"disturbance artifact" involved in the application of labeled precursors in the environment [185].

Mass spectrographic analysis enables the use of precursors labeled with stable isotopes to study the rate of synthesis of biochemical signature compounds in microbial communities. Stable isotopes
are superior to radioactive isotopes in that stable isotopes have

higher specific activities, include isotope markers for nitrogen, and can be efficiently detected using SIM techniques with the MS. The high specific activity enables the assay of critical reactions at substrate concentrations that are similar to the levels present in natural environments. Methods have been developed to measure ^{15}N incorporation into D-alanine in bacterial cell walls and to determine ^{13}C environment in PLFA [186,187].

VI. CONCLUSION

Chemical measures for the biomass, structure, nutritional status, and metabolic activity of microbial communities, based on the analyses of specific cell components, represent a quantitative and sensitive method for the assay of microorganisms in soils. Several studies have demonstrated that membrane lipids, such as phospholipids and ergosterol, are particularly useful as signatures for biomass. They are comparatively easy to extract, they have a rapid turnover, and estimates of biomass based on their content corresponds well with classic methods. The analysis of the ester-linked fatty acids in the phospholipids enables the detection of specific subsets of the microbiota. With this technique, shifts in the structure of the microbial community can be quantitatively assayed. Rates of formation of membrane lipids and turnover of endogenous storage polymers, such as PHB, provide insights into the nutritional status and metabolic activities of the community. The validation of these techniques has been reviewed [11]. Further applications of these techniques will provide both insights into the ecology of microorganisms in soil as well as further validations of the methods.

ACKNOWLEDGMENTS

This work was supported by grants from the Swedish Natural Science Research Council (to A. Tunlid), N00014-87-K0012 from the Office of Naval Research, and DE-FG05-88ER60643 from the Office of Health and Environmental Research, U.S. Department of Energy.

REFERENCES

1. Casida, L. E. 1968. Methods for the isolation and estimation of activity of soil bacteria, p. 97—122. *In* T. R. G. Gray and D. Parkinson (eds.), The ecology of soil bacteria. Liverpool University Press, Liverpool.

2. Faegri, A., V. Lid Torsvik, and J. Goksoyr. 1977. Bacterial and fungal activities in soil: separation of bacteria and fungi by a rapid fractionation centrifugation technique. Soil Biol. Biochem. 9:105–112.

3. Olsen, R. A., and L. R. Bakken. 1987. Viability of soil bacteria: optimization of the plate-count technique. Microb. Ecol. 13:59–74.

4. Millar, W. N., and L. E. Casida. 1970. Evidence for muramic acid in soil. Can. J. Microbiol. 16:299–304.

5. Jenkinson, D. S., and J. N. Ladd. 1981. Microbial biomass in soil: measurement and turnover, p. 415–471. *In* E. A. Paul and J. N. Ladd (eds.), Soil biochemistry, Vol. 5. Marcel Dekker, New York.

6. Soderstrom, B. E. 1977. Vital staining of fungi in pure culture and in soil with fluorescein diacetate. Soil Biol. Biochem. 9:59–63.

7. Lundgren, B. 1981. Fluorescein diacetate as a stain for metabolically active bacteria in soil. Oikos 36:17–22.

8. Ramsay, A. J. 1984. Extraction of bacteria from soil: efficiency of shaking or ultrasonication as indicated by direct counts and autoradiography. Soil Biol. Biochem. 16:475–481.

9. Baath, E. 1988. Autoradiographic determination of metabolically active fungal hyphae in forest soil. Soil Biol. Biochem. 20:123–125.

10. White, D. C. 1983. Analysis of microorganisms in terms of quantity and activity in natural environments, p. 37–66. *In* J. H. Slater, R. Whittenbury, and J. W. T. Wimpenny (eds.), Microbes in their natural environments, Soc. Gen. Microbiol. Symp. 34.

11. White, D. C. 1986. Validation of quantitative analysis for microbial biomass, community structure, and metabolic activity. Arch. Hydrobiol. Beih. Ergebn. Limnol. 31:1–18.

12. White, D. C. 1986. Quantitative physiochemical characterization of bacterial habitats, p. 177–203. *In* J. S. Poindexter and E. R. Leadbetter (eds.), Bacteria in nature, Vol. 2. Plenum Publishing, New York.

13. Holm-Hansen, O. 1973. The use of ATP determinations in ecological studies. Bull. Ecol. Res. Commun. 17:215–222.

14. Nannipieri, P. 1984. Microbial biomass and activity measurements in soil: ecological significance, p. 515–521. *In* M. J. Klug and C. A. Reddy (eds.), Current perspectives in microbial ecology. Proc. Third Int. Symp. Microb. Ecol. American Society for Microbiology, Washington, D.C.

15. Karl, D. M. 1980. Cellular nucleotide measurements and applications in microbial ecology. Microbiol. Rev. 44:739–796.

16. Conklin, A. R., Jr., and A. N. MacGregor. 1972. Soil adenosine triphosphate: extraction, recovery and half-life. Bull. Environ. Contam. Toxicol. 7:296–300.

17. Jenkinson, D. S., and J. M. Oades. 1979. A method for measuring adenosine triphosphate in soil. Soil Biol. Biochem. 11:193–199.

18. Chappelle, E. W., G. L. Piccolo, and J. M. Deming. 1978. Determination of bacterial content in fluids. Methods Enzymol. 57:65–72.

19. Chappelle, E. W., G. L. Piccolo, H. Okrend, R. R. Thomas, J. Deming, and D. A. Nibley. 1977. Significance of luminescence assays for characterizing bacteria, p. 611–629. *In* G. A. Borun (ed.), Second Biannual ATP Methodology Symposium. SAI Technology, San Diego.

20. Sparling, G. P., and F. Eiland. 1983. A comparison of methods for measuring ATP and microbial biomass in soils. Soil Biol. Biochem. 15:227–229.

21. Lee, C. C., R. F. Harris, J. D. H. Williams, D. E. Armstrong, and J. K. Syers. 1971. Adenosine triphosphate in lake sediments-I. Soil Sci. Soc. Am. Proc. 35:82–86.

22. Greaves, M. P., R. E. Wheatley, H. Shephard, and A. H. Knight. 1973. Relationship between microbial populations and adenosine triphosphate in a basin peat. Soil Biol. Biochem. 5:685–687.

23. Ausumus, B. S. 1973. The use of ATP assay in terrestrial decomposition studies. Ecol. Res. Commun. Bull. 17:223–234.

24. Eiland, F. 1979. An improved method for determination of adenosine triphosphate (ATP) in soil. Soil Biol. Biochem. 11:31–35.

25. Eiland, F. 1983. A simple method for quantitative determination of ATP in soil. Soil Biol. Biochem. 15:665–670.

26. Karl, D. M., and O. Holm-Hansen. 1976. Effects of luceriferin concentration on the quantitative assay of ATP using crude luciferase preparations. Anal. Biochem. 75:100–112.

27. Gray, T. R. G., R. Hisset, and T. Duxbury. 1974. Bacterial populations of litter and soil in a deciduous woodland. II. Numbers, biomass and growth rates. Rev. Ecol. Biol. Soil 11:15–26.

28. Karl, D. M., and P. A. LaRock. 1975. Adenosine triphosphate measurements in soil and marine sediments. J. Fish. Res. Board Can. 32:599–607.

29. Hodson, R. E., O. Holm-Hansen, and F. Azam. 1976. Improved methodology for ATP measurements in marine environments. Mar. Biol. 34:143–149.

30. Karl, D. M., and O. Holm-Hansen. 1978. Methodology and measurement of adenylate energy charge ratios in environmental samples. Mar. Biol. 48:185–197.

31. Davis, W. M., and D. C. White. 1980. Fluorometric determination of adenosine nucleotide derivatives as measures of the microfouling, detrital and sedimentary microbial biomass and physiological status. Appl. Environ. Microbiol. 40:539–548.

32. Nannipieri, P., R. L. Johnson, and E. A. Paul. 1978. Criteria for measurement of microbial growth and activity in soil. Soil Biol. Biochem. 10:223–229.

33. Oades, J. M., and D. S. Jenkinson. 1979. The adenosine triphosphate content of the soil microbial biomass. Soil Biol. Biochem. 11:201–204.

34. Jenkinson, D. S., S. A. Davidson, and D. S. Powlson. 1979. Adenosine triphosphate and microbial biomass in soil. Soil Biol. Biochem. 11:521–527.

35. West, A. V., G. P. Sparling, and W. D. Grant. 1986. Correlation between four methods to estimate total microbial biomass in stored, air dried and glucose-amended soils. Soil Biol. Biochem. 18:569–576.

36. Kates, M. 1964. Bcaterial lipids. Adv. Lipid Res. 2:17–90.

37. White, D. C., and A. N. Tucker. 1969. Phospholipid metabolism during bacterial growth. J. Lipid Res. 10:220–233.

38. White, D. C., W. M. Davis, J. S. Nickels, J. D. King, and R. J. Bobbie. 1979. Determination of the sedimentary microbial biomass by extractable lipid phosphate. Oecologia 40:51–62.

39. Tollefson, T. S., and R. B. McKercher. 1983. The degradation of ^{14}C-labelled phosphatidyl choline in soil. Soil Biol. Biochem. 15:145–148.

40. Tunlid, A., and G. Odham. 1986. Ultrasensitive analysis of bacterial signatures by gas chromatography/mass spectrometry, p. 447–454. *In* F. Megusar and G. Kantar (eds.), Perspectives in microbial ecology, Proc. Fourth Int. Symp. Microb. Ecol. Slovene Society for Microbiology, Ljubljana, Yugoslavia.

41. Bligh, E. G., and W. J. Dyer. 1959. A rapid method of total lipid extraction and purification. Can. J. Biochem. Physiol. 37:911–917.

42. King, J. D., D. C. White, and C. W. Taylor. 1977. Use of lipid composition and metabolism to examine structure and activity of estuarine detrital microflora. Appl. Environ. Microbiol. 33:1177–1183.

43. White, D. C., R. J. Bobbie, J. D. King, J. S. Nickels, and P. Amoe. 1979. Lipid analysis of sediments for microbial biomass and community structure, p. 87–103. *In* C. D. Lichtfield

and P. L. Seyfrieds (eds.), Methodology for biomass determinations and microbial activities in sediments. ASTM STP
673, American Society for Testing and Materials, Philadelphia.

44. Kowalenko, C. G., and R. B. McKercher. 1970. An examination of methods for extraction of soil phospholipids. Soil
Biol. Biochem. 2:269–273.

45. Gehron, M. J., and D. C. White. 1983. Sensitive measures
of phospholipid glycerol in environmental samples. J. Microbiol. Methods 1:23–32.

46. Bobbie, R. J., and D. C. White. 1980. Characterization of
benthic microbial community structure by high resolution gas
chromatography of fatty acid methyl esters. Appl. Environ.
Microbiol. 39:1212–1222.

47. O'Leary, W. M. 1982. Lipoidal contents of specific microorganisms. CRC Handb. Microbiol. 4:391–429.

48. Wassef, M. K. 1977. Fungal lipids. Adv. Lipid Res. 15:
159–232.

49. Lechevalier, M. P. 1977. Lipids in bacterial taxonomy — a
taxonomist's view. CRT. Rev. Microbiol. 7:109–210.

50. White, D. C., R. J. Bobbie, J. S. Herron, J. D. King, and
S. J. Morrison. 1979. Biochemical measurements of microbial biomass and activity from environmental samples, p.
69–81. *In* J. W. Costerton and R. R. Colwell (eds.), Native aquatic bacteria: enumeration, activity and ecology.
ASTM STP 695, American Society for Testing and Materials,
Philadelphia.

51. Nichols, P. D., J. M. Henson, C. P. Antworth, J. Parsons,
J. T. Wilson, and D. C. White. 1987. Detection of a microbial consortium including type II methanotrophs by use of
phospholipid fatty acids in an aerobic halogenated hydrocar-
bon degrading soil column enriched with natural gas. Environ.
Toxicol. Chem. 6:89–97.

52. Balkwill, D. L., F. R. Leach, J. T. Wilson, J. F. McNabb,
and D. C. White. 1988. Equivalence of microbial biomass
measures based on membrane lipid and cell wall components,
adenosine triphosphate, and direct counts in subsurface aquifer sediments. Microb. Ecol. 16:73–84.

53. Wilkinson, S. G. 1977. Composition and structure of bac-
terial lipopolysaccharides, p. 97–175. *In* I. Sutherland (ed.),
Surface carbohydrates of the prokaryotic cell. Academic Press,
London.

54. Saddler, J. N., and A. C. Wardlaw. 1980. Extraction, distribution and biodegradation of bacterial lipopolysaccharides
in estuarine sediments. Antonie Leeuwenhoek J. Microbiol.
46:27–39.

55. Perry, G. J., J. K. Volkman, and R. B. Johns. 1979. Fatty acids of bacterial origin in contemporary marine sediments. Geochim. Cosmochim. Acta 43:1715–1725.

56. Cranwell, P. A. 1981. The stereochemistry of 2- and 3-hydroxy acids in a recent lacustrine sediment. Geochim. Cosmochim. Acta 45:547–552.

57. Goossens, H. W. I. C. Rijpstra, R. R. Duren, J. W. de Leeuw, and P. A. Schenk. 1986. Bacterial contribution to sedimentary organic matter; a comparative study of lipid moieties in bacteria and recent sediments. Adv. Org. Geochem. 10:683–696.

58. Parker, J. H., G. A. Smith, H. L. Fredrickson, R. J. Vestal, and D. C. White. 1982. Sensitive assay, based on hydroxy fatty acids from lipopolysaccharide lipid A, for gram-negative bacteria in sediments. Appl. Environ. Microbiol. 44:1170–1177.

59. Levin, J., and F. B. Bang. 1964. A description of cellular coagulation in the Limulus. Bull. John Hopkins Hosp. 115:337.

60. Watson, S. W., J. Levin, and T. J. Novitsky (eds.). 1987. Detection of endotoxins with the Limulus amebocyte lysate test. Alan R. Liss, New York.

61. Westphal, O., and K. Jann. 1965. Bacterial lipopolysaccharide extraction with phenol-water and further applications of the procedure. Methods Carbohydr. Res. 5:83–91.

62. Ford, S. R., J. J. Webster, and F. R. Leach. 1985. Extraction of bacterial lipopolysaccharides from soil samples and their determination using the Limulus amebocyte reaction. Soil Biol. Biochem. 17:811–814.

63. Elin, R. J., and S. S. Wolf. 1973. Nonspecificity of the Limulus amoebocyte lysate test: positive reactions with polynucleotides and proteins. J. Infect. Dis. 128:349–352.

64. Suzuki, M., T. Mikami, T. Matsumoto, and S. Suzuki. 1977. Gelation of Limulus lysate by synthetic dextran derivatives. Microbiol. Immunol. 21:419–425.

65. Mikami, T., T. Nagase, T. Matsumoto, S. Suzuki, and M. Suzuki, 1982. Gelation of Limulus amebocyte lysate by simple polysaccharides. Microbiol. Immunol. 26:403–409.

66. Novitsky, T. J., P. F. Rolandsky, G. R. Siber, and H. S. Warren. 1985. Turbidometric method for quantifying serum inhibition of Limulus amebocyte lysate. J. Clin. Microbiol. 20:211–216.

67. White, D. C., G. A. Smith, M. J. Gehron, et al. 1983. The ground water aquifer microbiota: biomass, community structure and nutritional status. Dev. Ind. Microbiol. 24:201–211.

68. Smith, G. A., J. S. Nickels, B. D. Kerger, et al. 1985. Quantitative characterization of microbial biomass and community

structure in subsurface material: a prokaryotic consortium responsive to organic contamination. Can. J. Microbiol. 32: 104–111.

69. Odham, G., A. Tunlid, G. Westerdahl, L. Larsson, J. B. Guckert, and D. C. White. 1985. Determination of microbial fatty acid profiles at femtomolar levels in human urine and the initial marine microfouling community by capillary gas chromatography–chemical ionization mass spectrometry with negative ion detection. J. Microbiol. Methods 3:331–344.

70. Maitra, S. K., R. Nachum, and F. C. Pearson. 1986. Establishment of beta-hydroxy fatty acids as chemical marker molecules for bacterial endotoxin by gas chromatography–mass spectrometry. Appl. Environ. Microbiol. 52:510–514.

71. Sonesson, A., L. Larsson, G. Westerdahl, and G. Odham. 1987. Determination of endotoxins by gas chromatography: evaluation of electron-capture and negative-ion chemical-ionization mass spectrometry derivatives of beta-hydroxy-myristic acid. J. Chromatogr. 417:11–25.

72. Bryn, K., and E. Jantzen. 1986. Quantification of 2-keto-3-deoxyoctonate in (lipo)polysaccharides by methanolytic release, trifluoroacetylation and capillary gas chromatography. J. Chromatogr. 370:103–112.

73. McDonald, I. J., and G. A. Adams. 1971. Influence of cultural conditions on the lipopolysaccharide composition of *Neisseria sicca*. J. Gen. Microbiol. 65:201–207.

74. Wartenberg, K. W., W. Knapp, N. M. Ahmed, C. Widemann, and H. Mayer. 1983. Temperature-dependent changes in the sugar and fatty acid composition of lipopolysaccharides from *Yersinia enterocolitica* strains. Zentralbl. Bakteriol. Parasitenkd. Infektionskr. Abt. 1 Orig. 253:523–530.

75. Logel, D. M. 1988. Fungal lipids, p. 699–806. *In* C. Ratledge and S. G. Wilkinson (eds.), Microbial lipids, Vol. 7. Academic Press, London.

76. Nes, W. R. 1977. The biochemistry of plant sterols. R. Proletti and D. Kritchevsky (eds.). Lipid Res. 15:233–324.

77. Seitz, L. M., D. B. Sauer, R. Burroughs, H. E. Mohr, and J. D. Hubbard. 1979. Ergosterol as a measure of fungal growth. Phytopathology, 69:1220–1203.

78. Osswald, W. F., W. Höll, and E. F. Elstner. 1986. Ergosterol as biochemical indicator of fungal infection in spruce and fir needles from different sources. Z. Naturforsch. 41c: 542–546.

79. Grant, W. D., and A. W. West. 1986. Measurement of ergosterol, diaminopimelic acid and glucosamine in soil: evaluation as indicators of microbial biomass. J. Microbiol. Methods 6:47–53.

80. West, A. G., W. D. Grant, and G. P. Sparling. 1987. Use of ergosterol, diaminopimelic acid and glucosamine contents of soils to monitor changes in microbial populations. Soil Biol. Biochem. 19:607–612.

81. Newell, S. Y., T. L. Arsuffi, and R. D. Fallon. 1988. Fundamental procedures for determining ergosterol content of decaying plant material by liquid chromatography. Appl. Environ. Microbial. 54:1876–1879.

82. White, D. C., R. J. Bobbie, J. S. Nickels, S. D. Fazio, and W. M. Davis. 1980. Nonselective biochemical methods for the determination of fungal mass and community structure in estuarine detrital microflora. Bot. Mar. 23:239–250.

83. Matcham, S. E., B. R. Jordan, and D. A. Wood. 1985. Estimation of fungal biomass in a solid substrate by three independent methods. Appl. Microbiol. Biotechnol. 21:108–112.

84. Newell, S. Y., J. D. Miller, and R. D. Fallon. 1987. Ergosterol content of salt-marsh fungi: effect of growth conditions and mycelial age. Mycologia 79:688–695.

85. Schleifer, K. H., and O. Kandler. 1972. Peptidoglycan types of bacterial cell walls and their taxonomic implications. Bacteriol. Rev. 36:407–477.

86. Shields, J. A., E. A. Paul, W. E. Lowe, and D. Parkinson. 1973. Turnover of microbial tissue in soil under field conditions. Soil Biol. Biochem. 5:753–764.

87. Durska, G., and H. Kaszubiak. 1980. Occurrence of D,L-diaminopimelic acid in soil III. D,L-diaminopimelic acid as the nutritional component of soil microorganisms. Pol. Ecol. Stud. 6:201–206.

88. Nelson, D. W., J. P. Martin, and J. O. Erwin. 1979. Decomposition of microbial cells and components in soil and their stabilization through complexing with model humic acid-type phenolic polymers. Soil Sci. Soc. Am. J. 43:84–88.

89. Verma, L., J. P. Martin, and K. Haider. 1975. Decompostion of ^{14}C-labelled proteins, peptides, and amino acids; free and complexed with humic polymers. Soil Sci. Am. Proc. 39:279–283.

90. Durska, G., and H. Kaszubiak. 1983. Occurrence of bound muramic acid and D,L-diaminopimelic acid in soil and comparison of their contents with bacterial biomass. Acta Microbiol. Pol. 32:257–263.

91. Casagrande, D. J., and K. Park. 1978. Muramic acid levels in bog soils from the Okenfenokee swamp. Soil Sci. 125:181–183.

92. Gunnarsson, T., and A. Tunlid. 1986. Recycling of fecal pellets in isopods: microorganisms and nitrogen compounds as

potential food for *Oniscus asellus*. L. Soil Biol. Biochem. 18: 595–600.

93. Durska, G., and H. Kaszubiak. 1980. Occurrence of *D,L*-diaminopimelic acid in soil II. Usefulness of *D,L*-diamino-pimelic acid determination for calculations of the microbial biomass. Pol. Ecol. Stud. 6:195–199.

94. Durska, G., and H. Kaszubiak. 1980. Occurrence of *D,L*-diaminopimelic acid in soil I. The content of *D,L*-diamino-pimelic acid in different soils. Pol. Ecol. Stud. 6:189–193.

95. Steubing, L. 1970. Chemische Methoden zur Bewertung des Mengenmässigen Vorkommens von Bakterien und Algen im Boden. Zentralbl. Bakteriol. Parasitenkd. Abt. 2, 124:245–249.

96. Tunlid, A., and G. Odham. 1983. Capillary gas chroma-tography using electron capture detection or selected ion mon-itoring detection for the determination of muramic acid, di-aminopimelic acid and the ratio of *D/L*-alanine in bacteria. J. Microbiol. Methods 1:63–76.

97. Findlay, R. H., D. J. W. Moriarty, and D. C. White. 1983. Improved method of determining muramic acid from environ-mental samples. Geomicrobiology 3:135–150.

98. Hicks, R. E., and S. Y. Newell. 1982. An improved gas chromatographic method for measuring glucosamine and mu-ramic acid concentrations. Anal. Biochem. 128:438–445.

99. Fox, A., S. L. Morgan, J. R. Hudson, Z. T. Zhu, and P. Y. Lau. 1983. Capillary gas chromatographic analysis of alditol acetates of neutral and amino sugars in bacterial cell walls. J. Chromatogr. 256:429–438.

100. Moriarty, D. W. J. 1983. Measurement of muramic acid in marine sediments by high performance liquid chromatography. J. Microbiol. Methods 1:111–117.

101. Tunlid, A., and G. Odham. 1984. Diastereoisomeric deter-mination of *R*-alanine in bacteria using capillary gas chroma-tography and positive/negative ion mass spectrometry. Bio-med. Mass Spectrom. 11:428–434.

102. Ellwood, D. C., and D. W. Tempest. 1972. Effects of en-vironment on bacterial cell wall content and composition. Adv. Microb. Physiol. 7:83–117.

103. Moriarty, D. J. W. 1978. Estimation of bacterial biomass in water and sediments using muramic acid, p. 31–33. *In* M. W. Luotiti and J. A. R. Miles (eds.), Microbial Ecology. Springer-Verlag, Berlin.

104. Aronson, J. M. 1981. Cell wall chemistry, ultrastructure, and metabolism, p. 459–507. *In* G. T. Cole and B. Kendrick (eds.), Biology of conidial fungi, Vol. 2. Academic Press, New York.

105. Parsons, J. W. 1981. Chemistry and distribution of amino
     sugars in soils and soil organisms, p. 197—227. *In* E. A.
     Paul and J. N. Ladd (eds.), Soil biochemistry, Vol. 5. Mar-
     cel Dekker, New York.

106. Swift, M. J. 1973. The estimation of mycelial biomass by
     the determination of hexosamine content of wood tissue de-
     cayed by fungi. Soil Biol. Biochem. 5:321—332.

107. Swift, M. J. 1973. Estimation of mycelial growth during de-
     composition of plant litter. Ecol. Res. Commun. Bull. 17:
     323—328.

108. Frankland, J. C., D. K. Lindley, and M. J. Swift. 1978.
     A comparison of two methods for the estimation of mycelial
     biomass in leaf litter. Soil Biol. Biochem. 10:323—333.

109. Ride, J. P., and R. B. Drysdale. 1971. A chemical method
     for estimating *Fusarium oxysporum, F. lycopersici* in infected
     tomato plants. Physiol. Plant Pathol. 1:409—420.

110. Ride, J. P., and R. B. Drysdale. 1972. A rapid method
     for the chemical estimation of fungi in plant tissue. Physiol.
     Plant Pathol. 2:7—15.

111. Toppan, A., M. T. Esquerré-Tugayé, and A. Touzé. 1976.
     An improved approach for the accurate determination of fun-
     gal pathogens in diseased plants. Physiol. Plant Pathol. 9:
     241—251.

112. Becker, W. N., and J. W. Gerdemann. 1977. Colorimetric
     quantification of vesicular—arbuscular mycorrhizal infection
     in onion. New Phytol. 78:289—295.

113. Hepper, C. M. 1977. A colorimetric method for estimating
     vesicular—arbuscular mycorrhizal infection in roots. Soil
     Biol. Biochem. 9:15—18.

114. Haselwandter, K. 1979. Mycorrhizal status of ericaceous
     plants in alpine and subalpine areas. New Phytol. 83:427—
     431.

115. Plassard, C., D. Mousain, and L. Salsac. 1982. Dosage
     de la chitine sur des ectomycorhizes de pin maritime (*Pinus
     pinaster*) a *Pisolithus tinctorius*: evaluation de la masse my-
     celienne et de la mycorhization. Can. J. Bot. 61:692—699.

116. Vignon, C., C. Plassard, D. Mousain, and L. Salsac. 1986.
     Assay of fungal chitin and estimation of mycorrhizal infection.
     Physiol. Veg. 24:201—207.

117. Bethlenfalvay, G. J., R. S. Pacorvsky, and M. S. Brown.
     1981. Measurement of mycorrhizal infection in soybeans.
     Soil Sci. Am. J. 45:871—875.

118. Hurst, H. M., and G. H. Wagner. 1969. Decomposition of
     ^{14}C-labelled cell wall and cytoplasmatic fractions from hyaline
     and melanic fungi. Soil Sci. Soc. Am. Proc. 33:707—711.

119. Chen, G. C., and B. R. Johnson. 1982. Improved colorimetric determination of cell wall chitin in wood decay fungi. Appl. Environ. Microbiol. 46:13–16.

120. Aaronson, J. M. 1965. The cell wall, p. 49–76. *In* G. C. Ainsworth and A. S. Sussman (eds.), The fungi, Vol. 1. Academic Press, New York.

121. Hicks, R. E., and S. Y. Newell. 1984. A comparison of glucosamine and biovolume conversion factors for estimating fungal biomass. Oikos 42:355–360.

122. Sharma, P. D., P. J. Fisher, and J. P. Webster. 1977. Critique of the chitin assay technique for estimation of fungal biomass. Trans. Br. Mycol. Soc. 69:479–483.

123. Cochran, T. W., and J. R. Vercelloti. 1978. Hexosamine biosynthesis and accumulation by fungi in liquid and solid media. Carbohydr. Res. 61:529–543.

124. Collins, M. D. 1985. Isoprenoid quinone analyses in bacterial classification and identification, p. 267–287. *In* M. Goodfellow and D. E. Minnikin (eds.), Chemical methods in bacterial systematics. Academic Press, London.

125. Ross, H. N. M., W. D. Grant, and J. E. Harris. 1985. Lipids in archaebacterial taxonomy, p. 289–300. *In* M. Goodfellow and D. E. Minnikin (eds.), Clinical methods in bacterial systematics. Academic Press, London.

126. Tunlid, A., and D. C. White. 1989. Use of lipid biomarkers in environmental samples. *In* A. Fox and S. A. Morgan (eds.), Proceedings first international conference on analytical chemistry and microbiology. John Wiley & Sons, New York.

127. Vestal, J. R., and D. C. White 1989. Lipid analysis in microbial ecology. BioScience 535–541.

128. Guckert, J. B., and D. C. White. 1986. Phospholipid, ester-linked fatty acid analysis in microbial ecology, p. 455–459. *In* F. Megusar and G. Kantar (eds.), Perspectives in microbial ecology, Proc. Fourth Int. Symp. Microb. Ecol. Ljubljana. American Society for Microbiology, Washington, D.C.

129. Federle, T. W. 1986. Microbial distribution in soil – new techniques, p. 493–498. *In* F. Megusar and G. Kantar (eds.), Perspectives in microbial ecology. Proc. Fourth Int. Symp. Microb. Ecol., Ljubljana. American Society for Microbiology, Washington, D.C.

130. Federle, T. W., D. C. Dobbins, J. R. Thornton-Manning, and D. D. Jones. 1986. Microbial biomass, activity, and community structure in subsurface soils. Ground Water 24: 365–372.

131. Kroppenstedt, R. M. 1985. Fatty acid and menaquinone analysis of actinomycetes and related organisms, p. 173–199. *In* M. Goodfellow and D. E. Minnikin (eds.), Chemical methods in bacterial systematics. Academic Press, London.

132. Nichols, P. D., G. A. Smith, C. P. Antworth, R. S. Hanson, and D. C. White. 1985. Phospholipid and lipopolysaccharide normal and hydroxy fatty acids as potential signatures for the methane-oxidizing bacteria. FEMS Microbiol. Eco. 31:327–335.

133. Taylor, J., and R. J. Parkes. 1983. The cellular fatty acids of the sulfate reducing bacteria *Desulfobacter sp.*, *Desulfobulbus sp.*, and *Desulfovibrio desulfuricans*. J. Gen. Microbiol. 31:3303–3309.

134. Dowling, N. J. E., F. Widdel, and D. C. White. 1986. Phospholipid ester-linked fatty acid biomarkers of acetate-oxidizing sulfate reducers and other sulphide forming bacteria. J. Gen. Microbiol. 132:1815–1825.

135. Tunlid, A., H. A. J. Hoitink, C. Low, and D. C. White. 1989. Characterization of bacteria suppressive to *Rhizoctonia* damping-off in bark compost media by analysis of fatty acid biomarkers. Appl. Environ. Microbiol. 55:1368–1374.

136. Nichols, P. D., W. R. Mayberry, C. P. Antworth, and D. C. White. 1985. Determination of monounsaturated double bond positions and geometry in the cellular fatty acids of the pathogenic bacterium *Francisella tularensis*. J. Clin. Microbiol. 21:738–740.

137. Hensen, J. M., P. H. Smith, and D. C. White. 1985. Examination of thermophilic membrane-producing digesters by analysis of bacterial lipids. Appl. Environ. Microbiol. 50:1428–1433.

138. Guckert, J. B., C. B. Antworth, P. D. Nichols, and D. C. White. 1985. Phospholipid, ester-linked fatty acid profiles as reproducible assays for changes in prokaryotic community structure of estuarine sediments. FEMS Microbiol. Ecol. 31:147–158.

139. Gillan, F. T. 1983. Analysis of complex fatty acid methyl ester mixtures on non-polar capillary GC columns. J. Chromatogr. Sci. 21:293–297.

140. Traitler, H. 1987. Recent advantages in capillary gas chromatography applied to lipid analysis. Prog. Lipid Res. 26:257–280.

141. Ryhage, R., and E. Stenhagen. 1960. Mass spectrometry in lipid research. J. Lipid Res. 1:361–390.

142. Asselineau, C., and J. Asselineau. 1985. Fatty acids and complex lipids, p. 57–103. *In* G. Odham, L. Larsson, and P.-A. Maardh (eds.), Gas chromatography and mass spectrometry, applications in microbiology. Plenum Press, New York.

143. Nichols, P. D., J. B. Guckert, and D. C. White. 1986. Determination of monounsaturated fatty acid double bond position and geometry for microbial monocultures and complex consortia by capillary GC-MS of their dimethyl disulphide adducts. J. Microbiol. Methods 5:49–55.

144. McCloskey, J. A., and J. H. Law. 1967. Ring location in cyclopropane fatty acids by a mass spectrometric method. Lipids 2:225–230.

145. Tunlid, A., D. Ringelberg, T. J. Phelps, and D. C. White. 1989. Determination of phospholipid fatty acids from bacteria at picomolar sensitivities by gas chromatography and mass spectrometry. J. Microbiol. Methods 10:139–153.

146. Tunlid, A., B. H. Baird, M. B. Trexler, et al. 1985. Determination of phospholipid ester-linked fatty acids and poly-beta-hydroxybutyrate for the estimation of bacterial biomass and activity in the rhizosphere of the rape plant *Brassica napus* (L). Can. J. Microbiol 31:1113–1119.

147. Ringelberg, D. B., J. D. Davis, G. A. Smith, et al. 1989. Validation of signature polarlipid fatty acid biomarkers for alkane-utilizing bacteria in soils and subsurface materials. FEMS Microbiol. Ecol. 62:39–50.

148. Phelps, T. J., D. Ringelberg, D. Hedrick, J. Davis, C. B. Fliermans, and D. C. White. 1988. Microbial biomass and activities associated with subsurface environments contaminated with chlorinated hydrocarbons. Geomicrobiol. J. 6:157–170.

149. Mayberry, W. R., D. W. Lambe, Jr., and K. P. Ferguson. 1982. Identification of *Bacteroides* species by cellular fatty acid profiles. Int. J. Syst. Bacteriol. 32:21–37.

150. Goldfine, H., and P. O. Hagen. 1972. Bacterial plasmologens, p. 329–350. *In* E. Snyder (ed.), Ether lipids, chemistry and biology. Academic Press, New York.

151. DeRosa, M., A. Cambacorta, and A. Gliozzi. 1986. Structure, biosynthesis, and physical properties of archaebacterial lipids. Microbiol. Rev. 50:70–80.

152. Kamio, Y., S. Kanegasaki, and H. Takanashi. 1969. Occurrence of plasmalogens in anaerobic bacteria. J. Gen. Appl. Microbiol. 15:439–451.

153. Ward, D. M., S. C. Brassell, and G. Eglinton. 1985. Archaebacterial lipids in hot-spring mats. Nature 318:693–694.

154. Pauly, G. G., and E. S. Van Vleet. 1986. Acyclic archae-bacterial ether lipids in swamp sediments. Geochim. Cosmochim. Acta 50:1117–1125.

155. Mikell, A. T., Jr., T. J. Phelps, and D. C. White. 1986. Phospholipids to monitor microbial ecology in anaerobic digesters, p. 413–444. *In* W. R. Smith and J. R. Frank (eds.), Methane from biomass. A system approach. Elsevier Publishing, New York.

156. Martz, R. F., D. L. Sebacher, and D. C. White. 1983. Biomass measurement of methane-forming bacteria in environmental samples. J. Microbiol. Methods 1:53–61.

157. Mancuso, C. A., P. D. Nichols, and D. C. White. 1986. A method for the characterization of archaebacterial signature ether lipids. J. Lipid Res. 27:49–56.

158. Holzer, G., P. J. Kelly, and W. J. Jones. 1988. Analysis of lipids from a hydrothermal vent methanogen and associated vent sediment by supercritical fluid chromatography. J. Microbiol. Methods 8:161–173.

159. Hedrick, D. B., and D. C. White. 1986. Microbial respiratory quinones in the environment. I. A sensitive liquid chromatographic method. J. Microbiol. Methods 5:243–254.

160. Kroger, A., V. Dadak, M. Klingenberg, and F. Diemer. 1971. On the role of quinones in bacterial electron transport. Differential roles of ubiquinone and menaquinone in *Proteus rettgeri.* Eur. J. Biochem. 21:322–333.

161. Whistance, G. R., J. F. Dillon, and D. R. Threlfall. 1969. The nature, intergeneric distribution and biosynthesis of isoprenoid quinones and phenols in gram-negative bacteria. Biochem. J. 111:461–472.

162. Hollander, R., G. Wolf, and W. Mannheim. 1977. Lipoquinones of some bacteria and mycoplasmas, with considerations on their functional significance. Antonie Leeuwenhoek J. Microbiol. 43:177–185.

163. Gray, T. R. G., and S. T. Williams. 1971. Microbial productivity in soil. Symp. Soc. Gen. Microbiol. 21:255–286.

164. Dawes, E. A. 1984. Stress of unbalanced growth and starvation in micro-organisms, p. 19–43. *In* A. D. Russel and M. H. E. Andrew (eds.), Revival of injured microbes. Academic Press, London.

165. Dawes, E. A., and P. J. Senior. 1973. The role and regulation of energy reserve polymers in micro-organisms. Adv. Microb. Physiol. 10:135–266.

166. Gehron, M. J., and D. C. White. 1982. Quantitative determination of the nutritional status of detrital microbiota and the grazing fauna by triglyceride glycerol analysis. J. Exp. Mar. Biol. 64:145–158.

167. Findlay, R. H., and D. C. White. 1983. Polymeric beta-hydroxyalkanoates from environmental samples and *Bacillus megaterium*. Appl. Environ. Microbiol. 45:71–78.

168. Herron, J. S., and D. C. White. 1978. Recovery of poly-beta-hydroxybutyrate from estuarine microflora. Appl. Environ. Microbiol. 35:251–257.

169. Nickels, J. S., J. D. King, and D. C. White. Poly-beta-hydroxybutyrate accumulation as a measure of unbalanced growth of the estuarine detrital microbiota. Appl. Environ. Microbiol. 37:459–465.

170. Hanzlikova, A., A. Jandera, and F. Kunc. 1984. Formation of poly-3-hydroxybutyrate by a soil microbial community during batch and heterogenous cultivation. Folia Microbiol. 29:233–241.

171. Findlay, R. H., and D. C. White. 1987. A simplified method for bacterial nutritional status based on the simultaneous de-determination of phospholipid and endogenous storage lipid poly-beta-hydroxy alkanoate. J. Microbiol. Methods 6:113–120.

172. Odham, G., A. Tunlid, G. Westerdahl, and P. Marden. 1986. Combined determination of poly-beta-hydroxyalkanoic and cellular fatty acids in starved marine bacteria and sewage sludge by gas chromatography with flame ionization detector or mass spectrometric detection. Appl. Environ. Microbiol. 52:905–910.

173. Nichols, P. D., and D. C. White. 1989. Accumulation of poly-beta-hydroxybutyrate in a methane-enriched halogenated hydrocarbon-degrading soil column: implications for microbial community structure and nutritional status. Hydrobiologia 176/177:369–377.

174. Guckert, J. B., M. A. Hood, and D. C. White. 1986. Phospholipid, ester-linked fatty acid profile changes during nutrient deprivation of *Vibrio cholerae*: increases in the *trans*/*cis* ratio and proportions of cyclopropyl fatty acids. Appl. Environ. Microbiol. 52:794–801.

175. Moriarty, D. J. W. 1986. Measurement of bacterial growth rates in aquatic systems from rates of nucleic acid synthesis. K. C. Marshall (ed.). Adv. Microb. Ecol. 9:245–292.

176. Thomas, D. R., J. A. Richardson, and R. J. Ticker. 1974. The incorporation of tritiated thymidine into DNA as a measure of the activity of soil microorganisms. Soil Biol. Biochem. 6:293–296.

177. Christensen, H., and D. Funck-Jensen. 1989. Growth rates of rhizosphere bacteria measured by the tritiated thymidine incorporation technique. Soil Biol. Biochem. 21:113–117.

178. Thorn, P. M., and R. M. Ventullo. 1988. Incorporation of bacterial growth rates in subsurface sediments using the tritiated thymidine into DNA. Microb. Ecol. 16:3–16.

179. Phelps, T. J., D. B. Hedrick, D. Ringelberg, C. B. Fliermans, and D. C. White. 1989. Utility of radiotracer measurements for subsurface microbiology studies. J. Microbiol. Methods 9:15–27.

180. Phelps, T. J., E. G. Raione, D. C. White, and C. B. Fliermans. 1989. Microbial activities in deep subsurface environments. Geomicrobiol. J. 7:79–91.

181. White, D. C., R. J. Bobbie, S. J. Morrison, D. K. Oosterhof, C. W. Taylor, and D. A. Meeter. 1977. Determination of microbial activity of estuarine detritus by relative rates of lipid biosynthesis. Limnol. Oceanogr. 22:1089–1099.

182. Moriarty, D. J. W., D. C. White, and T. J. Wasenberg. 1985. A convenient method for measuring rates of phospholipid synthesis in seawater and sediments: its relevance to the determination of bacterial productivity and the disturbance artifacts introduced by measurements. J. Microbiol. Methods 3:321–330.

184. Findlay, R. H., P. C. Pollard, D. J. W. Moriarty, and D. C. White. 1985. Quantitative determination of microbial activity and community nutritional status in estuarine sediments: evidence for a disturbance artifact. Can. J. Microbiol. 31:493–498.

185. Tunlid, A., H. Ek, G. Westerdahl, and G. Odham. 1987. Determination of ^{13}C-enrichment in bacterial fatty acids using chemical ionization mass spectrometry with negative ion detection. J. Microbiol. Methods 7:77.

186. Tunlid, A., G. Odham, R. H. Findlay, and D. C. White. 1985. Precision and sensitivity in the measurement of ^{15}N enrichment in *D*-alanine from bacterial cell walls using positive/negative ion mass spectrometry. J. Microbiol. Methods 3:237–245.

8

Factors Affecting the Movement of Microorganisms in Soils

SHIMNA M. GAMMACK, ERIC PATERSON, JANE S. KEMP,
MALCOLM S. CRESSER, and KENNETH KILLHAM
University of Aberdeen, Aberdeen, Scotland

I. INTRODUCTION

The movement of microorganisms in soils has been studied for many years [e.g., 1–4]. Such studies provide information of value to a variety of aspects of soil biology (e.g., the spread of soil-borne plant pathogens through soil; the colonization of roots by symbiotic organisms, such as mycorrhizal fungi or root-nodulating rhizobacteria). However, the need is increasing for information on the movement of microorganisms in soils for the development and use of microbial inocula in the soil–plant system, both to assess the performance of an inoculum and to evaluate the environmental impact of the release of a particular inoculum.

Microbial inocula, involving indigenous or nonindigenous organisms, may be added to soils or agricultural soil–plant systems for several purposes. These include attempting to increase crop yield with microbial symbionts, such as N_2-fixing rhizobia [5,6] and mycorrhizal fungi; antagonistic control of soilborne plant pathogens [7,8]; the cloning of genes involved in the bacterial production of insecticidal toxins [9]; inocula for the production of plant growth regulators [10–12]; the detoxification of industrially contaminated soil [13]; and foliar treatments for plant protection against attack by frost [14,15] or by pests [16]. These applications depend on survival and establishment of the inocula in the chosen environment. In addition to the aforementioned deliberate inoculations, nonindigenous microorganisms may also enter the soil indirectly; for

example, in the use of cattle slurry or domestic sewage as fertilizers (in addition to nitrogenous compounds, these wastes contain
large numbers of bacteria, some of which may be pathogenic), by
accident, or through environmental transport. Therefore, inasmuch
as the ultimate fate of microbial inocula and fecal bacteria from organic wastes will often be the soil environment, it is timely to develop a predictive understanding of factors that affect the movement
of microorganisms through soil as well as the dispersal of soilborne
microorganisms between sites by external vectors.

To minimize the risks of release of potential pathogens and for
optimization of inoculum success, an understanding of the fate of
the introduced organism(s) in the environment is essential. If
nonindigenous organisms (including genetically engineered microorganisms; GEMs) are to be introduced, knowledge of the fate of
any foreign DNA they may carry is essential in risk assessment.
Hence, the dispersal [17], survival and growth, and possible gene
transfer [see 18–20] of microbial inocula in the environment should
be studied.

Information on the movement of microorganisms in soil can be
gained from studies on indigenous soil organisms, as well as from
studies with labeled or marked inocula. Inasmuch as sorptive interactions between microbial cells and soil particles are likely to
limit microbial movement through soil, this chapter begins with a
discussion of the factors affecting such interactions. The effects
of soil physical properties, such as particle size distribution and
pore size distribution, on the potential pathways available for microbial movement are reviewed. Active microbial movement, by motility or through cell growth, is discussed. The current knowledge
on the effects of plant roots and soil animals on movement through
soil and of the transport of microorganisms from soil by aboveground agents is reviewed. The chapter then identifies where information is lacking and suggests approaches to remedy this.

II. ADSORPTION OF MICROBIAL CELLS
ON SOIL PARTICLES

Microbial adsorption on particles is thought to be an important factor in determining microbial movement in soils [21] and, although
there is little direct evidence of adsorption in soil, the circumstan-
tial (or empirical) evidence is strong, in the form of retention of
cells in soil [see 22,23]. Retention may be explained by mechanisms other than adsorption of cells on soil surfaces (e.g., by filtration or entrapment within soil crumbs), but inasmuch as vigorous methods of soil disruption are often used to disperse soil before enumeration of microorganisms, it appears unlikely that such

mechanisms are entirely responsible for the retentitive properties of soil. Therefore, adsorption (physicochemical surface interactions) of cells with soil particulates must still be considered as a major mechanism for the observed retention of microbes in soil [22].

Cells adsorbed on soil particles will not move by their own motility or by Brownian motion (diffusion). However, the soil particles themselves may be moved by external agents, such as rainfall [24] or the action of other organisms.

In aquatic habitats, it can be shown that most microbial cells are associated with surfaces [25], and it is generally believed that this is also true in soil [26–28]. Direct evidence for this is scarce and difficult to obtain for soil [23].

In vitro studies with electrophoresis have indicated that, under certain circumstances (e.g., low pH or high concentrations of polyvalent cations), clay particles can adsorb on microbial cells and vice versa [29–31]. At typical soil pH values, however, the net electrical charges of clay and bacterial surfaces are negative [23,31,32]; therefore, they would be expected to repel each other by electrostatic forces.

Adsorption of microbial cells on soil particles is dependent on the following factors.

A. Soil Type

The movement of microorganisms through soil appears to vary with soil type [21,33]. This variation may, in part, be a result of variations in the adsorptive properties of different soils. Adsorption of microbes in different soils is often considered in terms of soil physical properties (e.g., percent sand, silt, clay, and organic matter), and the degree of adsorption between microorganisms and soil particles is broadly related to the surface area and surface charge properties of the particles [23,28]. The major soil components that affect bacterial adsorption are the organic matter [27] and clay fractions. Both these components possess large surface areas and consist of primarily negatively charged particles. Clay soils are termed "heavy" and sandy soils "light" [34].

Bashan and Levanony [35] found that about 90% of an added inoculum of the rhizosphere bacterium, *Azospirillum brasilense*, was retained in columns of either a light (0.33% organic matter, 5.7% clay) or two heavy (19% and 35.9% clay) textured soils when water (equivalent to 100 mm) or more of rainfall) was leached through the columns. However, only 5 to 6% of the added bacteria were retained by columns of natural quartz sand (from a quarry; <0.01% organic matter, 0.2% clay), which was characterized by a small surface area [91] (between 1.61 and 1.88 m^2 g^{-1}),

compared to the other soils used in the study (approximate surface
area of light-textured soil: 43 m^2 g^{-1}; heavy-textured soils: 143
m^2 g^{-1} and 270 m^2 g^{-1}). The authors attributed this retention to
adsorption of the bacteria on soil or sand in the columns, with
the lower degree of retention in the sand columns being attributed
to the relatively low surface area for adsorption and low contents
of clay and organic matter. The differences in retention may have
been a result of differences in mass flow or pore sizes and do not
directly prove that adsorption to clay and organic matter occurred.
However, by comparing bacterial numbers in washed soil with numbers in the supernatant, the authors demonstrated that 80 to 90%
of the added bacteria could not be removed by washing from (and
were, therefore, assumed to be adsorbed to) the light-textured
soil, whereas fewer than 1% remained in the natural quartz sand
after washing. In natural conditions, however, inert particles,
such as quartz, are often enveloped by skins or cutans of clay
and iron oxide, which may greatly increase adsorption [23,28,36].

B. Microbial Isoelectric Point

The *isoelectric point* (pI) of a microbial cell is defined as the pH
at which the net charge of the cell surface is zero. The net surface charge of a cell is negative at pH values above, and positive
at pH values below, the pI [37,38]. Most clay particles (exceptions are the variable-charged clays, which include allophane that
has a net positive charge at low pH, and kaolinite) have a net negative charge, irrespective of pH (in the absence of large polyvalent
cations, which adsorb to negatively charged clay surfaces). The
net negative charge is a result of isomorphous substitution in the
crystal structures of, for example, tetravalent Si^{4+} by trivalent
Al^{3+} or Al^{3+} by Mg^{2+}, which results in excess negative charge
[39]. Adsorption through coulombic interactions between clays and
microbial cells can occur only if the pH is below the pI of cells or
if polyvalent cations are present. This was borne out in pure
culture studies [38]. The pI of microbial cells in pure culture in
vitro (pH 2.5 to 3.5 [38]) is generally much lower than the bulk
pH of agronomic soils; therefore, adsorption between clays and microbial cells might not be expected to occur. However, under natural soil conditions, adsorption appears to occur (if it does, indeed,
occur [23]) between bacteria and soil particles at pH values above
the apparent pI of cells in vitro and in the presence of predominantly K^+, Na^+, Ca^{2+}, Mg^{2+}, and H^+, which do not appear to enhance adsorption [31]. It appears, therefore, that the pI of cells
in soil in vivo is higher than that in vitro. It must be mentioned,
however, that a *net* negative charge does not preclude the existence of local positively charged sites. For example, clay platelets

with a net negative charge may be positively charged at their edges at pH values below 7 [39], and edge-to-face adsorption of such small platelets may occur with negatively charged bacterial surfaces [29,40], and negatively charged bacteria may have cellular appendages, the tips of which are positively charged (see under Section II.F).

ᵩ A negatively charged surface in an aqueous environment will be neutralized by the accumulation of positively charged counterions [39]. The negative and positive charges together form the electrical double layer. Clay particles and bacteria will, at typical soil pH values, both have electrical double layers. Colloid particles, such as clays and bacteria, will tend to be attracted to each other through van der Waals forces, but their double layers will tend to repel each other and must be bridged, for example, by polyvalent cations, for adsorption to take place. The thickness of the double layer is dependent on the concentration of the electrolyte(s) and the cationic valency and is more compressed (i.e., the concentration gradient becomes steeper, so that the repulsion between particles is reduced) at higher concentrations and in the presence of polyvalent versus monovalent ions [23,31,32].

The pI of cells in soil may be influenced by the type of cations present (e.g., Fe^{3+} or Al^{3+} at a concentration of 3×10^{-4} M can increase the pI of bacteria from pH 2.7 to 7 [23,38]). Thus, as similar concentrations of such polyvalent cations are usually present in soil, microbes in situ may be positively charged, or at least have a lower net negative charge, than in vitro, which would facilitate adsorption with negatively charged clays [23].

C. Involvement of Cations

As discussed in the foregoing, cations affect the thickness of the electrical double layer and the pI of bacterial cells. The compression of the double layer is amplified with increasing valency, according to the Schulze—Hardy rule [see 39]. Cations also enhance adsorption between bacteria and clays by adsorbing to negatively charged surfaces, thus neutralizing the negative charge.

Santoro and Stotzky [31] demonstrated the influence of cations on the adsorption between clay particles and microbial cells. Peele [30] showed that various bacteria were adsorbed most strongly on soils saturated with trivalent cations, such as Fe^{3+} and Al^{3+}; there was 92 to 94% adsorption between microbes and a silt loam soil saturated with trivalent cations, whereas there was only 1 to 4% adsorption when monovalent cations were present.

These results can be explained by the adsorption and desorption equilibria of bacteria on the surface of the clay present in soil, which acts as an ion-exchange surface. Monovalent cations

in fertilizers could influence the ion-exchange surfaces in soil and
may cause a desorption effect between microbes and soil particles.

The involvement of polyvalent cations in adsorption has been
highlighted by Ruddick and Williams [41]. They suggested that
the results of Hepple [42], which demonstrated the inability of
spores of the fungus, *Mucor ramannianus*, to pass through the B
horizon of an iron—humus podzol (a zone characterized by deposi-
tions of fine clay, small amounts of organic matter, and iron oxide)
after they had moved through the strongly leached top soil of the
A horizon, were attributable to increased adsorption on clays in
the presence of Fe^{3+}.

D. Soil pH

Soil pH measurements give an average value for the bulk system.
However, soil is not a homogeneous medium, and there may be differences in pH between microhabitats, with the possibility that the
local pH may be below the pI of microbial cells. This would result
in a net positive cell surface charge [31], facilitating adsorption on
negatively charged soil particles.

The positive charge density on the edges of clay particles also
increases with decreasing pH [29,39], and this may also have a substantial effect on adsorption. Bashan and Levanony [35] attributed
the poor colonization of grass roots by *Azospirillum* spp. to this phenomenon, stating that the decrease in pH of the rhizosphere soil
(caused by proton extrusion by the growing root) resulted in increased adsorption on the soil, thereby preventing bacteria from
reaching the roots. A reduction in pH also increases the solubility
of some polyvalent cations (e.g., Fe^{3+}, Al^{3+}, Mn^{2+}), which, in turn,
may increase adsorption as described in the previous section.

E. Specificity of Adsorption of Microorganisms in Soil

Some studies have shown that different species of microorganism
show different rates of movement in soil [e.g., 43]. Differences
in adsorption may, in part, account for such differences in movement. The degree of adsorption of different bacteria on soils or
colloidal materials varies greatly; for a range of bacteria in a loamy
soil, adsorption varied from 95.5% for *Bacillus cereus* to 11.5% for
Escherichia coli [44]. As previously discussed, soil type, the concentration of polyvalent cations, and the pH of the microhabitat, all
have an effect on the degree of adsorption. However, adsorption
of microbes on soil particles is thought to be influenced also by
the action of the type of functional groups present on the surface
of the bacterial cells [40,45,46]. Rhizobial cells that contain car-
boxyl-type surfaces have been shown to adsorb more strongly on

Na^+-montmorillonite or Na^+-illite than cells that contain more complex amino-carboxyl surfaces [40,46]. Another indication of how bacterial species vary in their inherent sorptive properties comes from investigations of bacteria isolated from saline muds [47]. It was found that the muds contained bacteria, the degree of adsorption of which was higher than that of bacteria isolated from associated saline water. These results would suggest that these bacteria have adapted to surface-associated life in saline environments, where, as previously mentioned, monovalent salts tend to desorb bacteria from the soil matrix.

F. Active Microbial Processes

Inert colloidal particles, such as polystyrene latex spheres [48,49] or killed bacterial cells [31,50,51], have been shown to adsorb to surfaces through physicochemical interactions, such as hydrophobic, van der Waals, or coulombic (electrostatic) forces. These attractions can be predicted by mathematical models [e.g., 52], and microbial adsorption often follows these models. Where attachment deviates from the predictions of the models, active microbial processes are assumed to have a part in adsorption. These processes include an energy requirement for attachment [53], the production of adhesive exopolymers, and chemotaxis.

Production of Extracellular Polymers and Cellular Appendages

Soil microorganisms produce extracellular polysaccharides in pure culture [54–57], and the production of extracellular polymers has been associated with the attachment of cells (isolated from soil) to surfaces [58]. Restriction of microbial movement in soils through adsorption on soil particles [e.g., 21] may involve the production of extracellular polymers and appendages. Both gram-positive and gram-negative soil bacteria can produce extracellular polymers [59], and they are able to utilize water-soluble materials released from the rhizosphere of plants and to modify root mucilage to produce a polysaccharidic material, termed *mucigel* [56,60,61]. Polysaccharides produced by bacteria not only serve as a protective layer against predation and desiccation [62], but also they may have a vital role in mediating bacterial attachment [37], aggregate formation, and stability [63,64]. The forces of attraction between polysaccharides and clay particles are thought to involve hydrogen bonding, ionic bonding, and van der Waals forces [65]. Extracellular polymers have been implicated in binding of microbial cells to surfaces in aqueous environments and in pathogenesis [66,67], but there is little direct evidence for the formation of exopolymers in soil or for their involvement in binding of cells to soil particles [23].

Binding of bacteria by extracellular polymeric fibrils on soil particles is an active bacterial process that requires the expenditure of energy to produce fibrillar material and involves protein bridging [50,68,69]. Protein bridging is much weaker than ionic bonding and is greatly dependent on the natural status of the microhabitat, as highlighted by Bashan and Levanony [50], who, using the rhizosphere bacterium, *A. brasilense*, demonstrated that increasing the nutritional status of a sandy soil markedly increased bacterial attachment by increasing the production of polymeric fibrillar material. The authors cited Foster [70] to deny that fibrils are artifactual (as suggested by Fraser and Gilmour [71]) and proposed that fibrillar binding has an important role in sandy soils, as those soils are characterized by low cation-exchange capacities. In soils with greater amounts of clay and organic matter, and in the presence of a range of cations, other forces, such as ionic (coulombic) interactions, hydrogen bonding, and hydrophobic interactions, are thought to be of more importance [50].

Plant roots release different amounts and type of organic materials and, correspondingly, the rhizosphere population varies among plant species [37]. Thus, it would be expected that the production of extracellular polysaccharides involved in adsorption and aggregate stabilization also varies among plant species.

Many soil bacteria produce numerous proteinaceous cellular appendages [72], termed pili or fimbriae (there is widespread disagreement on the terminology of these cellular appendages [73], and for the purposes of this review, the term *pili* will be used to cover all nonflagellar appendages, as suggested by Brinton [74]). Pili appear to be largely confined to the gram-negative bacteria, including pseudomonads, enterobacteria, *Rhizobium*, and *Agrobacterium* [reviewed in Ref. 73]. It has been suggested [28] that pili are important in adsorption processes, as they substantially alter the net surface charge of cells [75,76]. Inasmuch as removal of the pili by mechanical agitation resulted in cells with the same surface charge as cells without pili [76], their effect may be due to the pili being positively charged at their tips. Because pili are thin and relatively long, they have the potential to bridge the electrical double layers and link negatively charged cells and clays [23]. It is believed that the tips of pili may be positively charged and aid in the adsorption of bacteria on negatively charged surfaces [77]. Adsorption of microorganisms on soil particulates by pili and, occasionally, by flagella, has been demonstrated by electron microscopy [78]. Inasmuch as flagella are far less numerous on most cells than are pili, the role of flagella in adsorption is not thought to be as important, but flagella have been implicated in the attachment of *Pseudomonas fluorescens* to glass in a liquid culture [51]. The bacteria were observed to be attached to glass slides at one end, apparently by their polar flagella, which continued to beat, causing rotation of the cell. The mechanism of flagellar attachment is probably similar to that of pili (i.e., their length and thinness allows them to penetrate the electrical double layers, and their tips may be positively charged, allowing adsorption to a negatively charged surface).

Active Microbial Movement

When chemical gradients are imposed in vitro on a population of motile bacteria, the cells migrate to, and accumulate in, that part of the gradient that provides an optimum concentration of the chemical. Accumulation of cells at sites with high nutrient concentrations is the result of positive chemotaxis [79,80]. With inhibitory chemicals, the optimum concentration would be approximately zero, and negative chemotaxis, away from the source of the inhibitor, occurs. In natural soil systems, nutrient gradients probably also exist. The interfaces between two different phases (e.g., solid—liquid, liquid—gas) possess a greater concentration of nutrients and may have more microbial activity than the surrounding bulk soil [81]. This intensified microbial activity may be the result of bacterial chemotactic responses. Inasmuch as transport of microbial cells to a surface is a prerequisite for adsorption onto the surface, chemotaxis may increase the rate of microbial adsorption on soil particles by attracting cells to the surface of the soil particle at a rate faster than that which would be achieved in the absence of a chemotactic response. Whether adsorption to soil particles is beneficial to the cell is the subject of much debate [reviewed in Refs. 23 and 82], as is the question of whether particle-associated effects (beneficial or otherwise) are the result of direct surface interactions or of indirect effects [see 23]. Chemotaxis to root exudates would also be expected to increase the attraction, and subsequent adsorption, of rhizoplane bacteria on the root surface. Although chemotaxis to increased concentrations of nutrients at a surface may be responsible for the initial colonization of that surface, the organisms at the surface will probably utilize the surface-concentrated nutrients faster than the nutrients can be replenished [83]; accordingly, enhanced concentrations of nutrients at surfaces may confer only short-term benefits to surface-associated cells.

For chemotaxis to occur, bacteria must be flagellated or otherwise motile (see Section IV). There is little evidence for this in nonsterile soil, but reports of apparent motility in sterilized soil are reviewed here and in Section IV. Thornton and Gangulee [4] observed motile forms in suspensions of sterilized soil 14 to 18 h after it had been inoculated with "*Bacillus radicicola.*"

Most studies on chemotaxis of soil bacteria have been conducted in vitro [e.g., 84], but chemotaxis of A. brasilense and P. fluorescens to root exudates in sterilized soil has been demonstrated [85]. Mellor et al. [86] observed no difference in root colonization in sterilized soil between motile and nonmotile rhizobia. Liu et al. [87] found that motility conferred little advantage in the nodulating competitiveness of Bradyrhizobium japonicum in a nonsterile sandy loam. Nevertheless, some studies have shown that motile rhizobia have a competitive advantage over nonmotile strains in root colonization and nodule formation [86–88], but only when the motile and nonmotile strains were inoculated together (i.e., the nonmotile strains were able to colonize roots effectively when inoculated in the absence of motile competitors). In addition, mutant strains of P. fluorescens that lacked flagella colonized roots in a soil system as effectively as wild-type strains that possessed flagella [89], suggesting that either (1) the wild-type strains are not flagellated in soil, but only in vitro, or (2) there is insufficient energy available in the soil for active flagellar movement to occur in the wild-type strains. In a nonsterile soil, P. putida and P. fluorescens were motile and showed chemotaxis toward soybean exudates [90]. More evidence is needed for the occurrence of chemotaxis in soils that have been neither sterilized nor amended with nutrients (i.e., untreated soils).

III. SOIL PHYSICAL PROPERTIES AND SOIL WATER IN RELATION TO MICROBIAL MOVEMENT

The physical structure of soil and its relation to the holding capacity and movement of water is of prime importance in determining how likely microorganisms are to move through soils.

A. Soil Texture and Structure

Soils are characterized in terms of the contribution to total mass of particles in selected size ranges, known as the sand, silt, and clay fractions. The particle size classes (diameter) adopted by the Soil Survey of England and Wales are: clay particles (< 2 µm); silt (2 to 60 µm); fine sand (60 to 200 µm); medium sand (200 to 600 µm); and coarse sand (600 to 2000 µm). The texture of a soil refers to its particle size distribution [e.g., 91]. In a system that consists of particles of the same size, the size of the pores is proportional to the size of the particles; for example, a coarse sand will contain larger pores than a fine sand. In mixed systems, the pore size distribution will be more complex, with smaller particles occupying the spaces between the larger particles. These generalities refer to unstructured systems, in which the primary particles (sand, silt, clay) are not held together in aggregates.

In natural soils, individual particles, such as clay and silt, adhere together to form complex aggregates of varying sizes and shapes [92—95]. The structure of a soil refers to the size, shape, and arrangement of its aggregates [91]. Soil structure and particle size distribution, in turn, determine the pore size distribution in soil and the way in which pores are interconnected. Soils vary in structure, depending on their particle size distribution and other factors, and it is to be expected that the degree of microbial migration will vary from one soil structural type to another.

B. Movement in Soils in the Absence of Percolating Water

As stated, structure is a fundamental property of soil that will influence bacterial movement. However, some of the studies relating to movement will have used artificial matrices or sands [96—100], which are unstructured and, therefore, more representative of subsoil than topsoil.

In saturated soil, the availability of continuous water pathways potentially facilitates bacterial movement (by passive diffusion or active motility) throughout the soil [101]. Movement of bacteria through active motility (speeds of 20 to 60 μm s^{-1} have been recorded in vitro [102]) will greatly exceed that through diffusion. However, diffusion is the only mechanism for movement of cells that are not actively motile in the absence of transporting agents, such as percolating water. Inasmuch as diffusion is a very slow process, bacterial dispersal in the absence of transporting agents is likely to be severely limited for cells that are not actively motile. Studies that demonstrate that microbial movement is limited in the absence of such transporting agents [43,103—105] suggest that active microbial motility is not reponsible for any substantial microbial movement in soil.

Bacterial movement can occur only where the pore neck sizes are not smaller than the bacterial dimensions. For a rod-shaped bacterium, such as *Pseudomonas aeruginosa*, a pore neck radius of 1 to 1.5 μm would severely restrict passage, whether by flagellar movement or Brownian motion [96]. In all soils, except some clay subsoils that are structureless and with pore necks too small to allow passage, some migration would, therefore, be expected.

Water is held in unsaturated soil at a negative pressure, which is referred to as the *matric potential* ("the negative pressure or suction experienced by soil water as a result of affinity for the soil matrix"; International Soil Science Society). This can be measured with a tensiometer [106], or the soil, after initial saturation, can be equilibrated at a given potential by either sucking water from the soil (on a tension table) or forcing water out of the soil under pressure (with a pressure plate apparatus) [107]. A graph of the matric potential (ψ_m in pascals [Pa]) versus soil water content is called the *water release characteristic* for that soil (Figure 1). At any given matric potential, pores with necks larger than a critical size will have drained and, therefore, be unavailable for microbial movement. The critical pore neck size is inversely proportional to the matric potential:

$$\psi_m = \frac{0.3}{d}$$

where ψ_m is the matric potential (in kPa) and d is the diameter of the largest water filled pore neck (in mm) (derived from equation [3] of Marshall [108]). Consequently, microbiological studies in unsaturated soil can be interpreted only if the soil matric potential is known; soil water content expressed as a percentage of water-holding capacity is meaningless unless the water release characteristic of the soil is known.

In experiments using structureless matrices, it was found that as matric potential was reduced (i.e., as the water content was reduced), there was quantitatively less movement of bacteria [96,98, 99]. This effect was more pronounced in matrices composed of large particles than in those consisting of smaller particles [96,98,99], which have a greater proportion of small pores. Inasmuch as smaller pores empty at a lower potential (i.e., more negative or greater suction, resulting in more water removal), fine-grained soils retain a higher percentage of their water content than do coarse-grained soils at any given matric potential. Therefore, in unstructured soils, at water contents below field capacity (i.e., the -33 kPa water tension, at which pores with neck diameters larger than 9.1 μm will be air-filled), there are likely to be more potential pathways for movement through fine-grained soils, as fine-grained soils will have a greater proportion of pores that are 9.1 μm or smaller in diameter, and air-filled pores in coarser soils will act as barriers to aqueous movement [91].

The situation is greatly complicated in natural structured soils, in which the pore size distribution is complex. The matric suction required to drain a pore is inversely proportional to the diameter of its pore neck (see Fig. 1). Therefore, as a soil drains (or dries), the largest pores lose their water first, followed by a descending range of smaller pores. In a structured soil (i.e., one containing aggregates), there are two distinct pore populations: those between aggregates (large) and those within aggregates (small). When the interaggregate pores are filled with water, the possibility for bacterial movement exists, either by active motility

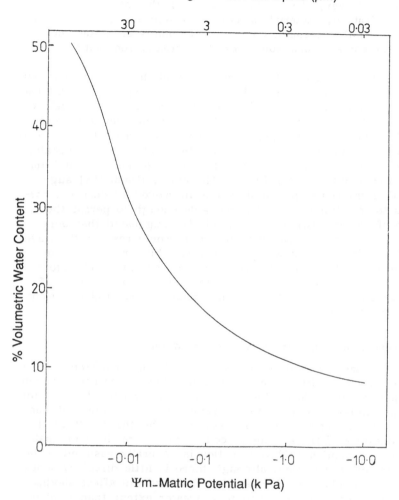

Figure 1 A soil moisture release characteristic (for a hypothetical soil), showing the relationship between volumetric water content, matric potential, and diameter of largest water-filled pore. The term *moisture release* refers to the loss of soil water through the drainage of progressively smaller pores with increasing matric stress.

(if indeed, active motility does occur in soil) or by Brownian movement (which is very slow). However, when the interaggregate pores are drained, bacterial movement can only occur from one aggregate to another through water films at points of contact between aggregates, which would result in greatly reduced bacterial dispersal.

Hamdi [98] found that the matric potential that limited movement of *Rhizobium trifolii* was dependent on the soil type. In a silt loam soil, the limiting matric potential was -39 kPa (i.e., below field capacity of -33 kPa), but in fine sand, the limiting matric potential was -24.5 kPa (i.e., only pores larger than 12 µm in diameter were drained of water), which corresponded to a water content greater than field capacity. At field capacity, pores smaller than 9.1 µm in diameter would be water-filled. Therefore, Hamdi [98] suggested that the matric potential that limits movement does not correspond to the draining of all pores wide enough to permit the passage of bacteria (Griffin and Quail [96] suggested that pore necks with a diameter smaller than 1.5 µm would restrict the movement of a bacterium that was 0.5-µm wide), but, rather, to the potential at which water pathways are discontinuous. Therefore, pore connectivity and pore size distribution are both aspects of soil structure that will determine the extent of bacterial movement in nonsaturated soils.

C. Movement in Soils with Mass Flow of Water

Movement in mass flow of water will be of more importance than diffusion (which is slow) or active motility (the occurrence of which in soil has not been demonstrated conclusively) when soils are subject to precipitation or irrigation [24,109–113]. Retention of bacteria in soil during mass flow is assumed to be the result of the combined effect of the sieving of cells that are retained in small pores and adsorption [21] (adsorption is generally assumed to be a major factor in retention, although there is little direct evidence for this in soil [23]; see Section II). Cell size will affect sieving, with larger cells being retained to a greater extent than smaller cells [214]. Bitton et al. [21] found that retention of *Klebsiella aerogenes* in soil columns increased as the clay content of the soil increased and that a larger encapsulated strain was retained to a greater extent than one that was nonencapsulated. However, it is not clear to what extent these results were caused by increased sieving, as the result of the presence of more pore necks restrictive to movement of bacteria, as opposed to increased adsorption; both are likely to be important, since an increased clay content will decrease pore size, but it will also provide a greater surface area that is potentially available for adsorption, if adsorption

of microbial cells on clay occurs. Hepple [42] showed that spores
of *M. ramannianus* (2 to 3 μm in diameter) could be washed through
the sandy A horizons of forest soil but that they were retained in
the more compact B_1 horizon. This observation was attributed to
sieving. Indirect evidence for the importance of sieving was found
by Lahav and Tropp [111] when investigating the movement of negatively charged synthetic microspheres in natural soils. They found
that in a clay soil, movement of larger microspheres (>8 μm) did not
occur.

The chemistry of adsorption has already been discussed (see
Section II). Several factors will also influence the extent of adsorption. Particles that are more dense than water deviate from
fluid streamlines as a result of sedimentation [111], and Brownian
motion causes bacteria to deviate randomly from the direction of
flow [114]. These two factors will be more important at low flow
rates. In small pores, a bacterium may experience a torque that
results from differences in flow velocity along the length of the
cell. This leads to rotation and movement of the cell perpendicular to the direction of flow [115]. Each of these three factors
makes bacterial contact with soil surfaces and, potentially, adsorption more likely.

In saturated soils, it is postulated that at high flow rates,
movement of water (e.g., driven by gravity and the suction
caused by transpiration in plants) is predominantly through large
(>50 μm in diameter [115]) interconnected pores; at lower flow
rates, a larger proportion of the water moves through smaller pores
[112]. It is to be expected that there will be less retention of
cells in soils at high flow rates, as the pores that are traversed
will have a smaller surface/volume ratio and the flow rates will reduce sedimentation [111,112].

Rainfall and irrigation can influence the physical structure of
soil [24,63,116]. The kinetic energy of falling raindrops may be
sufficient to disrupt aggregates at the soil surface and reduce the
infiltration rate. This reduction results from movement and accumulation of suspended solids released during the breakdown of soil
aggregates [24,117]. This is important in terms of bacterial migration, as the result of both the reduction in infiltration rate and
because bacteria may, themselves, be considered as suspended solids [115] that may be released during the breakdown of aggregates.
Bertrand and Sor [24] found that the disruption of aggregates was
directly related to rainfall intensity and that suspended soil particles could be carried to a depth of 7.5 cm.

Mass flow of water will tend to pass through channels offering
the least resistance. In well-structured soils, there may be channels that allow the flow to bypass the bulk of the soil matrix al-
together during periods of high rainfall. Cracks in dry soil can

facilitate rapid movement of water, suspended bacteria, and soil particles. The surfaces of plant roots [118] and tunnels produced by earthworms [119] and, presumably, other fauna can also provide pathways for bypassing flow; these aspects are discussed in later sections.

IV. ACTIVE MICROBIAL MOVEMENT

A. Active Microbial Motility

Microbial motility may be expected to have an important part in dispersal, but evidence in the literature is contradictory. Bowen and Rovira [120] stated that flagella were generally not considered to have a major influence in migration. Migration of *P. fluorescens* is not apparently affected by motility, for mutant strains lacking flagella colonized roots in a soil system as effectively as wild-type motile strains [89]. Stotzky and Post (Agron. Abstr., 1965; data shown in [38]) also observed that rates of bacterial spread were independent of bacterial characteristics, such as morphology and the presence of flagella. The general conclusion drawn from these studies was that microbial movement was more dependent on the mass movement of water in a soil system than on flagellar motion. However, some studies have shown that motility [99,121,122] increases movement of bacteria through sterilized (i.e., autoclaved) soil. Wong and Griffin [99] demonstrated greater migration of *P. fluorescens* and *Bacillus subtilis* in autoclaved soils when the soils were amended with nutrients, attributing this to greater active motility in the presence of nutrients. The fact that some authors have detected a difference in movement between motile and nonmotile strains, albeit in sterilized soils [e.g., 123], suggests that the motile strains are flagellated in soil, as do electron microscopic studies on soil [72,78]. That natural soils may not contain sufficient nutrients to support active flagellar motility has been highlighted by Catlow et al. [123], who showed that there was no significant difference between movement of motile and nonmotile strains of *R. trifolii* through an unsterilized, nutrient-poor sandy soil but that the motile strain showed considerably greater movement in steam-treated soil. They attributed this difference to nutrients for growth and flagellar movement being released by the steam treatment and, therefore, concluded that, in nutrient-poor soils, active motility may not be important and that studies on motility in steam-treated soil may lead to invalid conclusions about motility in situ. Sterilization of soil may also influence its structure and, thereby, the possible pathways for active motility, Brownian motion, or passive transport.

Wilkinson et al. [121] demonstrated greater movement (in terms of distance) of fungal zoospores, compared with cysts, during infiltration of water into dry, autoclaved soils, although the numbers of fungal propagules involved were not mentioned. This greater motility was attributed to the ability of motile forms to bypass restricting pore necks encountered during the mass flow of water. Thus, active movement may be important as a supplement to passive movement of some microbes in soil, especially in soils for which the pore size distribution dictates that size-sieving of microbes is a major factor.

B. Movement by Growth

Spread of bacteria through soil at the -33 kPa water tension [124] is too slow to be attributed to motility and is, therefore, considered to be the result of growth (i.e., cell division). Motility is thought to be inhibited by the forces of surface tension of the water films associated with clays in soil [23]. The hyphal mode of growth in fungi, coupled with their ability to translocate water and nutrients, enables them to cross pore spaces that are devoid of water [23,124,125]; accordingly, their movement through soil is less restricted at lower matric potentials than that of bacteria.

V. PLANT ROOTS

A. Bacterial Movement in the Rhizoplane

The *rhizoplane* is a narrow zone that constitutes the interface between the plant root and the soil. As such, it behaves as a distinct ecological niche that offers ecophysiological conditions that are different from the surrounding rhizosphere soil.

The presence of plant roots in soil has been demonstrated to increase the movement of bacteria [e.g., 104,105]. A number of root-mediated modes of dispersal have been suggested [120].

1. Active motility of bacteria suspended in a film of water on the root [100,126], perhaps especially in the grooves between epidermal cells [118].
2. Convection of bacteria in films of water along the root surface [127]. There is no experimental evidence for this, and it is more likely that colonies will be firmly anchored when actually on the root surface [128]; for example, being embedded in the mucigel layer [60,129,130], thereby greatly limiting such dispersal.

3. Carriage on apices of roots or extending fungal hyphae.
    Evidence for limited dispersal by extending roots has been
    obtained by Liddell and Parke [131].

In the presence of percolating water, a fourth root-mediated
mode of dispersal is possible: the preferential flow of water along
the root surface. Diurnal changes in root diameter of up to 20%
[132,133] may create gaps between the root and the surrounding
soil [118]. Percolating water has been shown to enhance root colonization by a number of species [110,118,131].

In the absence of percolating water, there is a minimum soil matric potential below which any substantial microbial dispersal by
developing roots does not occur [43,89,100,118]. The soil matric
potential that is limiting to microbial dispersal by developing roots
is not limiting to root growth [120]. Therefore, it is assumed that
at lower potentials, water films on the roots are too thin to sus-
tain movement of bacteria and, in very thin water films, the bacteria would be trapped by surface tension [23,100]. This suggests
that motility in water films on roots is of primary importance for
root-mediated transport. Further evidence for movement in water
films comes from direct microscopic examination of the bacterial
distribution on plant roots [118]: in soil with no percolating water, the colonies of bacteria were randomly scattered over the root
surface, whereas in the presence of percolating water, bacteria
frequently occurred as continuous longitudinal colonies extending
down the grooves between epidermal cells. These are the most
likely regions for thicker water films to exist [99] and, thus, the
distribution is interpreted as reflecting routes of dispersal along
the root. Another interpretation might be that this distribution is
the result of those sites being richer nutritionally (i.e., bacterial
aggregations resulting from short-range chemotaxis). Both inter-
pretations may be valid, with epidermal grooves being both nutritionally rich and constituting sites for migration along the root.

The colonization of roots by bacteria inoculated onto seed coatings demonstrates movement along the rhizoplane [89,126,131,134–
136]. This colonization may be the result of growth along the radicle, active motility, or passive transport.

A role for active motility in root-mediated transport has been
suggested by Catlow et al. [137], who showed that a motile strain
of *R. trifolii* was able to colonize and nodulate clover roots at a
considerably greater vertical distance from the inoculation point
(16.5 to 21.0 cm after 8 weeks; there was no significant difference after 6 weeks) that nonmotile strains (6.0 to 6.5 cm). The
nonmotile strains were indistinguishable from the motile strain in
terms of growth rate, so the results strongly suggest that active
motility aids in dispersal along roots. The study was conducted

in sterile soil; hence, it is possible that there were more nutrients available for active flagellar motility than there would be under natural conditions, but this may be less important in the rhizoplane, as root exudates may provide sufficient energy sources. This study shows that passive transport (e.g., in water films) on the root can disperse nonmotile organisms 6 to 6.5 cm in 8 weeks, but no conclusions can be drawn about the effect of motility on dispersal until motility has been demonstrated in nonsterile soil.

Fungal hyphae are also important in relation to dispersal of bacteria (as discussed in Section IV, the mycelial mode of growth is also important in fungal dispersal), both along the root surface and through soil generally [100]. As with bacteria, fungi are often associated with epidermal grooves on roots, as a result of growth responses of hyphae to these grooves (thigmodifferentiation) or chemotaxis. The relevance of this for microbial migration is that the hyphae can trap menisci of water between themselves and the plant roots that may allow passage of bacteria, provided the water films are thick enough for the bacteria to escape the surface tension. In addition, mycelial strands (aggregations of hyphae running in parallel) may hold sufficient water by the forces of surface tension to assist microbial dispersal in soils at -4.9 kPa, but not at -14.7 kPa [100]. The fungal hyphae also provide a source of nutrients, enabling bacteria to remain viable, and even to multiply, during dispersal. However, it is also necessary to consider the possible negative effects of fungal hyphae on bacteria. Wong and Griffin [100] state that migration was greater in the presence of dead hyphae than of live hyphae, an observation they attributed to the production of inhibitory substances by the living fungus, thus giving some evidence about the importance of the physiological state of the hyphae involved.

B. Rhizosphere Phenomena

Inasmuch as, by definition, the character of the rhizosphere is strongly influenced by the plant, it forms a distinct ecological niche [138]. When considering migration in the rhizosphere, it is first necessary to understand the special conditions that exist there.

An important influence of the root is in the alteration of the structure of rhizosphere soil. This is caused by root extension; during this process, the root displaces its own volume of soil and, in doing so, can have a compaction effect on the adjacent soil matrix [e.g., 139]. Another influence of roots relates to the uptake of mineral ions; X-ray fluorescence studies have shown that the breakdown of soil minerals by root exudates reduces the mean diameter of rhizosphere soil particles, compared with bulk soil [140]. Reduction in particle size in conjunction with compaction produces a cylinder of soil around the root in which pore diameters are considerably reduced (e.g., from 30 μm down to 10 μm).

In addition to the physical changes imposed on the soil by the plant root, there is a host of ecophysiological factors that result from the growth of the root. Under conditions of rapid transpiration, water potentials of -800 kPa may be generated at the soil-root interface [141], and these potentials act on rhizosphere microbes. Transpiration also causes movement of water in the rhizosphere, which may mediate microbial transport and cause cells to move toward the root, as suggested by Breitenbeck et al. for *B. japonicum* [33]. In short, water potentials in the rhizosphere are dependent on the plant and its rate of transpiration; thus, for a microbe to survive, it must be able to tolerate extremes of, and rapid changes in, water potential. Indeed, it has been demonstrated that osmotic tolerance is a beneficial character for biological control agents designed to inhabit the rhizosphere [135,142,143].

Ion concentrations and pH values in the rhizosphere may also be grossly different from those in the surrounding soil, as the result of specific uptake mechanisms of the root [140,144,145]. Moreover, release of organic compounds by the plant root system into the rhizosphere may enable prolific colonization by certain microbial species (e.g., pseudomonads). Total bacterial numbers in the rhizosphere can be up to 1000-fold higher than in surrounding soil, depending on soil type [146], and the enhanced microbial activity in the rhizosphere may result in increased movement by growth (see Section IV.B).

The compaction of soil by the growing root might have the effect of limiting microbial migration in the rhizosphere. In addition, it is likely that the bacteria in the rhizosphere will be adapted to stay there, as it is a nutritionally advantageous site. Such adaptations would include chemotaxis to root exudates [85], and the production of exopolymers [e.g., 58] and pili [35]. Therefore, once the bacteria have colonized the rhizosphere, they might be expected to show limited movement.

There is little or no evidence that the motility of rhizobacteria is important in root colonization in natural soil, with most or all of the studies having been conducted in sterilized soil [e.g., 137] or nutrient-amended artificial matrices [147]. However, there is evidence that external agents, such as percolating water [43,110,131] or earthworms [43], are necessary for significant dispersal. Growth of roots may contact soil microorganisms which colonize the root, as shown in vitro [126].

C.   Movement in the Endorhizosphere

In addition to movement on the rhizoplane, bacteria are carried inside the plant root. Endorhizosphere-inhabiting bacteria, such as

Azospirillium lipoferum [148,149], flourescent pseudomonads [150], *Enterobacter* spp. [212], and *Clavibacter* spp. [213], exist in the root cortex. Movement of microorganisms in roots is also particularly evident for those species that are adapted as pathogens and for those that may act as opportunists and enter through wound sites (e.g., *Erwinia* spp. [151]). Within the plant, some of these bacteria may move within the plant tissue and may later be released into the soil with sloughed-off plant cells or tissue [130]. This means of dispersal of pathogenic microorganisms will probably be limited in healthy plants, as a result of the host defenses, but it may be important in diseased plants or on crop residues, such as grain stubble [152—155].

VI. SOIL ANIMALS

A. Mesofauna

Substantial populations of animals (both meso- and microfauna) are commonly present in soil [156]. Of the mesofauna, nematodes, annelids (commonly, earthworms and potworms), and arthropods (commonly, centipedes, millipedes, woodlice, and mites) predominate in most temperate soils.

The influence of animals on microbial dispersal in soil has rarely been quantitatively studied. Rather, it is generally concluded that soil animals are of importance in dispersal [41,43,157]. Opperman et al. [158] found that presumptive coliform bacteria from cattle slurry added to the top 3 to 4 cm were moved to a depth of 7.5 to 17.5 cm in a sandy soil by the earthworm, *Eisenia foetida*, after 4 days. The following mechanisms for the influence of soil fauna on microbial dispersal have been proposed.

1. Soil animals are important, as they influence the structure of soil. A typical community of earthworms (mean biomass of 120 g m^{-2}) has an annual intake of soil of about 2640 g m^{-2} [159]. Their activity can reduce bulk density when soil is removed from lower depths and deposited in casts at or near the surface. Earthworm activity can, therefore, increase the proportion of large pores in soil and, hence, the number of potential routes for bacterial migration. Another related consequence of the activity of mesofauna in soil is the production of stable tunnels, which provide a route for bypassing the flow of percolating water [119,160,161] and a pathway for dispersal of bacteria [43]. Mesofauna, other than earthworms, also have burrowing activities in soil (e.g., cicadas [162]), but the effect of their burrows on water flow and microbial dispersal is not so well documented. The pattern of tunnel formation varies among earthworm species, but *Lumbricus terrestris*, for example, produces stable vertical burrows to a depth of 80 cm or

more [160]. Burrows in soil can provide rapid routes for microbial movement, not only in water flow, but also through microbes that are attached to soil particles being carried down into the burrow on falling soil crumbs.

2. The ingestion of soil and, consequently, of soil bacteria by mesofauna, in particular earthworms, is another likely means of dispersal. The extent of dispersal will depend on the length of time that the bacterium is within the earthworm and the distance traveled by the worm during that time. Parle [163] discovered that during tunneling, soil stays within the gut of *L. terrestris* for 12 h, and during feeding, for about 20 h. Individuals of *L. terrestris* can move distances of 3.8 to 19.3 m over the soil surface during a single foray [164], so lateral movement of ingested microbes in earthworm guts is of potential significance.

Another important factor is whether on passing through the gut, the microorganisms remain viable and are, consequently, able to colonize on being reintroduced into soil. Thornton [165] found viable propagules of soil-dwelling aquatic fungi (e.g., *Pythium* spp.) in the casts of earthworms, and Parle [163] provided evidence that bacteria would be viable after passing through an earthworm. Although fungal populations were reduced in the guts of *L. terrestris*, *Allolobophora caliginosa*, and *A. terrestris*, populations of bacteria and actinomycetes increased by up to 1000 times. The gut microbiota of *L. terrestris* did not differ substantially in species present from the populations in surrounding soil [163]. A similar enrichment of bacteria (10- to 100-fold) has been shown in the gut of the millipede, *Glomeris marginata* [166]. Thus, it seems that while within the gut, transported bacteria undergo a growth phase. However, for this work to be truly relevant to migration studies, an investigation into subsequent bacterial colonization of soil sites is essential.

3. In addition to transport through the gut, soil mesofauna may disperse microbes on their external surfaces. This type of dispersal has been demonstrated for actinomycete spores with soil mites as vectors [41] and for fungal spores with earthworms [165] as vectors. Spores with hydrophobic surfaces were most efficiently transported in this way [41] as a result of hydrophobic interactions between the two surfaces involved (i.e., the actinomycete spore surface and the mite cuticle). For strongly hydrophobic particles, animal-mediated transport may constitute the major means of dispersal, as such particles have been shown to be resistant to movement in soil water [41,97].

For spores (of both fungi and bacteria), there may be two distinct dispersal strategies through soil: either animal-mediated (hydrophobic spores) or soil water-mediated (hydrophilic spores).

It is important to consider the effect of earthworms on the transport of microbes that are introduced into the environment as foliar applications, as leaf litter (and its associated microbes) on the surface of soil is likely to be carried into the soil by the foraging activities of earthworms.

B. Influence of Microfauna

Microbial population dynamics are undoubtedly influenced by the activities of predators. For bacteria, the main predators are protozoa, but bacterial predators such as *Ensifer adhaerens* [167,168] and *Bdellovibrio* [e.g., 169] also exist. Stout and Heal [170] estimated that protozoa in an arable field consumed 150 to 900 g m^{-2} y^{-1} of bacteria, and several studies have concluded that protozoa are responsible for the decline in numbers of certain species of bacterial inoculants [171–175]. As protozoa typically constitute 1 to 6% of the total soil biomass [176], their grazing activity is likely to be an important factor in bacterial population dynamics.

In addition to reducing bacterial populations, protozoa may also contribute substantially to bacterial dispersal. The evidence for this comes from observations of protozoan feeding, which have shown that a notable number of bacteria that are ingested by protozoa are later egested in a viable condition (J. F. Darbyshire, personal communication). Thus, as protozoa are motile organisms, the period that a bacterium spends within a protozoan cell could be an important factor influencing spatial dispersal in soil. This type of dispersal would be largely dependent on the presence of water pathways suitable for movement of the protozoa.

VII. DISPERSAL BY HUMAN BEINGS AND OTHER ABOVEGROUND AGENTS

In addition to vertical and horizontal dispersal of microbial inocula within the soil–plant system, it is important to recognize that dispersal of microorganisms between sites can occur through the action of factors outside of the soil–plant system.

A. Agricultural Practices

Inasmuch as one of the main roles foreseen for microbial inocula is as biological control agents of plant diseases (see Section I), it is necessary to consider factors that are likely to influence spread of microbes in agricultural systems.

Residues of soil from an inoculated field will inevitably stick to surfaces of farm machinery, facilitating dispersal. The importance of this as a means of dispersal of bacterial plant pathogens has been studied [157]. It was found that almost every animal, insect, tool, or person passing through an infested field can disseminate pathogenic agents. The role of farm machinery in dispersal of soft rot *Erwinia* spp. has been reviewed by Pérombelon and Kelman [151].

There is a strong analogy between the dissemination of soilborne plant pathogens and that of microbial inocula. Therefore, for a particular GEM or other nonindigenous microbial strain (such as *Rhizobium* spp.) to be realistically confined to any particular area, sterilization of equipment would be required as standard practice.

Agricultural practices may also indirectly affect microbial dispersal through soil by altering the soil structure. For example, conventional tillage practices can affect earthworm populations and, therefore, the number of earthworm burrows [119]. The use of agrochemicals may alter the concentrations and valencies of ions in soil and, thereby, potentially affect adsorption of cells on particles (see Section II).

Another means of dispersal related to agricultural practices is the use of irrigation water. As discussed earlier, water is a requisite of most forms of dispersal through soil, and as such, irrigation may increase dispersal. When a field is irrigated for a long period, water that contains soil bacteria may run off the field into drainage ditches [177]. Consequently, soil microbes could be dispersed through water courses. If the runoff were collected in a pool and used to irrigate another field, large numbers of microorganisms may be transported in conditions conducive to their survival. According to Bashan [157], this type of sequential irrigation is common in arid regions. This means of dispersal could be important in risk assessment when nonindigenous microbial inocula are to be used. In Scotland, farmers use natural water courses for irrigation, and it is not difficult to see how, if microbial inocula entered the water course upstream, they could be spread by irrigation farther downstream.

B. Effects of Animals Above the Soil

In addition to the influences of human beings, it is possible to imagine natural means of dispersal over land by other animals. For example, a worm containing a quantity of microbes in its gut from an inoculated field could be eaten by a bird. The microbes could then later be released (in droppings) from the bird miles away from the original site. Soil organisms may also be ingested directly by some birds: for example, Canada geese sometimes ingest soil when feeding on roots [178]. Again, survival of the microorganisms in the avian gut would be necessary for this type of dispersal to occur. Five species of soil-dwelling aquatic fungi have been shown to survive and to be dispersed in this manner [179]. Human-pathogenic fungi found in soil, such as *Histoplasma capsulatum*, may also be dispersed in a similar manner by bats [23]. Small mammals, such as moles, voles, mice, and rabbits, are also likely to spread microbial inocula as a result of their activities in and around agricultural fields. Although such means of dispersal are certainly possible, the question is whether the scale at which they occur is large enough to provide a viable inoculum at a new site.

C. Aerosols

A more general and probably more frequent means of dispersal is in aerosols. Drops of rain break up and form a range of smaller droplets on hitting a solid surface [180], such as soil or leaves [181]. These smaller droplets form an aerosol above the surface, which can be dispersed by air currents [182]. The importance of this is that during the initial impact on the soil, bacterial cells or spores or fungal propagules may become incorporated into the aerosol and, consequently, be dispersed [183–185] subject to their survival when airborne [e.g., 186]. This type of dispersal has been demonstrated to be a method for carrying the spores of plant-pathogenic fungi by Jarvis [187], who found that the spores could commonly travel over a meter laterally and even over longer distances with significant air currents. Much of the literature on aerosols concerns dispersal of plant-pathogenic fungi [e.g., 185, 187] and bacteria [e.g., 181] from infected leaf surfaces, with little information available on aerosols formed from soil surfaces. However, a report by Lindemann et al. [188] suggests that aerosols can be formed above wet or dry soil by the action of wind alone, in the absence of rain.

Dispersal in aerosols may also be of relevance during the initial release of GEMs or nonindigenous microorganisms, if the organisms are applied by spraying [189]. Lindow et al. [190] spray-inoculated a plot with a strain of *Pseudomonas syringae*, and detected cells of that strain up to 27-m downwind of the target plot immediately after spraying.

D. Dust Dispersal

Another potential means of dispersal is in windblown dust from the topsoil of bare fields [180]. The limiting factor in this type of dispersal is the very low water potential of dust. Therefore, it is unlikely that most vegetative cells could survive such dispersal, although it is likely that a significant number of spores would. For example, viable *Bacillus* spores from a sandstorm near the Black Sea were detected in air in Sweden [191], and plant-pathogenic fungi have been spread by winds 2000 km across the Tasman Sea from Australia to New Zealand [192] and 4000 km across the Atlantic Ocean [193]. Viruses (some of which are used as inocula to control insect pests) may also be spread in airborne soil particles [194].

VIII. CONCLUSIONS

Knowledge of microbial dispersal through soil is of major importance in determining the fate of GEMs and other nonindigenous microorganisms. This is particularly important when the organisms could enter ground or drainage waters that contribute to drinking water supplies.

The soil system is in dynamic equilibrium, whereby numerous physicochemical and biological processes interact. Such processes have a great influence on the degree of microbial movement and dispersal in soil systems.

The inherent characteristics of soil, namely, its particle size distribution and chemical nature, together with its structure, will largely determine how microbes move through bulk soil. These features differ greatly between soils; therefore, it is difficult to predict from observations in one particular soil how a given bacterium will behave in another soil. Thus, any study that claims universitality must be concerned with microbial migration within a range of soils, using representative bacterial species, under conditions that are characteristic of those found in the field.

Common to all soil systems is that, in the absence of any physical or biological vectors, microbial movement is dependent on the presence of a continuous water pathway of requisite size, with migration being severely restricted in soils, the moisture content of which is below field capacity (i.e., -33 kPa). Thus, it would appear that appreciable movement of microbes in soil above the water table will probably occur only during, or immediately after, heavy rain, snow melts, or irrigation. Active microbial movement, whether by motility or cell division (growth), is strongly dependent on the nutrient status of the soil. As soil is generally considered to be an oligotrophic environment [195], passive transport of microorganisms by external agents will probably be more important than active movement in soil.

Microbial migration is strongly influenced by adsorption reactions between microorganisms and soil particles. The degree of adsorption depends upon the chemical properties of the soil and soil solution, namely: clay content, organic matter content, pH, and electrolyte type and concentration. Biological macromolecules, such as polysaccharides (of bacterial or plant origin), also affect adsorption.

Paradoxically, chemotaxis may have a limiting effect on microbial migration. Although it is commonly associated with directed movement (e.g., toward regions of high nutrient concentration), this movement may result in the organism being attracted to a solid surface and, consequently, adsorbing there, thereby preventing further movement. Chemotaxis may also be responsible for the retention of microorganisms in the rhizosphere region.

In general, there is the danger of studies of microbial migration that when considering processes that are basically physical (e.g., Brownian motion, movement in mass flow, adsorption on colloids), it is easy to overlook biological factors, which may be important in soils when the moisture content is below field capacity. Although the rhizosphere extends no more than approximately 1 cm from the plant root surface, it provides a niche that supports heavy microbial colonization, as a result of the presence of organic compounds that are released from the root. Root systems and fungal mycelia are important for migration, since they provide surfaces on which continuous water films may develop that allow active or passive movement of microorganisms. Passive migration (e.g., by convection or diffusion) in water films on biological surfaces is a physical process that occurs at a biological interface and is, therefore, likely to be influenced by biological factors, such as rhizodeposition (i.e., the release of carbon by roots) or the production of inhibitory substances by fungi.

Serious consideration must also be given to microbial dispersal by means of animal vectors. Soil fauna, both meso- and microfauna, are potential vectors for microbial movement, as are human beings and other animals that pass over or through soil. However, there is a distinct absence of quantitative data on the importance of these vectors in terms of distance moved and microbial numbers carried.

There is little quantitative information available concerning bacterial migration through nonsterile soil; hence, it is not known how genetically altered or other nonindigenous microorganisms would spread. The main reason for this gap in knowledge in this aspect of microbial ecology in soil is probably the difficulty of accurately assaying for particular microorganisms in soil. Traditional methods of isolation and incubation are difficult, as the result of the retentive properties of soil (i.e., the often-observed phenomenon that when a known number of cells are added to soil, 100% recovery of those cells in a viable state immediately after addition is very difficult [e.g., 196]) and of the complex nature of the soil ecosystem. Therefore, it is a priority in this type of research to have an armory of sensitive and specific detection methods [197–200]. Several methods now exist that fulfill these criteria; for example, the use of enzyme-linked immunosorbent assays (ELISA) [e.g., 201], specific fluorescent antibodies [202–204], and DNA probes [205,206] in conjunction with the polymerase chain reaction (PCR) [207]. The potential of genetic engineering is not only for altering organisms for inoculation and industrial purposes, but also for inserting genetic markers into organisms that are to be used in ecological research, for example, specific enzyme markers [208]. Of particular use for experimental purposes is the incorporation of a bioluminescence tag by insertion of the *lux* genes from marine vibrios into the genome of other bacteria of interest [209–211]. Once present, the luminescent tag provides relatively easy identification against a background of indigenous soil organisms in situ, in soil suspensions (by luminometry), and after cell extraction. These techniques can be used to study the effect of variables, such as soil structure, the addition of agrochemicals, soil pH, rainfall intensity, the presence and absence of roots or soil animals, and hydraulic gradients, on microbial movement and transport through nonsterile soil. However, as soil is a complex system, altering one variable may affect more than one aspect of the soil ecosystem, so that the reasons for any observed effects might be difficult to pinpoint [38].

Further research on the movement of microbes through soil must focus on mechanistic aspects, to enable the construction of predictive models that will form an integral part of future assessments of the dispersal of microorganisms from wastes, the epidemiology of soilborne pathogens, the performance of plant growth-promoting rhizobacterial inocula, and the risks of the release of GEMs and nonindigenous inocula into the environment.

ACKNOWLEDGMENTS

This work was conducted with a grant from the Department of the Environment of the United Kingdom. Thanks are expressed to Dr. Chris E. Mullins (Plant and Soil Science Dept., University of Aberdeen) for helpful comments during the preparation of this manuscript.

REFERENCES

1. Kellerman, K. F., and E. H. Fawcett. 1907. Movement of certain bacteria in soils. Science 25:806.

2. Ball, O. M. 1909. A contribution to the life history of *Bacillus* (Ps.) *radicicola* Beij. Centralbl. Bakteriol. Abt. 2 23:47–59.

3. Frazier, W. C., and E. B. Fred. 1922. Movement of legume bacteria in soil. Soil Sci. 14:29–35.

4. Thornton, H. G., and N. Gangulee. 1926. The life-cycle of the nodule organism, *Bacillus radicicola* (Beij.), in soil and its relation to the infection of the host plant. Proc. R. Soc. Lond. B 99:427–451.

5. Maier, R. J., and Brill, W. J. 1978. Mutant strains of *Rhizobium japonicum* with increased ability to fix nitrogen for soybean. Science 201:448–450.

6. Paau, A. S. 1989. Improvement of *Rhizobium* inoculants. Appl. Environ. Microbiol. 55:862–865.

7. Burr, T. J., M. N. Schroth, and T. Suslow. 1978. Increased potato yields by treatment of seedpieces with specific strains of *Pseudomonas fluorescens* and *P. putida*. Phytopathology 68:1377–1383.

8. Schippers, B. 1988. Biological control of pathogens with rhizobacteria. Philos. Trans. R. Soc. Lond. B 318:283–293.

9. Obukowicz, M. G., F. J. Perlak, K. Kusano-Kretzmer, E. J. Mayer, S. L. Bolten, and L. S. Watrud. 1986. Tn5-mediated integration of the delta-endotoxin gene from *Bacillus thuringiensis* into the chromosome of root-colonizing pseudomonads. J. Bacteriol. 168:982–989.

10. Jackson, R. M., M. E. Brown, and S. K. Burlingham. 1964. Similar effects on tomato plants of *Azotobacter* inoculation and application of gibberellins. Nature 203:851.

11. Merriman, P. R., R. D. Price, J. F. Kollmorgen, T. Piggott, and E. H. Ridge. 1974. Effect of seed inoculation with *Bacillus subtilis* and *Streptomyces griseus* on the growth of cereals and carrots. Aust. J. Agric. Res. 25:219–226.

12. Kloepper, J. W., D. J. Hume, F. M. Scher, C. Singleton, B. Tipping, M. Laliberté, K. Frauley, T. Kutchaw, C. Simonson, R. Lifshitz, I. Zaleska, and L. Lee. 1988. Plant growth-promoting rhizobacteria on canola (rapeseed). Plant Dis. 72: 42–46.

13. Jain, R. K., and Sayler, G. S. 1987. Problems and potential for in situ treatment of environmental pollutants by engineered microorganisms. Microbiol. Sci. 4:59–63.

14. Lindow, S. E. 1983. Methods of preventing frost injury caused by epiphytic ice-nucleation-active bacteria. Plant Dis. 67:327–333.

15. Lindow, S. E., D. C. Arny, and C. D. Upper. 1983. Biological control of frost injury. II. Establishment and effects of an antagonistic *Erwinia herbicola* isolate on corn in the field. Phytopathology 73:1102–1106.

16. Blakeman, J. P., and N. J. Fokkema. 1982. Potential for biological control of plant diseases on the phylloplane. Annu. Rev. Phytopathol. 20:167–192.

17. Andow, D. A. 1986. Fate and movement of microorganisms in the environment. Part 2: Dispersal of microorganisms with emphasis on bacteria. Environ. Manage. 10:470–487.

18. Stotzky, G., and H. Babich. 1986. Survival of, and genetic transfer by, genetically engineered bacteria in natural environments. Adv. Appl. Microbiol. 31:93–138.

19. Beringer, J. E., and M. J. Bale. 1988. The survival and persistence of genetically-engineered micro-organisms, p. 29–46. *In* M. Sussman, C. H. Collins, F. A. Skinner, and D. E. Stewart-Tull (eds.), The release of genetically engineered micro-organisms. Academic Press, London.

20. Levy, S. B., and B. M. Marshall. 1988. Genetic transfer in the natural environment, p. 61–76. *In* M. Sussman, C. H. Collins, F. A. Skinner, and D. E. Stewart-Tull (eds.), The release of genetically-engineered micro-organisms. Academic Press, London.

21. Bitton, G., N. Lahav, and Y. Henis. 1974. Movement and retention of *Klebsiella aerogenes* in soil columns. Plant Soil 40:373–380.

22. Stotzky, G., and R. G. Burns. 1982. The soil environment: clay–humus–microbe interactions, p. 105–133. *In* R. G. Burns and J. H. Slater (eds.), Experimental microbial ecology. Blackwell Scientific Publications, Oxford.

23. Stotzky, G. 1986. Influence of soil mineral colloids on metabolic processes, growth, adhesion, and ecology of microbes and viruses, p. 305–428. *In* P. M. Huang and M. Schnitzer (eds.), Interactions of soil minerals with natural organics and microbes. Soil Science Society of America, Madison, Wisconsin.

24. Bertrand, A. R., and K. Sor. 1962. The effects of rainfall intensity on soil structure and migration of colloidal materials in soils. Soil Sci. Soc. Am. Proc. 26:297–300.

25. Harvey, R. W., and L. Y. Young. 1980. Enumeration of particle-bound and unattached respiring bacteria in the salt marsh environment. Appl. Environ. Microbiol. 40:156–160.

26. Balkwill, D. L., T. E. Rucinsky, and L. E. Casida, Jr. 1977. Release of microorganisms from soil with respect to transmission electron microscopy viewing and plate counts. Antonie Leeuwenhoek J. Microbiol. Serol. 43:73–87.

27. Gray, T. R. G., P. Baxby, I. R. Hill, and M. Goodfellow. 1967. Direct observation of bacteria in soil, p. 171–192. *In* T. R. G. Gray and D. Parkinson (eds.), The ecology of soil bacteria. Liverpool University Press, Liverpool.

28. Marshall, K. C. 1971. Sorptive interactions between soil particles and microorganisms, p. 409–445. *In* A. D. McLaren and J. Skujins (eds.), Soil biochemistry, Vol. 2. Marcel Dekker, New York.

29. Lahav, N. 1962. Adsorption of sodium bentonite particles on *Bacillus subtilis*. Plant Soil 17:191–208.

30. Peele, T. C. 1936. Adsorption of bacteria by soils. Cornell Agric. Exp. Stat. Mem. 197:1–18.

31. Santoro, T., and G. Stotzky. 1968. Sorption between microorganisms and clay minerals as determined by the electrical sensing zone particle analyzer. Can. J. Microbiol. 14: 299–307.

32. Dikusar, M. M. 1940. Adsorption of bacteria and its effect on microbial processes. Mikrobiologiya 9:895–908.

33. Breitenbeck, G. A., H. Yang, and E. P. Dunigan. 1988. Water-facilitated dispersal of inoculant *Bradyrhizobium japonicum* in soils. Biol. Fertil. Soils 7:58–62.

34. Leeper, G. W. 1964. Introduction to soil science, 4th ed. Cambridge University Press, London.

35. Bashan, Y., and H. Levanony. 1988. Adsorption of the rhizosphere bacterium *Azospirillum brasilense* Cd to soil, sand and peat particles. J. Gen. Microbiol. 134:1811–1820.

36. Brewer, R. 1976. Fabric and mineral analysis of soils. Robert E. Krieger Publishing, New York.

37. Lynch, J. M., and E. Bragg. 1985. Microorganisms and soil aggregate stability. Adv. Soil Sci. 2:133–171.

38. Stotzky, G. 1972. Activity, ecology and population dynamics of microorganisms in soil. Crit. Rev. Microbiol. 2:59–137.

39. van Olphen, H. 1977. An introduction to clay colloid chemistry for clay technologists, geologists and soil scientists, 2nd ed. Wiley-Interscience, New York.

40. Marshall, K. C. 1968. Interaction between colloidal montmorillonite and cells of *Rhizobium* species with different ionogenic surfaces. Biochim. Biophys. Acta 156:179–186.

41. Ruddick, S. M., and S. T. Williams. 1972. Studies on the ecology of actinomycetes in soil. V. Some factors influencing the dispersal and adsorption of spores in soil. Soil Biol. Biochem. 4:93–103.

42. Hepple, S. 1960. The movement of fungal spores in soil. Trans. Br. Mycol. Soc. 43:73–79.

43. Madsen, E. L., and M. Alexander. 1982. Transport of *Rhizobium* and *Pseudomonas* through soil. Soil Sci. Soc. Am. J. 46:557–560.

44. Chudiakow, N. N. 1926. Adsorption of bacteria in the soil and its influence on the microbiological soil processes. Centralbl. Bakteriol. Abt. 2 68:345–358.

45. Busscher, H. J., and A. H. Weerkamp. 1987. Specific and non-specific interactions in bacterial adhesion to solid substrata. FEMS Microbiol. Rev. 46:165–173.

46. Marshall, K. C. 1969. Studies by microelectrophoretic and microscopic techniques of the sorption of illite and montmorillonite to rhizobia. J. Gen. Microbiol. 56:301–306.

47. Rubentschik, L., M. B. Roisin, and F. M. Bieljansky. 1936. Adsorption of bacteria in salt lakes. J. Bacteriol. 32:11–31.

48. Rutter, P., and R. Leech. 1980. The deposition of *Streptococcus sanguis* NCTC7868 from a flowing suspension. J. Gen. Microbiol. 120:301–307.

49. Uyen, H. M., H. C. van der Mei, A. H. Weerkamp, and H. J. Busscher. 1988. Comparison between the adhesion to solid substrata of *Streptococcus mitis* and that of polystyrene particles. Appl. Environ. Microbiol. 54:837–838.

50. Bashan, Y., and H. Levanony. 1988. Active attachment of *Azospirillum brasilense* Cd to quartz sand and to a light textured soil by protein bridging. J. Gen. Microbiol. 134:2269–2279.

51. Meadows, P. S. 1971. The attachment of bacteria to solid surfaces. Arch. Mikrobiol. 75:374–381.

52. van Loosdrecht, M. C. M., J. Lyklema, W. Norde, and A. J. B. Zehnder. 1989. Bacterial adhesion: a physicochemical approach. Microb. Ecol. 17:1–15.

53. Feldner, J., W. Bredt, and S. Razin. 1981. Possible role of ATP and cyclic AMP in glass attachment of *Mycoplasma pneumoniae*. FEMS Microbiol. Lett. 11:253–256.

54. Finch, P., M. H. B. Hayes, and M. Stacey. 1971. The biochemistry of soil polysaccharides, p. 257–319. *In* A. D. McLaren and J. Skujins (eds.), Soil biochemistry, Vol. 2. Marcel Dekker, New York.

55. Forsyth, W. G. C., and D. M. Webley. 1949. The synthesis of polysaccharides by bacteria isolated from soil. J. Gen. Microbiol. 3:395–399.

56. Martin, J. P. 1945. Microorganisms and soil aggregation I. Origin and nature of some of the aggregating substances. Soil Sci. 59:163–174.

57. Webley, D. M., R. B. Duff, J. S. D. Bacon, and V. C. Farmer. 1965. A study of polysaccharide-producing organisms occurring in the root region of certain pasture grasses. J. Soil Sci. 16:149–157.

58. Keen, G. A., and J. I. Prosser. 1988. The surface growth and activity of *Nitrobacter*. Microb. Ecol. 15:21–39.

59. Sutherland, I. W. 1977. Bacterial exopolysaccharides—their nature and production, 27–96. *In* I. Sutherland (ed.), Surface carbohydrates of the prokaryotic cell. Academic Press, London.

60. Jenny, H., and K. Grossenbacher. 1963. Root—soil boundary zones as seen in the electron microscope. Soil Sci. Soc. Am. Proc. 27:273—276.

61. Oades, J. M. 1978. Mucilages at the root surface. J. Soil Sci. 29:1—16.

62. Dudman, W. F. 1977. The role of surface polysaccharides in natural environments, p. 357—414. *In* I. Sutherland (ed.), Surface carbohydrates of the prokaryotic cell. Academic Press, London.

63. Helalia, A. M., and J. Letey. 1988. Cationic polymer effects on infiltration rates with a rainfall simulator. Soil Sci. Soc. Am. J. 52:247—250.

64. Mehta, N. C., H. Streuli, M. Muller, and H. Deuel. 1960. Role of polysaccharides in soil aggregation. J. Sci. Food Agric. 11:40—47.

65. Martin, J. P. 1971. Decomposition and binding action of polysaccharides in soil. Soil Biol. Biochem. 3:33—41.

66. Costerton, J. W., G. G. Geesey, and K. J. Cheng. 1978. How bacteria stick. Sci. Am. 238:76—85.

67. Costerton, J. W., R. T. Irvin, and K. J. Cheng. 1981. The bacterial glycocalyx in nature and disease. Annu. Rev. Microbiol. 35:299—324.

68. Bashan, Y., H. Levanony, and E. Klein. 1986. Evidence for a weak active external adsorption of *Azospirillum brasilense* Cd to wheat roots. J. Gen. Microbiol. 132:3069—3073.

69. Marshall, K. C., and G. Bitton. 1980. Microbial adhesion in perspective, pp. 1—5. *In* G. Bitton and K. C. Marshall (eds.), Adsorption of microorganisms to surfaces. John Wiley & Sons, New York.

70. Foster, R. C. 1986. The ultrastructure of the rhizoplane and the rhizosphere. Annu. Rev. Phytopathol. 24:211—234.

71. Fraser, T. W., and A. Gilmour. 1986. Scanning electron microscopy preparation methods: their influence on the morphology and fibril formation in *Pseudomonas fragi* (ATCC 4973). J. Appl. Bacteriol. 60:527—533.

72. Nikitin, D. I. 1964. Use of electron microscopy in the study of soil suspensions and cultures of microorganisms. Soviet Soil Sci. 1964:636—641.

73. Ottow, J. C. G. 1975. Ecology, physiology, and genetics of fimbriae and pili. Annu. Rev. Microbiol. 29:79—108.

74. Brinton, C. C. 1965. Structure, function, synthesis, and genetic control of bacterial pili and a molecular model for DNA and RNA transport in gram negative bacteria. Trans. N.Y. Acad. Sci. 27:1003—1054.

75. Brinton, C. C., Jr., and M. A. Lauffer. 1959. The electrophoresis of viruses, bacteria, and cells, and the microscope method of electrophoresis, p. 427—492. *In* M. Bier

(ed.), Electrophoresis: theory, methods, and applications, Vol. 1. Academic Press, New York.

76. Brinton, C. C., Jr., A. Buzzell, and M. A. Lauffer. 1954. Electrophoresis and phage susceptibility studies on a filament-producing variant of the *E. coli* B bacterium. Biochim. Biophys. Acta 15:533–542.

77. Marshall, K. C. 1976. Interfaces in microbial ecology. Harvard University Press, Cambridge.

78. Zvyagintsev, D. G., A. F. Pertsovskaya, V. I. Duda, and D. I. Nikitin. 1969. Electron-microscopic study of the adsorption of microorganisms on soil and minerals. Mikrobiologiya 38:1091–1095.

79. Adler, J. 1966. Chemotaxis in bacteria. Science 153:708–716.

80. Adler, J. 1975. Chemotaxis in bacteria. Annu. Rev. Biochem. 44:341–356.

81. Marshall, K. C. 1975. Clay mineralogy in relation to survival of soil bacteria. Annu. Rev. Phytopathol. 13:357–373.

82. van Loosdrecht, M. C. M., J. Lyklema, W. Norde, and A. J. B. Zehnder. 1990. Influence of interfaces on microbial activity. Microbiol. Rev. 54:75–87.

83. Ellwood, D. C., C. W. Keevil, P. D. Marsh, C. M. Brown, and J. N. Wardell. 1982. Surface-associated growth. Philos. Trans. R. Soc. Lond. B 297:517–532.

84. Barak, R., I. Nur, and Y. Okon. 1983. Detection of chemotaxis in *Azospirillum brasilense*. J. Appl. Bacteriol. 53:399–403.

85. Bashan, Y. 1986. Migration of the rhizosphere bacteria *Azospirillum brasilense* and *Pseudomonas fluorescens* towards wheat roots in the soil. J. Gen. Microbiol. 132:3407–3414.

86. Mellor, H. Y., A. R. Glenn, R. Arwas, and M. J. Dilworth. 1987. Symbiotic and competitive properties of motility mutants of *Rhizobium trifolii* TA1. Arch. Microbiol 148:34–39.

87. Liu, R., V. M. Tran, and E. L. Schmidt. 1989. Nodulating competitiveness of a nonmotile Tn7 mutant of *Bradyrhizobium japonicum* in nonsterile soil. Appl. Environ. Microbiol. 55:1895–1900.

88. Ames, P., and K. Bergman. 1981. Competitive advantage provided by bacterial motility in the formation of nodules by *Rhizobium meliloti*. J. Bacteriol. 148:728–729.

89. Howie, W. J., R. J. Cook, and D. M. Weller. 1987. Effects of soil matric potential and cell motility on wheat root colonization by fluorescent pseudomonads suppressive to take-all. Phytopathology 77:286–292.

90. Scher, F. M., J. W. Kloepper, and C. A. Singleton. 1985. Chemotaxis of fluorescent *Pseudomonas* spp. to soybean seed exudates in vitro and in soil. Can. J. Microbiol. 31:570–574.

91. Marshall, T. J., and J. W. Holmes. 1988. Soil physics, 2nd ed. Cambridge University Press, Cambridge.
92. Martin, J. P., W. P. Martin, J. B. Page, W. A. Raney, and J. D. DeMent. 1955. Soil aggregation. Adv. Agron. 7:1.
93. Chesters, G., O. J. Attoe, and O. N. Allen. 1957. Soil aggregation in relation to various soil constituents. Soil Sci. Soc. Am. Proc. 21:272–277.
94. Emerson, W. W. 1959. The structure of soil crumbs. J. Soil Sci. 10:235–244.
95. Allison, F. E. 1968. Soil aggregation—some facts and fallacies as seen by a soil microbiologist. Soil Sci. 106:136–143.
96. Griffin, D. M., and E. Quail. 1968. Movement of bacteria in moist, particulate systems. Aust. J. Biol. Sci. 21:579–582.
97. Burges, A. 1950. The downward movement of fungal spores in sandy soil. Trans. Br. Mycol. Soc. 33:142–147.
98. Hamdi, Y. A. 1971. Soil–water tension and the movement of rhizobia. Soil Biol. Biochem. 3:121–126.
99. Wong, P. T. W., and D. M. Griffin. 1976. Bacterial movement at high matric potentials (I)—in artifical and natural soils. Soil Biol. Biochem. 8:215–218.
100. Wong, P. T. W., and D. M. Griffin. 1976. Bacterial movement at high matric potentials (II)—in fungal colonies. Soil Biol. Biochem. 8:219–223.
101. McCoy, E. L., and C. Hagedorn. 1979. Quantitatively tracing bacterial transport in saturated soil systems. Water Air Soil Pollut. 11:467–479.
102. Gooday, G. W. 1979. The potential of the microbial cell, p. 5–21. *In* J. M. Lynch and N. J. Poole (eds.), Microbial ecology: a conceptual approach. Blackwell Scientific Publications, Oxford.
103. Postma, J., J. A. van Veen, and S. Walter. 1989. Influence of different initial soil moisture contents on the distribution and population dynamics of introduced *Rhizobium leguminosarum* biovar *trifolii*. Soil Biol. Biochem. 21:437–442.
104. Trevors, J. T., J. D. van Elsas, L. S. van Overbeek, and M.-E. Starodub. 1990. Transport of a genetically engineered *Pseudomonas fluorescens* strain through a soil microcosm. Appl. Environ. Microbiol. 56:401–408.
105. Bashan, Y., and H. Levanony. 1987. Horizontal and vertical movement of *Azospirillum brasilense* Cd in the soil and along the rhizosphere of wheat and weeds in controlled and field environments. J. Gen. Microbiol. 133:3473–3480.
106. Cassell, D. K., and A. Klute. 1986. Water potential: tensiometry, p. 563–596. *In* A. Klute (ed.), Methods of soil analysis, Part 1, Physical and mineralogical methods, 2nd ed. American Society of Agronomy, Madison, Wisconsin.

107. Mullins, C. E. 1990. Matric potential, p. 75–109. *In* K. A. Smith and C. E. Mullins (eds.), Soil analysis: physical methods. Marcel Dekker, New York.

108. Marshall, T. J. 1959. Relations between water and soil. Commonwealth Agricultural Bureaux, Farnham Royal, England.

109. Bahme, J. B., M. N. Schroth, S. D. Van Gundy, A. R. Weinhold, and D. M. Tolentino. 1988. Effect of inocula delivery systems on rhizobacterial colonization of underground organs of potato. Phytopathology 78:534–542.

110. Chao, W. L., E. B. Nelson, G. E. Harman, and H. C. Hoch. 1986. Colonization of the rhizosphere by biological control agents applied to seeds. Phytopathology 76:60–65.

111. Lahav, N., and D. Tropp. 1980. Movement of synthetic microspheres in saturated soil columns. Soil Sci. 130:151–156.

112. Wollum, A. G. II, and D. K. Cassell. 1978. Transport of microorganisms in sand columns. Soil Sci. Soc. Am. J. 42:72–78.

113. Zyman, J., and C. A. Sorber. 1988. Influence of simulated rainfall on the transport and survival of selected indicator organisms in sludge-amended soils. J. Water Pollut. Control Fed. 60:2105–2110.

114. Cridland, J. V., and P. C. Thonemann. 1984. Dispersal of motile bacteria from a plane layer. Biophys. J. 46:781–786.

115. Vinten, A. J. A., and P. H. Nye. 1985. Transport and deposition of dilute colloidal suspensions in soils. J. Soil Sci. 36:531–541.

116. Coutts, J. R. H., M. F. Kandil, and J. Tinsley. 1968. Use of radioactive ^{59}Fe for tracing soil particle movement. Part II. Laboratory studies of labelling and splash displacement. J. Soil Sci. 19:325–341.

117. Vinten, A. J. A., U. Mingelgrin, and B. Yaron. 1983. The effect of suspended solids in wastewater on soil hydraulic conductivity: II. Vertical distribution of suspended solids. Soil Sci. Soc. Am. J. 47:408–412.

118. Parke, J. L., R. Moen, A. D. Rovira, and G. D. Bowen. 1986. Soil water flow affects the rhizosphere distribution of a seed-borne biological control agent, *Pseudomonas fluorescens.* Soil Biol. Biochem. 18:583–588.

119. Edwards, W. M., M. J. Shipitalo, L. B. Owens, and L. D. Norton. 1990. Effect of *Lumbricus terrestris* L. burrows on hydrology of continuous no-till corn fields. Geoderma 46:73–84.

120. Bowen, G. D., and A. D. Rovira. 1976. Microbial colonization of plant roots. Annu. Rev. Phytopathol. 14:121–144.

121. Wilkinson, H. T., R. D. Miller, and R. L. Millar. 1981.
      Infiltration of fungal and bacterial propagules into soil.
      Soil Sci. Soc. Am. J. 45:1034–1039.
122. Soby, S., and K. Bergman. 1983. Motility and chemotaxis
      of *Rhizobium meliloti* in soil. Appl. Environ. Microbiol. 46:
      995–998.
123. Catlow, H. Y., A. R. Glenn, and M. J. Dilworth. 1990.
      The use of transposon-induced non-motile mutants in as-
      sessing the significance of motility of *Rhizobium legumino-
      sarum* biovar *trifolii* for movement in soils. Soil Biol. Bio-
      chem. 22:331–336.
124. Stotzky, G. 1965. Replica plating technique for studying
      microbial interactions in soil. Can. J. Microbiol. 11:629–
      636.
125. Stotzky, G. 1973. Techniques to study interactions be-
      tween microorganisms and clay minerals in vivo and in vitro.
      Bull. Ecol. Res. Comm. (Stockh.) 17:17–28.
126. Leben, C. 1983. Assocation of *Pseudomonas syringae* pv.
      lachrymans and other bacterial pathogens with roots. Phy-
      topathology 73:577–581.
127. Bandoni, R. J., and R. E. Koske. 1974. Monolayers and
      microbial dispersal. Science 183:1079–1080.
128. Dart, P. J. 1971. Scanning electron microscopy of plant
      roots. J. Exp. Bot. 22:163–168.
129. Darbyshire, J. F., and M. P. Greaves. 1970. An improved
      method for the study of the interrelationships of soil micro-
      organisms and plant roots. Soil Biol. Biochem. 2:63–71.
130. Old, K. M., and T. H. Nicolson. 1975. Electron micro-
      scopical studies of the microflora of roots of sand dune
      grasses. New Phytol. 74:51–58.
131. Liddell, C. M., and J. L. Parke. 1989. Enhanced coloniza-
      tion of pea taproots by a fluorescent pseudomonad biocontrol
      agent by water filtration into soil. Phytopathology 79:1327–
      1332.
132. Huck, M. G., B. Klepper, and H. M. Taylor. 1970. Di-
      urnal variations in root diameter. Plant Physiol. 45:529–
      530.
133. Faiz, S. M. A., and P. E. Weatherley. 1982. Root contrac-
      tion in transpiring plants. New Phytol. 92:333–343.
134. Weller, D. M. 1984. Distribution of a take-all suppressive
      strain of *Pseudomonas fluorescens* on seminal roots of winter
      wheat. Appl. Environ. Microbiol. 48:897–899.
135. Suslow, T. V., and M. N. Schroth. 1982. Rhizobacteria of
      sugar beets: effects of seed application and root colonization
      on yield. Phytopathology 72:199–206.

136. Scher, F. M., J. S. Ziegle, and J. W. Kloepper. 1984. A method for assessing the root-colonizing capacity of bacteria on maize. Can. J. Microbiol. 30:151–157.
137. Catlow, H. Y., A. R. Glenn, and M. J. Dilworth. 1990. Does rhizobial motility affect its ability to colonize along the legume root? Soil Biol. Biochem. 22:573–575.
138. Foster, R. C., and G. D. Bowen. 1982. Plant surfaces and bacterial growth: the rhizosphere and rhizoplane, p. 159–185. *In* M. S. Mount and G. H. Lacey (eds.), Phytopathogenic prokaryotes, Vol. 1. Academic Press, New York.
139. Greacen, E. L., D. A. Farrell, and B. Cockcroft. 1968. Soil resistance to metal probes and plant roots. Trans. Int. Soc. Soil Sci. (Adelaide) 1:769–778.
140. Sarkar, A. W., D. A. Jenkins, and R. S. Wyn-Jones. 1979. Modifications to mechanical and mineralogical composition of soil within the rhizosphere, p. 125–136. *In* J. L. Harley and R. Scott Russell (eds.), The soil–root interface. Academic Press, London.
141. Weatherley, P. E. 1979. The hydraulic resistance of the soil–root interface—a cause of water stress in plants, p. 275–286. *In* J. L. Harley and R. Scott Russell (eds.), The soil–root interface. Academic Press, London.
142. Loper, J. E., C. Haack, and M. N. Schroth. 1985. Population dynamics of soil pseudomonads in the rhizosphere of potato (*Solanum tuberosum* L.). Appl. Environ. Microbiol. 49:416–422.
143. Weller, D. M. 1988. Biological control of soil-borne plant pathogens in the rhizosphere with bacteria. Annu. Rev. Phytopathol. 26:379–407.
144. Jackson, P. C., and H. R. Adams. 1963. Cation–anion balance during potassium and sodium adsorption by barley roots. J. Gen. Physiol. 46:369–386.
145. Smiley, R. W. 1979. Wheat-rhizoplane pseudomonads as antagonists of *Gaeumannomyces graminis*. Soil Biol. Biochem. 11:371–376.
146. Newman, E. I., and A. Watson. 1977. Microbial abundance in the rhizosphere: a computer model. Plant Soil 48:17–56.
147. Hunter, W. J., and C. J. Fahring. 1980. Movement by *Rhizobium* and nodulation of legumes. Soil Biol. Biochem. 12:537–542.
148. Patriquin, D. G., and J. Döbereiner. 1978. Light microscopy observations of tetrazolium-reducing bacteria in the endorhizosphere of maize and other grasses in Brazil. Can. J. Microbiol. 24:734–742.
149. Bentjen, S. A., J. K. Fredrickson, P. Van Voris, and S. W. Li. 1989. Intact soil-core microcosms for evaluating the fate

and ecological impact of the release of genetically engineered organisms. Appl. Environ. Microbiol. 55:198–202.

150. van Peer, R., H. L. M. Punte, L. A. de Weger, and B. Schippers. 1990. Characterization of root surface pseudomonads in relation to their colonization of roots. Appl. Environ. Microbiol. 56:2462–2470.

151. Pérombelon, M. C. M., and A. Kelman. 1980. Ecology of the soft rot erwinias. Annu. Rev. Phytopathol. 18:361–387.

152. Burr, T. J., and M. N. Schroth. 1977. Occurrence of soft rot *Erwinia* spp. in soil and plant material. Phytopathology 67:1382–1387.

153. McIntyre, J. L., D. C. Sands, and G. S. Taylor. 1978. Overwintering, seed disinfestation, and pathogenicity studies of the tobacco hollow stalk pathogen, *Erwinia carotovora* var *carotovora*. Phytopathology 68:435–440.

154. Jardine, D. J., C. T. Stephens, and D. W. Fulbright. 1988. Potential sources of initial inoculum for bacterial speck in early planted tomato crops in Michigan: debris and volunteers from previous crops. Plant Dis. 72:246–249.

155. Gilbertson, R. L., R. E. Rand, and D. J. Hagedorn. 1990. Survival of *Xanthomonas campestris* pv. *phaseoli* and pectolytic strains of *X. campestris* in bean debris. Plant Dis. 74:322–327.

156. Kevan, D. K. McE. 1965. The soil fauna—its nature and biology, p. 33–51. *In* K. F. Baker and W. C. Snyder (eds.), Ecology of soil-borne plant pathogens: prelude to biological control. John Murray, London.

157. Bashan, Y. 1986. Field dispersal of *Pseudomonas syringae* pv. *tomato*, *Xanthomonas campestris* pv. *vesicatoria* and *Alternaria macrospora* by animals, people, birds, insects, mites, agricultural tools, aircraft, soil particles, and water sources. Can. J. Bot. 64:276–281.

158. Opperman, M. H., L. McBain, and M. Wood. 1987. Movement of cattle slurry through soil by *Eisenia foetida* (Savigny). Soil Biol. Biochem. 19:741–745.

159. Satchell, J. E. 1967. Lumbricidae, p. 259–322. *In* A. Burges and F. Raw (eds.), Soil Biology, Academic Press, London.

160. Ehlers, W. 1975. Observations on earthworm channels and infiltration on tilled and untilled loess soil. Soil Sci. 119:242–249.

161. Bouma, J., C. F. M. Belmans, and L. W. Dekker. 1982. Water infiltration and redistribution in a silt loam subsoil with vertical worm channels. Soil Sci. Soc. Am. J. 46:917–921.

162. Hughie, V. K., and H. B. Passey. 1963. Cicadas and their effect upon soil genesis in certain soils in southern Idaho,

northern Utah, and northeastern Nevada. Soil Sci. Soc. Am. Proc. 27:78—82.

163. Parle, J. N. 1963. Micro-organisms in the intestines of earthworms. J. Gen. Microbiol. 31:1—11.

164. Mather, J. G., and O. Christensen. 1988. Surface movements of earthworms in agricultural land. Pedobiologia 32: 399—405.

165. Thornton, M. L. 1970. Transport of soil-dwelling aquatic phycomycetes by earthworms. Trans. Br. Mycol. Soc. 55: 391—397.

166. Anderson, J. M., and D. E. Bignell. 1980. Bacteria in the food, gut contents and faeces of the litter-feeding millipede *Glomeris marginata* (Villers). Soil Biol. Biochem. 12:251—254.

167. Casida, L. E., Jr. 1980. Bacterial predators of *Micrococcus luteus* in soil. Appl. Environ. Microbiol. 39:1035—1041.

168. Casida, L. E., Jr. 1982. *Ensifer adhaerens* gen. nov., sp. nov.: a bacterial predator of bacteria in soil. Int. J. Syst. Bacteriol. 32:339—345.

169. Germida, J. J. 1987. Isolation of *Bdellovibrio* spp. that prey on *Azospirillum brasilense* in soil. Can. J. Microbiol. 33:459—461.

170. Stout, J. D., and O. W. Heal. 1967. Protozoa, p. 149—195. *In* A. Burges and F. Raw (eds.), Soil biology. Academic Press, New York.

171. Habte, M., and M. Alexander. 1975. Protozoa as agents responsible for the decline of *Xanthomonas campestris* in soil. Appl. Microbiol. 29:159—164.

172. Danso, S. K. A., S. O. Keya, and M. Alexander. 1975. Protozoa and the decline of *Rhizobium* populations added to soil. Can. J. Microbiol. 21:884—895.

173. Habte, M., and M. Alexander. 1977. Further evidence for the regulation of bacterial populations in soil by protozoa. Arch. Microbiol. 113:181—183.

174. Sardeshpande, J. S., R. H. Balasubramanya, J. H. Kulkarni, and D. J. Bagyaraj. 1977. Protozoa in relation to *Rhizobium* S-12 and *Azotobacter chroococcum* in soil. Plant Soil 47:75—80.

175. Ramirez, C., and M. Alexander. 1980. Evidence suggesting protozoan predation on *Rhizobium* associated with germinating seeds and in the rhizosphere of beans (*Phaseolus vulgaris* L.). Appl. Environ. Microbiol. 40:492—499.

176. Rutherford, P. M., and N. G. Juma. 1989. Dynamics of microbial biomass and soil fauna in two contrasting soils cropped to barley (*Nordum vulgare* L.). Biol. Fertil. Soils 8:144—153.

177. Evans, M. R., and J. D. Owens. 1973. Soil bacteria in land-drainage water. Water Res. 7:1295–1300.

178. Goodman, D. C., and H. I. Fisher. 1962. Functional anatomy of the feeding apparatus in waterfowl (Aves: Anatidae). Southern Illinois University Press, Carbondale, Illinois.

179. Thornton, M. L. 1971. Potential for long-range dispersal of aquatic phycomycetes by internal transport in birds. Trans. Br. Mycol. Soc. 57:49–59.

180. Gregory, P. H. 1973. The microbiology of the atmosphere, 2nd ed. John Wiley & Sons, New York.

181. Venette, J. R., and B. W. Kennedy. 1975. Naturally produced aerosols of *Pseudomonas glycinea*. Phytopathology 65:737–738.

182. Graham, D. C., C. E. Quinn, and L. F. Bradley. 1977. Quantitative studies on the generation of aerosols of *Erwinia carotovora* var *atroseptica* by simulated raindrop impaction on blackleg-infected potato strains. J. Appl. Bacteriol. 43:413–424.

183. Lloyd, A. B. 1969. Dispersal of streptomycetes in air. J. Gen. Microbiol. 57:35–40.

184. Graham, D. C., and M. D. Harrison. 1975. Potential spread of *Erwinia* spp. in aerosols. Phytopathology 6:739–741.

185. Fitt, B. D. L., H. A. McCartney, and P. J. Walklate. 1989. The role of rain in dispersal of pathogen inoculum. Annu. Rev. Phytopathol. 27:241–270.

186. Anderson, J. D., and C. S. Cox. 1967. Microbial survival, p. 203–226. *In* P. H. Gregory and J. L. Monteith (eds.), Airborne microbes, Cambridge University Press, Cambridge.

187. Jarvis, W. R. 1962. Splash dispersal of spores of *Botrytis cinerea* Pers. Nature 193:599.

188. Lindemann, J., H. A. Constantinidou, W. R. Barchet, and C. D. Upper. 1982. Plants as sources of airborne bacteria, including ice nucleation-active bacteria. Appl. Environ. Microbiol. 44:1059–1063.

189. Knudsen, G. R. 1989. Model to predict aerial dispersal of bacteria during environmental release. Appl. Environ. Microbiol. 55:2641–2647.

190. Lindow, S. E., G. R. Knudsen, R. J. Seidler, M. V. Walter, V. W. Lambou, P. S. Amy, D. Schmedding, V. Prince, and S. Hern. 1988. Aerial dispersal and epiphytic survival of *Pseudomonas syringae* during a pretest for the release of genetically engineered strains into the environment. Appl. Environ. Microbiol. 54:1557–1563.

191. Bovallius, A., B. Bucht, R. Roffey, and P. Ånäs. 1978. Long-range air transmission of bacteria. Appl. Environ. Microbiol. 35:1231–1232.

192. Close, R. C., N. T. Moar, A. I. Tomlinson, and A. D. Lowe. 1978. Aerial dispersal of biological material from Australia to New Zealand. Int. J. Biometeorol. 22:1–19.

193. Nagarajan, S., and D. V. Singh. 1990. Long-distance dispersion of rust pathogens. Annu. Rev. Phytopathol. 28: 139–153.

194. Olofsson, E. 1988. Dispersal of the nuclear polyhedrosis virus of *Neodiprion sertifer* from soil to pine foliage with dust. Entomol. Exp. Appl. 46:181–186.

195. Lynch, J. M. 1982. Limits to microbial growth in soil. J. Gen. Microbiol. 128:405–410.

196. Ramsay, A. J. 1984. Extraction of bacteria from soil: efficiency of shaking or ultrasonication as indicated by direct counts and autoradiography. Soil Biol. Biochem. 16:475–481.

197. Atlas, R. M., and G. S. Sayler. 1988. Tracking microorganisms and genes in the environment, p. 31–45. *In* G. S. Omenn (ed.), Environmental biotechnology: reducing risks from environmental chemicals through biotechnology. Plenum Press, New York.

198. Colwell, R. R., C. Somerville, I. Knight, and W. Straube. 1988. Detection and monitoring of genetically-engineered micro-organisms, p. 47–60. *In* M. Sussman, C. H. Collins, F. A. Skinner, and D. E. Stewart-Tull (eds.), The release of genetically-engineered micro-organisms. Academic Press, London.

199. Ford, S., and B. H. Olson. 1988. Methods for detecting genetically engineered microorganisms in the environment. Adv. Microb. Ecol. 10:45–79.

200. McCormick, D. 1986. Detection technology: the key to environmental biotechnology. Biotechnology 4:419–423.

201. Levanony, H., Y. Bashan, and Z. E. Kahana. 1987. Enzyme-linked immunosorbent assay for specific identification and enumeration of *Azospirillum brasilense* Cd. in cereal roots. Appl. Environ. Microbiol. 53:358–364.

202. Eren, J., and D. Pramer. 1966. Application of immunofluorescent staining to studies of the ecology of soil microorganisms. Soil Sci. 101:39–45.

203. Bohlool, B. B., and E. L. Schmidt. 1980. The immunofluorescent approach in microbial ecology. Adv. Microb. Ecol. 4:203–241.

204. Brayton, P. R., and R. R. Colwell. 1987. Fluorescent antibody staining method for enumeration of viable environmental *Vibrio cholerae* O1. J. Microbiol. Methods 6:309–314.

205. Holben, W. E., J. K. Jansson, B. K. Chelm, and J. M. Tiedje. 1988. DNA probe method for the detection of specific microorgansms in the soil bacterial community. Appl. Environ. Microbiol. 54:703–711.

206. Zeph, L. R., and G. Stotzky. 1989. Use of a biotinylated DNA probe to detect bacteria transduced by bacteriophage P1 in soil. Appl. Environ. Microbiol. 55:661–665.

207. Steffan, R. J., and R. M. Atlas. 1988. DNA amplification to enhance detection of genetically engineered bacteria in environmental samples. Appl. Environ. Microbiol. 54:2185–2191.

208. Drahos, D. J., B. C. Hemming, and S. McPherson. 1986. Tracking recombinant organisms in the environment: β-galactosidase as a selectable non-antibiotic marker for fluorescent pseudomonads. Biotechnology 4:439–444.

209. Shaw, J. J., and C. I. Kado. 1986. Development of a *Vibrio* bioluminescence gene-set to monitor phytopathogenic bacteria during the ongoing disease process in a non-disruptive manner. Biotechnology 4:560–564.

210. Meighen, E. A. 1988. Enzymes and genes from the *lux* operons of bioluminescent bacteria. Annu. Rev. Microbiol. 42:151–176.

211. Rattray, E. A. S., J. I. Prosser, K. Killham, and L. A. Glover. 1990. Luminescence-based nonextractive technique for in situ detection of *Escherichia coli* in soil. Appl. Environ. Microbiol. 56:3368–3374.

212. Kleeberger, A., H. Castorph, and W. Klingmüller. 1983. The rhizosphere microflora of wheat and barley with special reference to gram-negative bacteria. Arch. Microbiol. 136:306–311.

213. Misaghi, I. J., and C. R. Donndelinger. 1990. Endophytic bacteria in symptom-free cotton plants. Phytopathology 80:808–811.

214. Gannon, J. T., V. B. Manilal, and M. Alexander. 1991. Relationship between cell surface properties and transport of bacteria through soil. Appl. Environ. Microbiol. 57:190–193.

9

Interactions Between Soil Minerals and Microorganisms

MICHEL ROBERT and CLAIRE CHENU *Station de Science du Sol, Institut National de la Recherche Agronomique, Versailles, France*

I. INTRODUCTION

Soils are complex living media and contain large amounts of microorganisms (e.g., 10^7 bacteria, 10^6 actinomycetes, 10^5 fungi, 10^5 protozoans, and 10^4 algae per gram of oven dried soil). Where are these microorganisms located? What are their relations with soil minerals? What are the main limiting factors in relation to these constituents that govern the development and growth of the microorganisms? These are questions that are very difficult to answer, both for microbiologists and for soil scientists. However, answers to these questions are needed, both for fundamental knowledge in microbial ecology and for applications in agronomy, forestry, and plant pathology. These questions also need to be answered if microorganisms are to be introduced into soils.

In this chapter, we aim to introduce some elements of discussion for soil scientists and to indicate new trends in research for the future. This review is not exhaustive, and we focus on only some of the interactions between soil minerals and microorganisms.

Interaction, as used in this chapter, means an action of the physical and chemical environment on the growth and development of microorganisms (e.g., supplying oxygen, water, nutrients, or toxins; Figure 1). It also means an action of the microorganisms on soil constituents, namely, weathering, release of elements, aggregation, and transformation of soil organic matter.

In the first section, new information is presented about soil constituents, their organization, and their reactivity, which is necessary

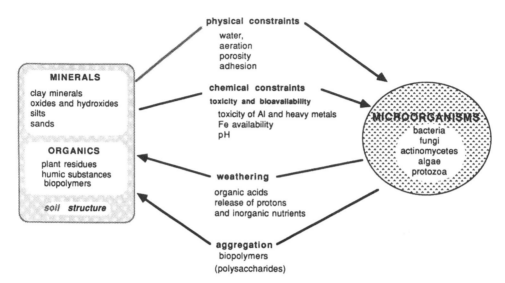

Figure 1 Some interactions between soil constituents and micro-organisms.

to better understand both the results of experimental studies and the phenomena that occur in the field. In the second section, selected interactions are presented. Two important environmental factors have been chosen to represent stresses for microbial life: a physical one (i.e., the role of water potential); and a chemical one (i.e., metal availability and its toxicity). The action of microorganisms on soil constituents through their role in aggregation and "weathering" processes will also be considered. Interactions between microorganisms and soil organic matter are not considered, as reviews exist on this subject [e.g., 1].

II. SOIL MINERAL CONSTITUENTS

A distinction is made between (1) the soil constituents considered as individual components (soil texture), which, for the finer and more active ones, corresponds to what is called the *colloidal level*; and (2) the natural association of these constituents (soil structure), which results in porosity, the *aggregate level*.

Most studies on interactions between minerals and microbes, whether under laboratory or field conditions, have been conducted at the colloidal level. The aggregate level, which appears to be predominant in several interactions, such as the resistance of microbes to drying, has to be viewed with special attention and as being considerably more complex than the colloidal level. Aggregate organization is more complicated than the model of Hattori [2], which was the first one proposed.

A. Constituents as Individuals: The Colloidal Level

Fractionation of Soil Constituents

The best way to determine the relative importance of soil constituents is to separate them according to their size. Therefore, mechanical analyses have been extensively developed for the study of the mineral constituents of soil, and these methods have been recently applied to the other constituents.

The classic granulometric methods, the purpose of which is to determine the true size of individual mineral particles, use chemical reagents to remove organic matter (OM) and amorphous compounds (H_2O_2 and reductive agents) and to disperse the clay minerals (K^+, Na^+, NH_4^+). If chemical reagents are avoided, the particle size separation of both mineral and organic constituents can be achieved by the addition of mechanical energy only.

For mineral constituents, separation can be achieved between the >2-μm fraction, which is composed of sand and silt, and the <2-μm fraction, which is the "granulometric" clay (Table 1). The former fraction represents the "inert" skeleton of soil, and the latter fraction represents the active fraction, especially if sorptive interactions are considered.

An important part of the <2-μm fraction is composed of micro-divided primary minerals (e.g., calcite, quartz, feldspars), which do not have the structure and properties of clay minerals. The "mineralogical" clay consists of phyllosilicate minerals, with specific structures and small particle sizes that confer specific chemical and physical properties. Oxides or hydroxides of Al, Fe, Ti, and Mn, with either a long- or a short-range crystalline structure, may also modify the clay properties.

Size fractionation demonstrates the very small size of clays that occur naturally in soil. The finer the fraction, the greater the surface area and swelling properties and the less the crystallinity [3].

The fractionation can be carried further on the basis of size, density, or magnetic properties. Size fractionation, combined with density, permits more accurate separation of oxides, clays, and OM. High field magnetic separation [4] enables the separation of clays on the basis of their chemical composition. Hence, classic fractionation methods, combined with new methods, such as high-resolution transmission electron microscopy (HRTEM), offer a new trend in research.

Table 1   Size Fractionation of Soil Constituents

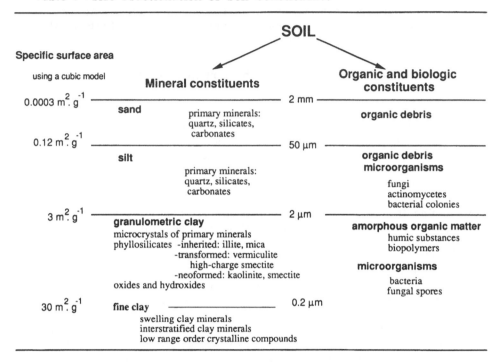

Organic and biological constituents can also be fractionated according to size after dispersion by mild agitation [5], ultrasonic treatment [6,7], or ion exchange on Na-resins [5,8,9]. Ramsay [10] used agitation or ultrasonication for the enumeration of bacteria by direct count and autoradiography. It must be emphasized that none of the size fractionation methods based on mechanical disruption results in the complete dispersion of the constituents. This indicates the presence of stable aggregates and the presence of organisms located inside or outside the aggregates [2].

Organization and Physicochemical
Properties of Soil Constituents

Structure, Size, and Organization of Minerals. The sand and silt fractions are composed mainly of primary minerals, such as quartz and feldspars. When phyllosilicates are present, they are monocrystals (micas and chlorites) with low surface area; only vermiculites and zeolites, which are very rare, can have specific chemical properties (e.g., high cation-exchange capacity; CEC). Gypsum and calcite are an important source of Ca, as the result of their solubility (3 g L^{-1} for $CaSO_4 \cdot 2H_2O$ and 20—100 mg L^{-1} for calcite). Depending on the size and shape of primary minerals, their arrangement can result in different overall porosities and range of pore sizes.

Since the beginning of the 20th century, X-ray diffraction has been the main method for the identification of clay phyllosilicates through the evaluation of the elementary layer size and the apparent displacement of the layer spacing by water or polyalcohols [11] (Table 2). This kind of identification, based on the layer and interlayer space, can help predict chemical properties, but it is important to know more about the organization of the clay, particularly at the particle level. The organization of clay phyllosilicates has been the subject of major studies in the last 10 years [12,13], based on the characterization of wet samples or samples the fabric of which is preserved in its natural state of humidity [14]. The development of new methodologies (e.g., low-angle X-ray scattering and electron microscopy) has made it possible to evaluate the number of layers per particle, their lateral extension, their mode of association [11], and to provide information on what is called the "texture" of the clay [15].

On the basis of such criteria, it is possible to better understand the behavior of the smectites in comparison with those of all the other clays. Smectites have an organization with variable geometry, the variation being dependent on the crystallochemical composition, the type of charge-compensation cations, the water content, the osmotic pressure (see under Section III.A), and the history of the clay. Only alkaline cations (e.g., Na, K, Li) at low concentration result in the development of a diffuse layer between the clay plates and, thus, in gel formation (layers separated by 2—15 nm) or in complete dispersion of tactoids composed of one to five layers, which correspond to the fundamental particles [16]. Other cations (e.g., Ca, Mg) give a quasicrystal organization [17], in which the interlayer space goes from an upper limit of 1 nm (four layers of water), which is the normal state under soil conditions, to 0.4 nm, which corresponds to air-drying (about 40% relative humidity; RH), or to even complete collapse under vacuum. Layers are associated in a subparallel manner (face-to-face), and their number ranges from 50 to more than 200, with decreasing water potential (decreasing water content) [18]. Such particles are highly deformable, and they delimit a three-dimensional porosity, the size of which depends on the size of the particles (0.5—2 μm) (Figures 2 and 3). With Ca or Mg cations, each drying leads to an irreversible increase in the number of layers.

Table 2 Classification of Clay Minerals and Major Surface Properties

Type of clay mineral	Layer type (nm)	Layer spacing Ca^a (nm)	Amount of charge (negative)[b]	Cation-exchange capacity (mEq Kg^{-1})	Surface area (m^2/g)	Dominant type of charge
Kaolinite	0.7	0.7	No charge or very low	10–200 silanol groups	5–200 External	Variable
Halloysite		1				
Illite (I)	1	1	0.6–0.9 (K)	100–400	50–100 Ext. + int.	Permanent (−)
Interstratified (I:S)	1	1–1.7	0.9–0.3 K–Ca	100–800	100–800 Internal	Permanent (−)
Smectite (S)	1	1–1.7	0.25–0.6 Ca	800–1300	800 Internal	Permanent (−)
Vermiculite	1	1–1.4	0.6–0.9 Ca	1300–1800	800 Internal	Permanent (−)
Hydroxy-Al-vermiculite	1	10–1.4	0.6–0.9 [Al(OH)n]	100–400 100–400	Partially internal	Variable and permanent
Chlorite	1 + 0.4	1.4	No charge	10–200	5–20 External	Variable

aWith water on polyalcohol.
bFor Si_4O_{10} in equivalent.

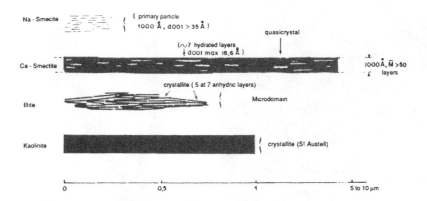

Figure 2 Organization of clay particles. From Ref. 12.

Other clays present a more stable organization in which a definite number of layers are built into rigid particles or crystallites (see Figures 2 and 3). If reference minerals derived from quarries are considered, illites and kaolinites have more than ten layers, and their lateral extension is important (Figure 2). Recent studies [3] have shown that in clays of soils and sediments, very small particles (one to ten layers) with limited lateral extension can occur (see Figure 3). Such clay can often be considered as *Texturally interstratified*, that is, superposition of very small particles with different layers that may be collapsed or opened. Minerals with 1-nm spacing (illites) are very abundant in temperate regions and correspond to two types of clay crystals: thick crystals of micromicas with several packs of layers; or very small particles with fewer than five layers, which are generally associated face-to-face to form domains. These are quite similar to small aggregates, but they correspond to more oriented clays.

Soil smectites that have mainly a beidellitic composition [19,20] are also quite different from reference minerals used in the laboratory, as these soil particles have a very small lateral extension and have always some collapsed layers [3]. A main characteristic of soil clays is a relatively high ratio of external (interparticle) to internal (interlayer) surface area. This ratio has to be considered in interactions of clays with polymers and microorganisms.

The difference observed in particle organization between smectites and other clays may contribute to the explanation of the specific

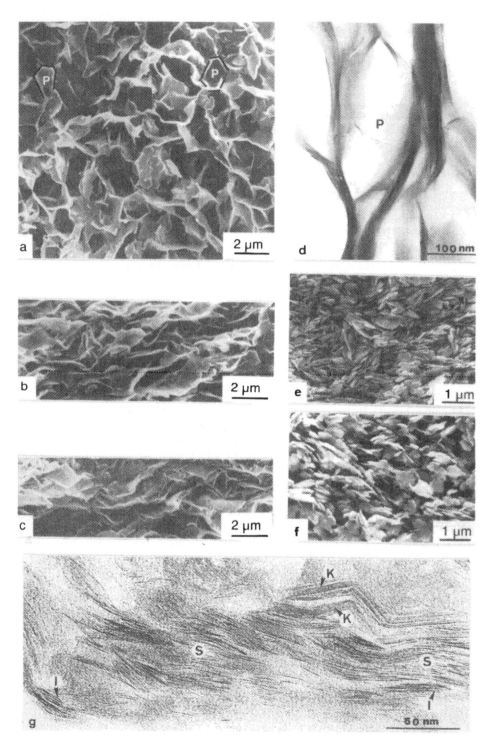

314

or even unique role of montmorillonite [1]. The properties of other clays, which are rigid, are more like those of silt particles, but the finer the size, the greater will be the surface properties and possible reactions with microbes.

Surface and Exchange Properties of Soil Minerals. The physicochemical properties of clay are governed by two main characteristics: surface area and electrical charge. Both are of major importance to biological activity.

Two kinds of surface area with different reactivities have to be distinguished: external surface, which corresponds to the surface of crystallites or particles; and internal surface, which corresponds to interlayer spaces.

External surface area decreases with an increase in particle size (see Table 1). Particles smaller than 2 μm have surface areas that exceed a few square meters per gram. External surface is the main surface for kaolinites, illites, and chlorites. Small particles of soil clays can have values up to 100 m^2 g^{-1}. For smectite, the determination of external surface area is very difficult and has no relevance if the clay is dried before the determination (which is the normal way when N_2 is used).

The internal surface area increases in interstratified illite—smectite, and becomes very large in smectites and vermiculites (700–800 m^2 g^{-1}) (see Table 2). Calcium-saturated smectites have an interlayer space limited to 0.8 nm, which can accommodate small polymers (e.g., polymerized Al, proteins, polyalcohols, polysaccharides). If the diffuse double layer is active (e.g., saturation with monovalent cation), the size of the interlayer has no limit and can even accommodate enzymes (>5 nm). However, such conditions will be scarce in soils, and interparticle sites are certainly the most common for polymers.

Figure 3 Clay microorganizations characterized by low-temperature scanning electron microscopy (a–c,e,f) and high-resolution transmission electron microscopy (d,g), at different water potentials. (a) Ca-Wyoming montmorillonite, −0.0032 MPa; (b) Ca-Wyoming montmorillonite, −0.1 MPa; (c) Ca-Wyoming montmorillonite, −1 MPa; (d) Ca-Wyoming montmorillonite, −0.0032 MPa; (e) St. Austell kaolinite, −0.1 MPa; (f) St. Austell kaolinite, −0.0032 MPa; (g) textural interstratified soil clay mineral. (P, pores; S, smectite layer; K, kaolinite layer; I, illite layer). (Photographs courtesy J. Berrier and A. M. Jaunet.) (a)–(f) from Ref. 12, (g) M. Robert, unpublished.

Two different types of electrical charge also need to be distinguished: a permanent negative charge and a variable charge. The first type is present mainly in 2:1 phyllosilicates, although it can exist in 1:1 minerals. It originates from isomorphic substitutions by trivalent cations (e.g., Al^{3+}, Fe^{3+}) in the tetrahedral sheet or by divalent cations (e.g., Mg^{2+}, Fe^{2+}) in the octahedra. Values of negative charge range from one charge in mica for each Si_4O_{10} unit cell (where the charge is blocked by nonexchangeable K) to 0.3 charge in smectites (see Table 2). If some negative permanent charge occurs in samples of kaolinite and halloysite, the presence of some 2:1 phyllosilicate layers has to be suspected, especially when CEC values are high [20].

Organic matter also has important exchange properties. Once the COOH or OH groups are dissociated, the charge can be considered as permanently negative, especially if the pH remains above the pKa. This charge ranges in extent similar to that for mineral constituents (1 to more than 2 mEq g^{-1}). The respective role of clay and organic matter in soil chemical properties will depend on their relative proportions.

The variable charge of minerals has received increasing attention in recent years. It originates from broken edges on the crystals and the dissociation Si-OH (silanols) or $Fe(OH)_3$ and $Al(OH)_3$. Such dissociation of $Al(OH)_3$ or $Fe(OH)_3$ depends on the pH, with the general reaction, $M(OH)_2^+ \longleftrightarrow M(OH) \longleftrightarrow MO^- + H_2O$, where M is the metal. Variable charges are found at the surfaces of oxides, hydroxides, and the edges of phyllosilicates, especially kaolinites. The smaller the particle size, the higher the charge. This explains the important role of short-range crystalline compounds that consist of particles of 5–10 nm for allophanes or even less for ferrihydrite or Al polymers. Fulvic and humic colloids have similar sizes. All these compounds can exist independently, but most often they are associated as coatings on clay surfaces (Figure 4) [22–24] or as clay–OM complexes, which appear as more diffuse (see Figure 4). At a pH lower than the zero point of charge value (ZPC), the overall charge is positive, but it changes to negative above this value. Hence, soils with variable charge will most often have a positive charge. This normal behavior can be modified by the specific adsorption of anions, such as SiO_4, PO_4^{3-}, or organic matter onto these positive charges. Coordination bonds can develop and can decrease the overall ZPC value of the complex and give a high-energy bond to the adsorbed compounds (e.g., PO_4^{3-}). Fixation of Al or Fe polymers on 2:1 phyllosilicates surfaces can result in a complicated situation, with a variable charge on the external surface and a permanent negative charge on the internal surface.

Soils with a permanent negative charge are generally formed in temperate regions (except for vertisols), whereas soils with a variable charge are generally formed in tropical areas and on volcanic rocks (except for podzols). In the case of soils with variable charge, upper horizons rich in OM can have a net negative charge, and B horizons, a net positive charge at the same soil pH [25].

Although the main mineral constituents determine general soil properties, the minor compounds are very important. For example, the presence of OM and 2:1 layer clays in kaolinitic soils is very important for cation fixation and, especially, for K fixation [26], and fixation of polyvalent cations on 2:1 minerals can change many physical and chemical properties [27,28].

Usually, there is a good correlation between the CEC and total surface as determined by ethylene glycol monoethyl ether (EGME), and such characteristics are good indicators of soil behavior [29]. These surface and exchange properties of clays are very important in relation to biological processes. Until recently, the importance of interlayer spaces was emphasized. However, essentially only cations and small polymers can be fixed on interlayer surfaces [30], and larger polymers (e.g., humic compounds), most enzymes, and microorganisms adsorb only on the external surfaces [31]. This is especially true for soil clay particles that are very short and thin. In the majority of soils developed in temperate regions, microorganisms, which are negatively charged colloids [32], will be in contact with negative clay surface, unless Al or Fe polymers are present at the interface. In soils with variable charge, exposed surfaces will be mostly positively charged.

In chemical properties (e.g., CEC, surface area), smectite and vermiculite have similar exceptional properties, but these clays differ mainly in physical properties. Such differences could be useful in devising experiments aimed at determining the relative importance of chemical and physical factors in biological processes.

Physical Properties of Soil Minerals. Definitions and Methods. The status of water in soil is defined by the concept of water potential. This potential indicates the energy by which water is retained by soils or clays. Hence, it is relevant for organisms, as they have to provide equivalent energy to utilize and extract the water. It covers two terms: the *matric* and the *osmotic* potential. The former is predominant in most soils, as the solute concentration of the soil solution is generally $<10^{-2}$ M. Water retention is related to water potential and is increasingly expressed in the percentage volume, rather than the percentage weight [12]. The *water ratio* is the ratio of the volume of water to the volume of solid, and the *void ratio* (e) is the ratio of the volume of voids to the volume of solid [33].

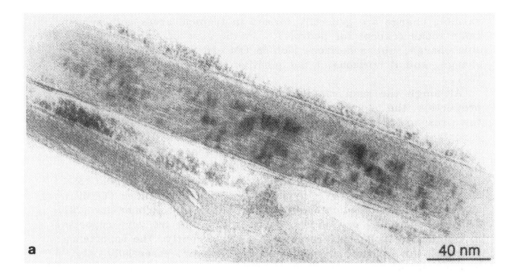

Figure 4 Clay coatings observed by high-resolution transmission electron microscopy. (a) Fe coating on Wyoming montmorillonite; (b) organic coating on smectite in a Vertisol (thin section fixed with OsO_4, and stained with lead citrate and uranyl acetate); (c) aggregates of Fe polycations-montmorillonite associations. (Photographs courtesy J. Berrier and A. M. Jaunet.) (a) and (c) from Ref. 24; (b) M. Robert, unpublished.

The monitoring of water potential is very important to better understand clay–microorganism interactions. However, true control of water potential is difficult to maintain and can very often alter biological functions. The method most generally used in laboratory experiments to establish a matric water potential is to add specific quantities of water to soil samples, this is correct unless the moisture release curve has been previously determined [34,35]. When desiccators and saline solutions are used (i.e., isopiestic control) [e.g., 36], it is difficult to renew the nutrients.

The use of polyethylene glycol is possible over only a short range of water potential, and it generates problems of aeration [37]. Suction or pressure plates are widely used [e.g., 38,39] and cover a −2 to −0.032 MPa water potential range. However, sterility of the plates is not easy to attain, although it is possible.

For high water potentials (which are probably the most important in biology) Tessier and Berrier [40] have developed a very

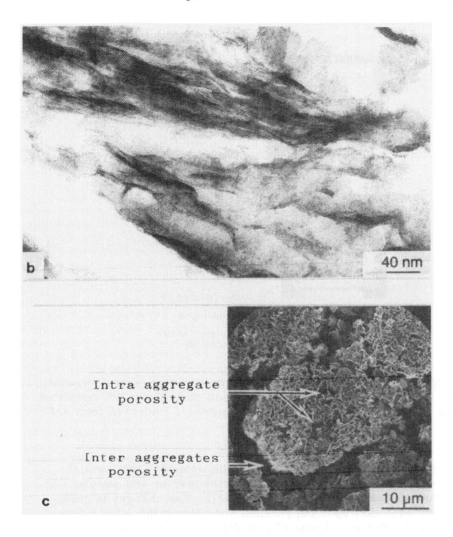

convenient technique (Figure 5). It consists of a filtration device that can be made of glass or polyvinyl chloride (PVC). The clod of soil or sample of clay is placed on a 0.2-μm filter, which assures the maintenance of sterility. The sample is in equilibrium with a solution that can be water, a nutritive solution, or a solution adjusted to a given osmotic pressure. The matric potential is applied by pneumatic air pressure in the range of −0.01 to −0.1 MPa. In such a system, the problem of aeration does not arise.

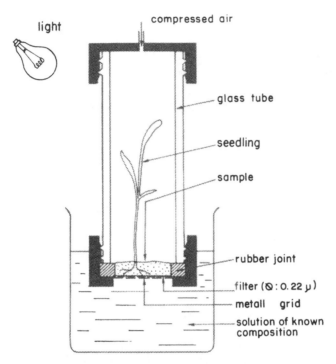

light

compressed air

glass tube

seedling

sample

rubber joint

filter (∅ : 0.22 μ)

metall grid

solution of known composition

Figure 5 Filtration device for the control of high water potentials. The sample can be a clay paste or a soil clod, it can be inoculated with microorganisms, or plants can be grown. Adapted from Ref. 40.

By measuring both the weight of the water at 110°C and the volume (e.g., with kerosene [41]) of the clay or soil sample, water and air indices can be deduced [13]. The changes in these indices' evolution upon dehydration and rehydration give information on the different types of porosity and their filling by air or water. These are certainly among the best parameters that can be measured on unperturbed samples (clods of a few cubic centimeters) to characterize the physical environment of microorganisms. Other methods, such as Hg or N porosimetry, give good values for the different types of pores (both quantity and diameters from 100 mm to 10 nm), but the measurements are made on dry samples with high vacuum, thereby causing strong modifications in many soil types.

Water Retention and Porosity of Clays. In most soils, clays have the predominant role in the water retention of soils. The organization

of the clay itself depends on the water content. Thus, the entire
clay—water system needs to be characterized [13]. All clays with
rigid particles or crystallites (e.g., kaolinite—illites) have a simple
behavior (Figure 6). At high water potential (−0.1 MPa), they retain relatively large amounts of water in a porosity that is less than
1 μm between particles. This water is readily available to microorganisms. With further increases in water potential, particles come
in contact, and the porosity is then filled with air. The quantity
of water retained, the energy of water retention, and the point of
air entry depend on the size of the particles or domains. The water retention properties of smectites are different in intensity from
those of other clay minerals, which, relative to microbial activity
are mainly the following (Figures 6 and 7):

1. The amount of water retained is very high at low suctions
    (i.e., at low pF and high water potential): as much as
    seven times their weight in water for Ca-montmorillonite
    and 15 times for Na-montmorillonite.
2. Most of the water, especially for Ca-montmorillonite, occurs
    in pores between the quasicrystals or tactoids, rather than
    in the interlayer spaces. With −1.6 MPa (pF 4.2) as a refer-
    ence, the interlayer water of Ca-montmorillonite is not avail-
    able to microorganisms, as a suction lower than −1 MPa
    (pF 4.5) is needed to extract the first two layers of water
    from Ca-montmorillonite (see Figure 7).
3. The interparticle porosity remains saturated under most hy-
    dric conditions in soil (see Figure 7). Upon desiccation,
    there is a continuous loss of water that corresponds to the
    collapse of these pores and in an equivalent loss of the ap-
    parent volume of the clay. These pores are highly deform-
    able.
4. Hysteresis occurs on rehydration of Ca-smectites, as the
    result of the aggregation of quasicrystals during desicca-
    tion [12]. Sodium-smectites, which are widely used in lab-
    oratory experiments, are very infrequent in natural soils,
    whereas Ca-smectites are widespread.

Organic Constituents. Because microorganisms grow primarily within or in the vicinity of the organic or organomineral phases, which
provide many of their basic nutrients as well as energy, some chemical and physical properties of organic constituents will be discussed.

Plant residues, present in the >50-μm fraction, are relatively inert chemically. They are characterized physically by high porosity,
which is unsaturated at most water potentials, and they have an
elastic behavior upon rehydration (Figure 8). On the other hand,
the small-sized fractions of OM are composed of the small debris of

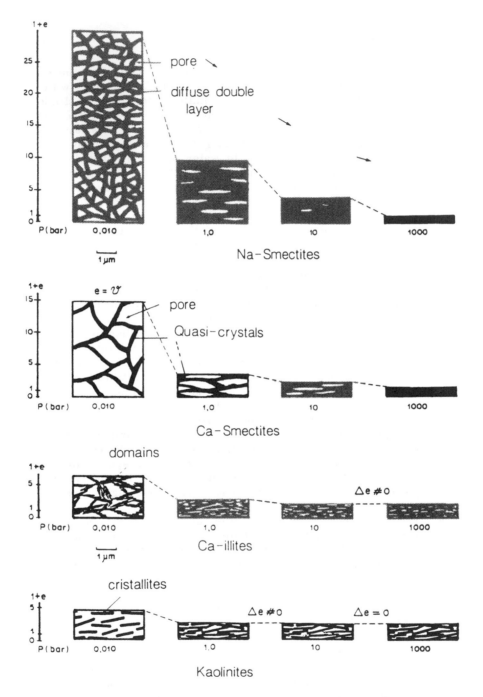

Figure 6   Variation in clay microorganizations and porosities with matric potential.   e is the void ratio and ν is the water ratio (in $cm^3 \ cm^{-3}$).   From Ref. 12.

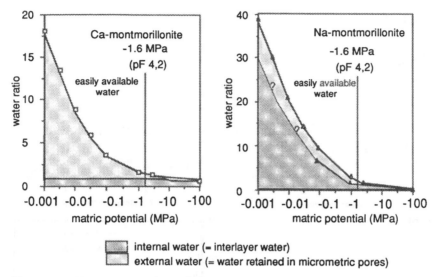

Figure 7 Water retention of Ca- and Na-montmorillonite in relation to matric potential. Water ratios (ν) are in cm^3 cm^{-3}. Adapted from Ref. 12.

plants and microbes at different stages of decomposition. Most small-sized fractions of OM are amorphous and consist of humic materials or chemically defined polymers, such as polysaccharides or proteins, which may be particulate or soluble. Some organic compounds, such as polysaccharides, have high water retention capacities, generally with considerable hysteresis upon rehydration (see Figure 8). Knowledge about the structure and organization of humic compounds has advanced much less than that of clays. The initial configuration of OM was thought to be spherical, with a radius of 6–8 nm and a mean molecular weight of 80,000. However, such particles were seen only under vacuum and an electron beam. More recent studies have led to a concept of linear polymers that exist in solution as random coils, which can be more-or-less tightly coiled, and result in ellipsoid particles that are cross-linked or swollen, depending on the ionic strength, pH, and charge-compensating cations [42,43].

B. Associations of Constituents: The Aggregate Level

The most widely accepted definition of a *soil aggregate* is that proposed by Martin et al. [44], who defined it as "a naturally occurring cluster or group of soil particles in which the forces holding the particles together are much stronger than the forces between

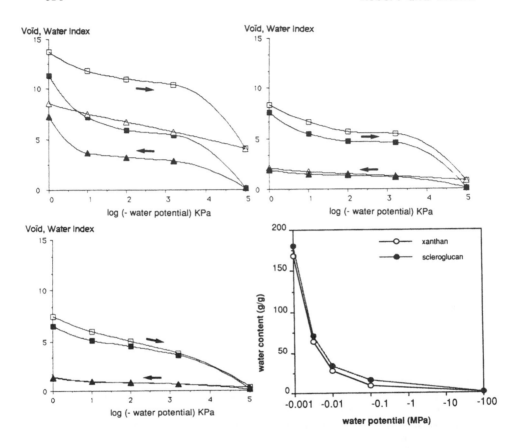

Figure 8   Water retention properties of different kinds of organic matter: (a) sphagnum peat; (b) herbaceous peat; (c) very decomposed wood peat; (d) bacterial and fungal extracellular polysaccharides.   Symbols in a, b, c stand for: void ratios (□) and water ratios (△) in desiccation, and void ratios (■) and water ratios (▲) in rehydration.   (a, b, c) From Valat, B., C. Jouany, and L. M. Rivière. 1991. Characterization of the wetting properties of air dried peats. Soil Sci., in press; (d) adapted from Ref. 124 and C. Chenu, unpublished data.

adjacent aggregates." Such aggregates are the basic units of soil structure; when they do not exist, structure results only from the juxtaposition of individual constituents. Between aggregates or constituents are pores, the volume of which depends on the size, shape, and stability of the aggregates or constituents.

Study of Soil Structure

Three main methods can be used to study the structure and aggregation of soil: (1) soil morphology at different scales; (2) fractionation of soil aggregates and indirect assessment of their stability; and (3) measurements of soil porosity. These methods can also be used to locate microbes and characterize their microhabitats in soil.

Direct Study of Soil Morphology. Soil macroorganization, with horizontal (soil horizons) or vertical zonation (prisms, fissures, tongues), will not be discussed here. On more microscopic levels, micromorphological methods are relevant to the study of interactions between microorganisms and minerals. One of the main problems, already mentioned for soil constituents, is to preserve the natural organizations that correspond to the state of humidity that prevails in the natural soils. A dissecting microscope can be brought to the field to study fresh clods [45], as well as to assess the microbial activity in soil. The same authors noted the importance of seasonal fungal development in A_2 Podzol horizons.

For scanning electron microscopy (SEM; $10^{-8}-10^{-3}$ m), critical-point drying or cryofixation of the samples for low-temperature SEM (LTSEM) is of particular interest. The soil aggregates around fungal hyphae and bacterial colonies can be observed after cryofixation [46,47]. Light microscopy ($10^{-3}-10^{-7}$ m) and transmission electron microscopy (TEM) ($10^{-4}-10^{-10}$ m) can now be performed on thin sections of soils obtained without too much perturbation of the samples by the use of fixative and hardening agents, and staining can be performed afterward [48]. Light microscopy allows the direct visualization of aggregates, pores, and main soil constituents, including roots, hyphae, and bacterial colonies [49,50]. Some quantification of the porosity can be achieved with light microscopy (quantimetry) [51,52], and it is easier if pores have been made fluorescent under ultraviolet (UV) radiation by a special dye. Pore quantification is also possible with SEM using cathode luminescence and backscattered electrons [53]. Microaggregates, microorganisms, biological remnants, and even amorphous organic matter can be visualized with TEM (see Figure 4).

To visualize all the different levels of soil organization and pores, it is necessary to combine these different methods. For example, combined light microscopy and TEM studies may be used to characterize porosities from several millimeters to nanometers in size. These observations are in good correlation with other porosity measurements.

Darbyshire et al. [54] have recently applied such methods to characterizing the microenvironment of soil microorganisms in undisturbed samples. Both UV light microscopy and SEM with backscattered electrons were used to estimate the soil pore network available to protozoa. Such a fluorescent technique was also used for viewing fungi in the presence of soil particles [55]. A new promising field of research seems open, but we must emphasize that, for all of these morphological studies, the biggest problem is not to disturb the organization of the soil from the sample collection for observation of the thin, or ultrathin, sections.

Fractionation of Aggregates. As indicated in Section I.A, the separation of aggregates implies the use of less destructive methods than those required for the separation of individual soil constituents. Physical fractionation in water allows stable tropical soil micropeds or "pseudosand" aggregates, which are composed of associations of clay particles with iron oxides or hydroxides, to be separated [43,56,57]. Organomineral aggregates resistant to dispersion are present in the 0.2- to 20-μm fraction of soils [6,7,58]. Conversely, methods of size fractionation are used to assess naturally occurring levels of organization in soils and their mechanical resistance [59,60].

Fractionation of soil aggregates and assessment of their stability classically involved wet sieving [61], simulated rainfall, and swelling and dispersion in electrolytes [62]. Ultrasonic dispersion methods were subsequently developed [63,64].

Methods for the fractionation of aggregates are also interesting in soil microbiology because they allow the differentiation and separation of microbial cells that are outside or inside the aggregates (e.g., the washing–sonification method of Hattori [2]). The main limitations are that (1) microorganisms adhering to the outside can stay on the aggregates upon washing, and (2) physical techniques are very often insufficient to separate biomass and metabolites from inorganic components. Such findings were illustrated with ATP determination by Ahmed and Oades [65], and by the fractionation of particle-attached and unattached bacteria [66].

Porosity Measurements. Porosity is one of the most important characteristics of both the physical properties of soil and the development and mobility of microorganisms. One of the major techniques to evaluate soil porosity is to measure apparent density, either in the field or in the laboratory, on an unperturbed sample. Such a measurement takes into account both the density of the constituents and all the voids in the sample.

On small clods, it is possible to measure, by the methods described in Section I.A, the volume of solid, air, and liquid at each water potential value (Figure 9). The Laplace law expresses the maximum diameter of water-saturated pores according to the suction pressure (Table 3).

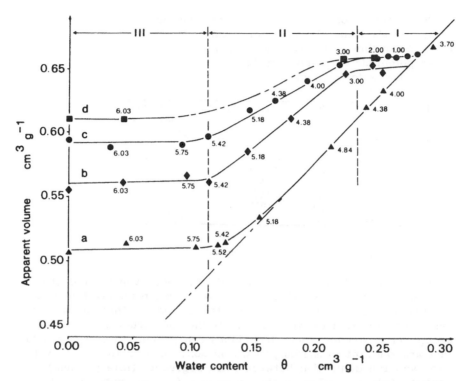

Figure 9 Shrinkage curves in a clayey soil (Terres d'Aubues) for samples of micrometric (a), millimetric (b), centrimetric (c), and decimetric (d) size. The areas separating the various curves correspond to the different levels of porosity, i.e., pores corresponding to the stacking of clay particles (below curve a), tubular pores (between curves a and b), and millimetric or centrimetric cracks (between curves b and d). Phase I correspond to the draining of pores with unchanged geometry, phase II to both shrinkage and draining of pores, and phase III to the shrinkage limit [68].

The Laplace law:

$$d = \frac{2\Gamma \cos \theta}{P \rho\, g}$$

where

d is the pore diameter   P is the suction pressure
Γ is the surface tension of water ρ is the volumetric mass of water
θ is the contact angle of water   g is gravity

Table 3 Water Potential Expressed with Various
Units and Maximum Size of Water-Saturated
Pores According to Laplace Law. Low
water potentials correspond to high
pF and low water contents.

pF	Ψ (MPa)	Ψ (bars)	a_w	maximum diameter of water-saturated pores (μm)
1	-0.001	-0.01	0.999993	300
2	-0.01	-0.1	0.99993	30
3	-0.1	-1	0.9993	3
4	-1	-10	0.9927	0.3
5	-10	-100	0.927	0.03
6	-100	-1000	0.484	0.003

It is also possible to measure directly the pore volume and the
equivalent pore radius using mercury or nitrogen penetration into
the sample. The main problem with this method is that the sample
has to be air-dried or lyophilized. If the soil structure is stable
enough after drying, these techniques can provide interesting data
that are relevant to microorganisms in soil, such as total pore volume, and the distinction between intra-aggregate (interparticle)
porosity (which is often in the range of a few nanometers) and
interaggregate (Figure 10) porosity (which ranges from a few to
more than 10 μm). The range of porosity can be defined from
mega-, through macro- and micro-, to nanoporosity.

The size and arrangement of individual particles, or the presence of aggregates, determine the retention and circulation of water in soil. If only texture is considered, clayey soils have mainly
hydric properties that characterize pure clay minerals (e.g., high
water retention and large variations of apparent volume upon desiccation and rehydration). However, differences exist between smectitic and kaolinitic soils (Figure 11) [67]. In contrast, sandy soils
have low water reserves, and there is no change in apparent volume during gain or loss of water (see Figure 11). Silty soils have
intermediate properties. General relationships between soil texture
and porosity (amount and size of the pores) exist.

Each kind of porosity has its specific function (e.g., drainage
of excess water and aeration for big pores, and retention of water
with higher and higher energy when the pore radius is decreasing
[68]). Recent emphasis has focused on the importance of such

determinations for microorganisms, for example, water-filled pore space (WFPS) has been demonstrated to regulate microbial activity, especially relative aerobic versus anaerobic processes [69—71].

Examples of Aggregated Soils

By using the techniques described in Section I.B, it has been demonstrated that many soils have specific aggregation levels, which are related to their constituents and conditions of genesis. For example, in brown acid soils, two different types of microaggregates (<250 µm) were recognized: the first was the result of microbial activity (see Section IV.B), whereas the second was related to chemical Al coatings on the clay [72]. In mollisols, the formation of stable microaggregates in the range of 0—20 µm is probably caused by organic matter [43]. The nature of the clay minerals also appears to have a role [e.g., larger aggregates (>250 µm) are formed with OM when montmorillonite is the dominant clay].

In vertisols, clay aggregation occurs only in the subsurface. In the profile, the main levels of porosities are the micrometric pores of the smectite quasicrystal network, together with fissures formed upon drying and some biological pores resulting from roots and fungi (Cabidoche, 1990, personal communication). Both types of porosities are revealed bimodally by water retention curves (see Figure 11). In these soils, bacteria were demonstrated to live mostly in the micrometer-scale pores [73], rather than on the walls of the fissures [59].

In clayey tropical soils, two very different structures were distinguished [56,74]. Red tropical soils are well structured: clay particles are linked together into 100- to 500-µm aggregates by Fe and Al cements (see Figure 10). In yellow soils (i.e., degraded red tropical soils), soil aggregation is absent, and clay particles occur as individuals. Porosities are different in the two types of soil and result in different hydric properties.

In andosols, two main scales of organization have also been shown: nanoaggregates (1—10 nm) resulting from the association of elementary mineral particles of allophanes, imogolites, or halloysites (which are very small: 1—10 nm); and microaggregates (100—1000 nm) resulting from drying or faunal activity [75].

In addition, the soil management or action of macrofauna can result in the formation of microaggregates and clods which are loose associations of microaggregates or individual constituents, with larger pores between them [58,60]. All these levels of organization are of great importance for both soil properties and microbial life.

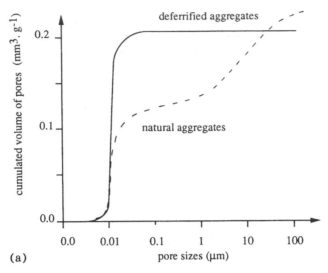

(a)

Figure 10 Levels of aggregation in an Oxisol: (a) porosity curves for natural aggregates and deferrified aggregates: absence of a 10-nm aggregated level; (b) schematic representation of the different levels of organization [57].

Conclusion. Some important properties of soil constituents have been discussed. The chemical properties have certainly received the most attention relative to microbial functions [1]. These chemical properties are directly related to the colloidal level, especially to the type of layer and interlayer of the clay minerals and to the kind of primary structure for OM. However, it is also necessary to consider the importance of the interparticle surface area of clay minerals and the extremely small size of clay particles in soils; as well as variable charges, which exist in half of the soils of the world, relative to coatings and low-range crystalline compounds [76].

The physical properties of soil that can affect microbes have been considered less than the chemical properties for various reasons. Soil physical properties involve the tertiary and quaternary structure of OM and the various levels of soil organization, from the colloidal to the aggregate level. Until recently, clay and soil organizations were not well understood. However, the development of new methods, based on morphological studies at different scales, have helped solve this problem. Moreover, until recently, physical constraints were more difficult to control and to experiment with than were chemical constraints. Another reason is the difficulty

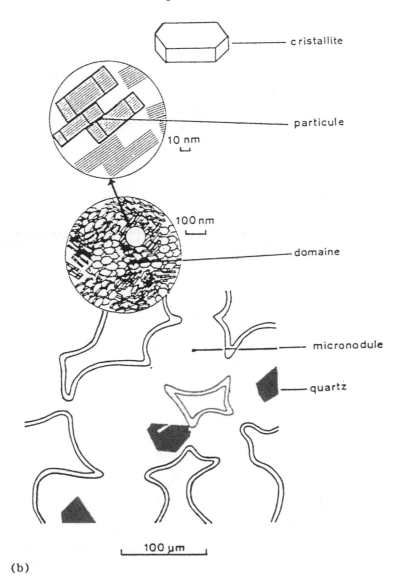

(b)

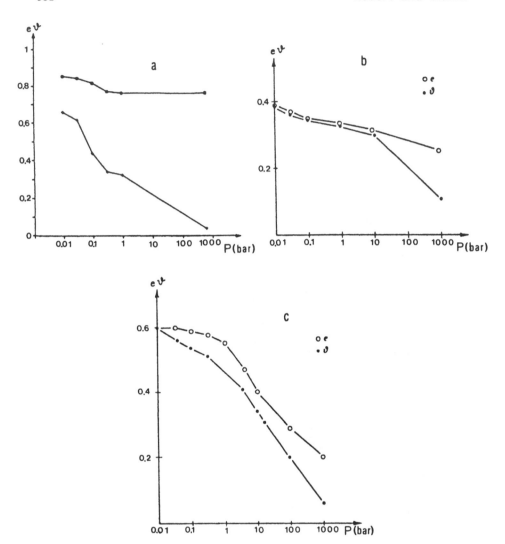

Figure 11 Shrinkage curves in soils of sandy (a), loamy (b), or clayey texture (c). ν is the water ratio and e the void ratio (both are in cm³ cm⁻³ [67].

in locating and quantifying accurately microorganisms in soil samples.

In the second part, some data on interactions between soil minerals and microbes, as influenced by these properties will be presented.

III. INFLUENCE OF THE MINERAL ENVIRON-
MENT ON MICROORGANISMS

Most of the research on the effects of soil minerals on the activity of microbes has been conducted at the colloidal level (i.e., the effect of clay minerals on microbes) rather than at the aggregate level.

Microbial functions, such as respiration, growth, reproduction, spore germination, nutrient uptake, can be affected by clay minerals (see [1], for a comprehensive review). The effects of clay minerals, particularly of the most active ones (i.e., smectites) appear to be very important and very complex [1,77—80]. For example, the addition of small amounts of clay, especially of montmorillonite, stimulates the respiration of fungi [81]. Further increasing the amount of clay (> 4%) gives the reverse effect. The first phenomenon was explained by a chemical action (i.e., buffering the pH of the medium), whereas the second one was mainly physical (i.e., the result of increased viscosity and reduced O_2 diffusion into the medium). The latter phenomenon can be explained by a better understanding of smectite organization (see Section II.A).

Adhesion of the microorganisms on clay minerals has been hypothesized to modify several aspects of their biological functions, although most effects of clay minerals on the activity of microorganisms do not necessarily imply adhesion [1]. More detailed aspects of the consequences of adhesion on microbial activity will not be discussed here, as comprehensive and recent reviews discuss the topic [e.g., 1].

The general effect of clays on microorganisms involves chemical aspects (e.g., the adsorption properties, charge, cation-exchange capacity, and cation retention by clays) as well as physical aspects (e.g., water retention, porosity, viscosity, and diffusion of O_2). The effects of clay on microorganisms have been studied mostly in pure liquid cultures [1,82]. Such an approach emphasizes the chemical effects. It should also be noted that in these suspension conditions, the clay surfaces are highly accessible to microorganisms or organic materials, whereas accessibility is probably lower in soils. This might be very important for interactions in which the clay minerals are involved as adsorbents. It is likely that in soils the physical effects of clay minerals will, in many situations, be as important as, or even overwhelm, the chemical effects.

Several actions of the soil mineral constituents on microbial activities occur at the aggregate level. Soil minerals and their organization influence, for example, the availability of water and the water potential conditions in which microbes live, the diffusion of oxygen and the relative occurrence of aerobic versus anaerobic conditions, and the accessibility of organic substrates for microbes to feed on.

An example of a major physical effect of soil minerals is how soil constituents and their organization affect microbial activities through water potential. Chemical factors will then be considered with the example of various elements that can cause major stresses for microbes: iron, through its availability to microorganisms and aluminum by its toxicity.

A. Physical Stress in Relation to Soil Minerals: Water Potential

Soils are porous media with variable water contents and water status. The most important physical aspect of microbial ecology in soils is related to the water potential. It is one of the main factors that governs both survival and activity of microorganisms and the competition between them [83,84]. The definition of water potential and its role in clay and soil organization were presented in the first part of this chapter. In most soils, the salinity remains lower than 0.2 g L^{-1} and thus, the matric potential is the major component of the soil water potential, and this discussion will mainly focus on it. However, with the development of irrigated agriculture, soil salinity has become an ever-increasing problem, and salinities as high as $1.5-3$ g L^{-1} can be measured [85]. Hence, more knowledge of the sensitivity of microbial populations, especially root nodule microbes, to osmotic stress is needed. However, it is frequently accepted that there is a good correlation between the effect of matric and osmotic potential [86–88].

In this review, the direct effects of variations in water potential on the activity of microbes are considered first. However, most aspects in which soil minerals are involved in the activity of microbes usually involve indirect effects of the clays on water potential.

General Effects of Water Potential

Disregarding, for the moment, very high water potentials, i.e., high water content, high water activities a_w (Table 3), there is a general correlation between water potential and the number and activity of microbes, such as numbers of bacteria (Figure 12) [89], hyphal growth [87,90], and respiration [83]. Maximum rates of microbial respiration, nitrification, and mineralization occur at the highest water content (lowest Ψ) at which soil aeration remains nonlimiting [91,92]. For many soils, this corresponds to -0.3 to -0.01 MPa, that is, to water contents of $10-35\%$, or 60% of the water-holding capacity, of the soil (Figure 13) [70,92,93]. Within the range of -0.03 to -1.5 MPa, the activity decreases linearly with the water potential [92], and the activity of microorganisms becomes very

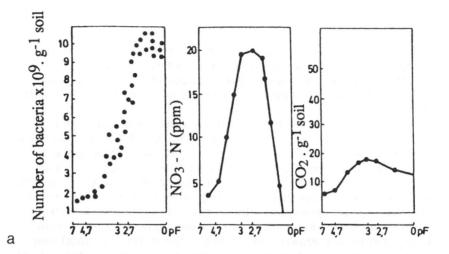

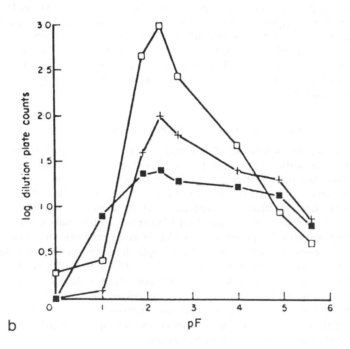

Figure 12   Dependence of the number of microbes and activities on water potential (a) number of bacteria, nitrification, and CO_2 production in a calcareous soil [89]. (b) Numbers of actinomycetes (+), bacteria (□) and fungi (■) in a Al soil horizon [90].

much less below −10 MPa (pF 5) (Figures 12 and 13). Dommergues
[94] found a water potential limit for microbial activities at about
−40 MPa (pF 5.6). Soil water potential conditions usually range
from saturation to −4.5 MPa in the upper meter of cultivated soils
[39].

The effect of water potential differs widely with the different
kinds of microorganisms. It has been established for decades that
bacteria are more sensitive to low water potentials than are actinomycetes or fungi (see Figure 12). Most fungi can grow without
problem above −1.5 MPa [95], and in this range of water potential,
the fungi have a competitive advantage over bacteria when the stress
is increased. Independent of these general trends, each microbe,
especially each fungal species, has its own optimum and minimum
water potential for growth [84]. Thus, the matric potential can
determine some general ecological distributions, for example, of soil-
borne pathogens, which can grow in dry (e.g., *Fusarium gramin-
earum culmorum* and *F. roseum graminis*) or wet soils (e.g., *Gaeumannomyces graminis*, various species of *Pythium*). These pathogenic fungi have higher resistance to dryness than do their plant
hosts. Changes in water potential can confer a relative competitive
advantage to some microorganisms [96], which provides the possibility of biological control. For example, *F. roseum* has a relative
advantage below −1 MPa (pF > 4) when bacteria that can lyse their
hyphae are absent.

Direct Effect of Water Potential on Microbial Cells

The internal water potential of microorganisms is in equilibrium with
the outside soil water potential [97]. Passive equilibration of the
cell potential can occur through cellular plasmolysis and a decrease
in internal water potential. However, this does not allow the physiological activities to function and the cell to survive. Many microorganisms can actively decrease their internal water potential by
accumulating solutes in response to external decreases in either
the matric or osmotic potential. This topic has received considerable recent attention [98,99], advancing to the level of genetic control [99,100]. The production of osmoprotectant solutes seems to
be general among soil microbes [97]. Microbes devoid of any osmoprotectant solutes would not be able to survive water potential
stresses, with decreases in water potential being immediately followed by cell plasmolysis and dehydration.

Osmoregulation. Osmoprotectant solutes can be more-or-less compatible with physiological activities; for example, compatible solutes in
procaryotes are, in general, amino acids and quaternary ammonium
compounds. Gram-negative bacteria (e.g., *Pseudomonas aeruginosa*;

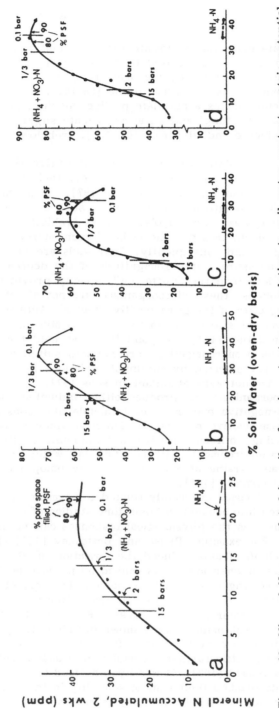

Figure 13 Mineral N accumulated in soils of various texture in relation to soil water content, matric potential, and percentage of water-filled pore space. (a) Sandy loam, (b) silty, (c) loamy, and (d) clayey soils. From Ref. 92.

rhizobia) accumulate potassium glutamate [98,101], and proline and aminobutyrate can also be accumulated (e.g., in species of *Serratia* or *Klebsiella* [98,102]). Gram-positive bacteria tend more to accuumulate neutral amino acids, such as proline [98,102]. Actinomycetes such as streptomycetes accumulate proline, alanine, and glutamine [103]. Eucaryotic fungi and algae accumulate sugars and polyols, such as glucose and glycerol [104–108], trehalose [109], or sucrose [110].

The osmoprotectant solutes can be either constitutive (i.e., normally accumulated by the cells) or inducible (i.e., their production can be induced under water potential stress) [97]. Some microorganisms accumulate constitutive solutes only, such as potassium glutamate, in procaryotes, and polyols, such as arabitol or manitol, in eucaryotes. When unstressed, the cells of such microbes have high internal turgor, and consequently, their walls are very strong to withstand this pressure. If the magnitude of the decrease in water potential is less than the "buffering" capacity provided by the constitutive solutes, these microorganisms are not affected. For example, the growth of the gram-positive bacteria *Arthrobacter crystallopoites*, was not affected when the Ψ > −1.5 MPa in relation to glutamate accumulation [111]. Beyond this threshold the cell plasmolyzed, and such microorganisms are not well suited to soil habitats that are characterized by continual variations in water potential, and they are not very abundant in soils [97].

Most soil microorganisms can produce either inducible solutes (e.g., most gram-negative bacteria), or inducible plus constitutive solutes (e.g., gram-positive bacteria, which have thick walls) [98]. Eucaryotes producing both constitutive and inducible solutes are species adapted to environments with constant low water potentials (e.g., Xero-tolerant strains of *Saccharomyces* or halophilic algae such as *Dunaliella viridis*) [104].

In soil, osmotic stresses are mainly related to permeating solutes, which must be taken into account in laboratory studies of osmoregulation of soil microbes, as the physiological responses can be different from that in soil. For example, Busse and Bottomley [112] showed that *Rhizobium meliloti* strains induced the formation of glycine betaine and K^+ formation when low Ψ was caused by permeating NaCl but not with a nonpermeating solute such as polyethylene glycol.

Physiological responses of microbes can also differ according to the magnitude of the water potential stress. For example *Streptomyces griseus* and *S. californicus* were shown to induce the production of amino acids upon osmotic stress. Beyond a threshold of 0.75 M NaCl there was little further accumulation of amino acids, but intracellular K^+ increased [103]. It was hypothesized that the accumulated amino acids reached their limit of solubility in the cell or that a higher concentration could be inhibitory to enzyme function.

The desiccation process in soil is, in general, slow enough to allow the microorganisms to accumulate osmoprotectants. On the contrary, as discussed by Kieft et al. [113], the most rapid changes in water potential occur when a dry soil is wetted. The difference between the soil Ψ and the higher internal Ψ of the microbe causes an influx of water into the organism. This influx of water results in high turgor pressure, release of internal solutes, and can lead to cell lysis (plasmoptosis) and bursting of the cell wall [106,114]. Factors of resistance to dilution shock include the thickness and cell wall on the ability to release or polymerize internal solutes without a loss in viability, such as glycerol in lichen [105] or sugars in the alga *Rivularia atra* [109].

Forms of Resistance to Drying. One of the main ecological advantages of fungi and actinomycetes is their ability to persist as dormant propagules, such as sclerotia, oospores, or chlamydospores [115]. These forms are of particular importance to soil-borne plant pathogens, as they have great resistance to extreme desiccation for long periods and can give rise to hyphae and other organs of reproduction in the presence of hosts or in appropriate moisture conditions.

The effects of water potential on reproduction are complex and depend on which step of the reproduction cycle is affected [87]. Spore germination can be less sensitive to low water potential than growth, whereas sporulation is more sensitive than growth (e.g., in *F. roseum*) [116], and Ψ values corresponding to soil saturation are needed for the discharge of zoospore. Hence, to better understand the competition between microorganisms in soil, knowledge of their resistance to low water potentials during all steps of reproduction, growth, and survival is necessary.

Role of Extracellular Polysaccharides. It is frequently suggested that extracellular polysaccharides of soil bacteria may efficiently protect them from desiccation [117,118]. Extracellular polysaccharides have been hypothesized, because they are very hygroscopic, to provide a source of water under dry environmental conditions [117] and to reduce the rates of drying and wetting of the microbial cell, thereby facilitating the rehydration of the cells without deplasmolysis [119,120]. However, the presumed beneficial effects of extracellular polysaccharides on desiccation lack convincing experimental evidence, and this concept needs careful evaluation.

The concept of protection against desiccation by polysaccharides comes mostly from microscopic observations of bacteria in soils: soil bacteria are frequently surrounded by a capsule or a loose slime that is polysaccharide in nature [121,122]. In soil aggregates subjected to one—six months of severe desiccation, gram-negative bacteria were observed to be embedded in a thick layer of polysaccharides [123]. The cells had well-preserved cell structures, indicating the viability of the bacteria. The authors concluded that extracellular polysaccharides were a major factor in the survival of gram-negative bacteria subjected to desiccation. However, no attempt was made, in this study, to correlate the presence of extracellular polysaccharide with the preservation of cell structures.

The assumption of a protective role for polysaccharides is also based on the hygroscopic properties of many microbial polysaccharides [117]. However, the water retention curve of microbial extracellular polysaccharides indicates that high water retention is restricted to high water potentials and not to low water potentials (e.g., <−0.1 MPa) (see Figure 8) [124].

Pena-Cabriales and Alexander [125] observed that the addition of purified exopolysaccharides to a *Rhizobium* strain in soil samples allowed somewhat better survival upon desiccation. However, in all other studies in which comparisons of the survival upon drying in soils of mucoid (i.e., polysaccharide-producing) versus nonmucoid strains of bacteria, were reported (e.g., *Klebsiella aerogenes* [126], *Pseudomonas solanacearum* [127], and various rhizobia [125,128–131], no significant difference could be recorded between the polysaccharide-producing or nonproducing strains. Thus, the weight of the evidence is against a protective role for polysaccharides. On the other other hand, in support for a possible protective role for polysaccharides, it may be noted that (1) when a pseudomonad isolated from soils undergoing severe desiccation was subjected to high water potentials, the production of its extracellular polysaccharide was enhanced [132]. (2) Polysaccharide gels are used as inoculant carriers for microorganisms; the desiccation of the gel causes severe mortality among the microorganisms, but the survival remains better than without any embedding agent [133–136].

If they are not involved in resistance to drying, what is the role of extracellular polysaccharides? They must have some survival advantages, as it is unlikely that, under highly competitive and oligotrophic conditions, bacteria would produce functionless substances requiring substrate and energy for their synthesis. Other possible roles have been reviewed by Dudman [118] (e.g., storage functions, ionic barriers, virulence agents, protectants against predation, or adhesive agents). Some of these will be discussed later in this chapter (see Section IV.B).

Indirect Effects of Water Potential Involving Soil Minerals

High Water Potentials and Aeration. At very high water potentials, the matric and osmotic potential become negligible and the main factor is the excess of water, which fills all the pores, even those larger than 10 μm (see Section II.B; Table 3). Hence, gas diffusion is reduced: in water the diffusion rate is about 1/10,000 of that in air [137]. For example, the diffusion coefficient of O_2 in air is 0.189 cm^2 sec^{-1} and 2.56 10^{-5} cm^2 sec^{-1} in water [138]. The low diffusion coefficients of O_2 become critical for aerobic microbial processes such as respiration, nitrification, and sulfur oxidation. For example, an optimum Ψ for nitrification is approximately -0.1 MPa [89], and water potential is directly or indirectly a limiting factor both at high ($-0.0032-0$ MPa) and low water potentials (below -1 MPa) (see Figures 12 and 13). Anaerobic conditions are characterized by the predominance of specific bacteria and specific chemical processes (e.g., production of CH_4, N_2, H_2, and H_2S).

The separation of anaerobic from aerobic conditions is assumed to correspond to Ψ values of -0.02 to -0.01 MPa [139] or to 70% of the water-holding capacity [69,140]. The threshold between aerobic and anaerobic conditions depends on the texture and structure of the soil, and neither the water content nor the water potential are particularly good indicators of aerobic versus anaerobic conditions [93] (see Figure 13). However, a meaningful factor to distinguish aerobic from anaerobic conditions is the proportion of soil pore space filled with water (water-filled pore space; WFPS) [70]. Aerobic microbial processes, such as respiration, increase linearly with the water content between 30 and 60% water-filled pore space, and declines beyond 60–70% (Figure 14) [70,71]. This relation is true for a wide range of soil textures. On the other hand, an anaerobic process, such as denitrification, occurs only beyond 70% WFPS [71,92,141]. The significance of an overall determination of the water-filled pore space when applied to microbial habitats depends on the organization and pore size distribution of the soil. Two of the soils considered by Doran et al. [71] exhibited microbial activity versus WFPS relations that were different from the other soils (i.e., their curves were shifted toward higher WFPS; see Figure 14). These soils were both weathered Hawaiian soils, with very fine granular structure, and it is probable that the active microbes were located in the interaggregate porosity, whereas the intra-aggregate pores, which were taken into account in the WFPS, were water-saturated and biologically inactive.

Clay Minerals and the Resistance of Microbes to Drying. With the purpose of inoculating soils with root nodule-forming bacteria, many studies have been conducted on the survival of rhizobia exposed to desiccation in soils, and clay minerals have been reported to protect bacteria from death associated with drying. The survival of rhizobia tended to increase with an increase in the clay content of the soil [142,143]. When clay minerals were added to sand or sandy soils, the survival of bacteria varied widely, depending on the clay type, amount added, and how the clay was added (as a dried powder

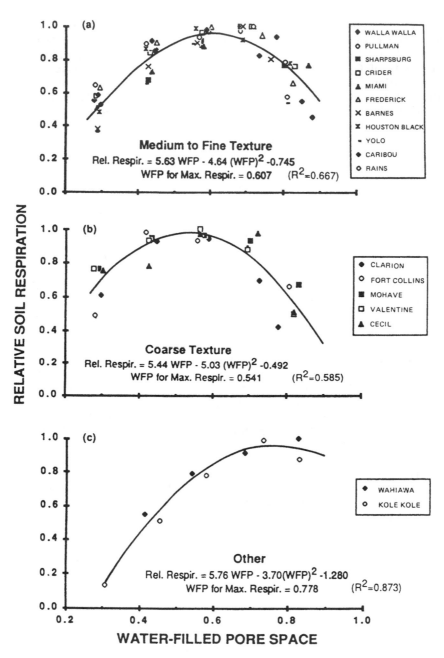

Figure 14   Relative soil respiration as a function of water-filled pore space (WFPS) for soils of various textures [71].

or as a suspension) [128,142]. Additions of kaolinite decreased the survival of *Rhizobium japonicum* in sand, whereas additions of illite had either no effect [142] or improved survival [17,131]. The addition of montmorillonite had, in general, a positive effect [126,128, 142], although no or negative effects were reported for some strains [128,142]. The protecting effect of clays was greater with strains less tolerant to desiccation [128,142].

A factor of major importance in these experiments was the rate of drying: for example, Bushby and Marshall [128] reported that the fast-growing rhizobia were more susceptible to desiccation than the slow-growing species under conditions of rapid drying. This result was confirmed by Jansen van Rensburg and Strijdom [144], although these authors found opposite patterns under slow desiccation: greater survival of fast-growing rhizobia than of slow-growing ones. They indicated that slow drying is probably more representative of drying conditions in the field than rapid drying, as in the Marshall procedure. The slower the soil dries, the greater the bacterial survival [145–148]. The protecting role of clay minerals might be to lower the rates of drying of the soil, which has been demonstrated [126,131]. The beneficial effect of clays can then be expected to depend on the rate of drying.

The numbers of survivors were directly related to the final water content of the soil matrix [126,142]. Clay was hypothesized to provide more available water to microbes upon desiccation [126], although other authors have considered, to the contrary, that the high affinity of montmorillonite for water caused the bacterial cells to be dried more thoroughly [142,144]. Indeed, for a wide range of microbial species, better viabilities are ensured at very low humidities, rather than at intermediate ones [118,133,142]. Bushby and Marshall [149] and Al-Rashidi et al. [143] proposed a similar hypothesis based on water adsorption isotherms of rhizobia: the more desiccation-sensitive strains were those of which freeze-dried cells retained more water.

With reference to the water retention properties of the different clay minerals (see Section II.A: Figures 6 and 7), the effectiveness in protecting microbes can be expected to be kaolinite < illite and interstratified clays < montmorillonite, which is consistent with most of the above-quoted results. It is worth noting that in some experiments, the clay minerals had a detrimental effect [142], which occurred when the clay was added as a suspension, rather than as a powder [128]. Thus, such a lethal effect of clay minerals occurs when the bacteria and clay can be expected to be the more intimately associated.

The same mechanisms were hypothesized to explain the protective effect of clay minerals and of extracellular polysaccharides on the death of microorganisms associated with drying (i.e., both have

very high affinities for water). In thin sections of soils, both are
frequently associated; that is, bacteria or bacterial colonies are embedded in polysaccharides, which are surrounded by a layer of clay
minerals [48,121,123] (Figure 15). As discussed by Kilbertus [150]
this "encapsulation" might be of ecological significance for gram-
negative bacteria that are devoid of forms of resistance such as
spores.

The protective effect of clay minerals (see Figure 15) has found
application in their use as inoculant carriers, either for root nodule bacteria [130] or for the biological control of soil pathogens by
selected microorganisms [151—153].

Mechanical Effects Associated with Water Potential and Clays. Recent studies have pointed out a new mechanism by which water potential can act on microorganisms through minerals. *Pseudomonas
solanacearum* is an important soil-borne plant pathogenic bacterium
that causes severe wilt of solanaceous crops in subtropical and tropical areas. In the French West Indies (Guadelupe), vertisols were
shown to be suppressive to the disease, whereas oxisols are conducive [154]. The suppressiveness seems to be related to the dominant clays in these soils (i.e., montmorillonite in vertisols, halloysite and kaolinite in oxisols). Schmit and Robert [127] studied the
survival of a mucoid (exopolysaccharide-producing) and a nonmucoid
strain of *P. solanacearum* in the presence or absence of pure clay
minerals (i.e., montmorillonite or kaolinite). Clay slurries were
mixed thoroughly with the bacterial suspensions to a ratio of 2×10^{11} bacteria per gram of dry clay. The water potential was accurately controlled using the methodology presented in Section I.A.
[40]. In these tropical soils, the water potential generally fluctuates
between −0.01 and −2.5 MPa, so the matric potential was maintained
constant at either −0.01, −1, or −2.5 MPa, or alternated from −0.01
to −2.5 MPa, and the survival of *P. solanacearum* was monitored by
periodic plate counts. At high water potential (−0.01 MPa), the
populations of bacteria were maintained close to the initial level,
but low water potential caused a rapid decline in both strains (Figure 16). A comparison with the reference samples of bacteria without clay indicated that desiccation of the bacterial cells could not
entirely account for the mortality in the clay at low water potential.
As observed with SEM and TEM, the bacteria were located in the mi-
cropores between clay particles (see Figure 15). Such pores are
closed between −0.1 and −1.0 MPa [12]. It is, therefore, assumed
that the mortality was due to mechanical stresses as the pores sizes
decrease below a limit lethal for cells. Both montmorillonite and
kaolinite had lethal effect on cells. Pores in the micron range are
abundant in vertisols, and the mortality of *P. solanacearum* below
−0.1 MPa was confirmed in vertisols and in purified clays from

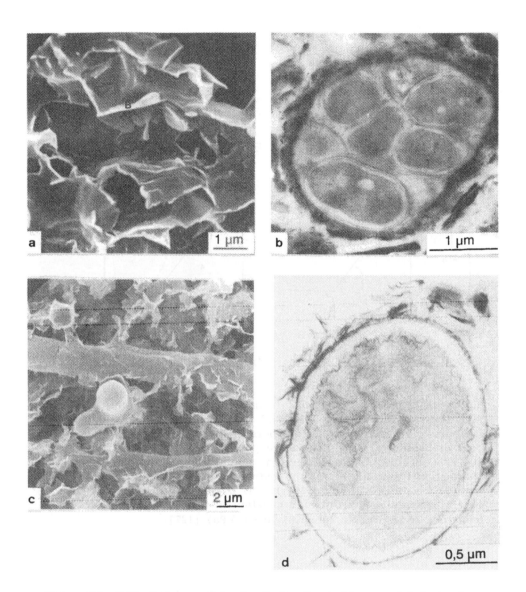

Figure 15 Soil microorganisms in their microhabitats: (a) bacteria
(*P. solanacearum*) in the micrometric porosity in montmorillonite,
low-temperature scanning electron microscopy [127]. (b) Bacterial
microoaggregates in the 0.2- to 2-μm fraction of a soil, transmission
electron microscopy observations on thin sections prefixed with OsO_4
and stained with uranyl acetate and lead citrate (C. Chenu, unpublished). (c) and (d) Fungi encapsulated in montmorillonite for biological control; (c) low-temperature scanning electron microscopy
and (d) transmission electron microscopy. From J. Fargues, unpublished, and Ref. 152.

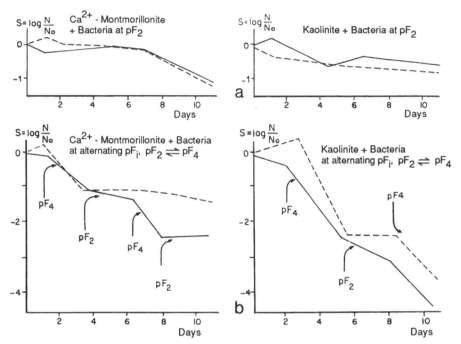

Figure 16 *Pseudomonas solanacearum* survival upon different water potential conditions in kaolinite and montmorillonite. Survival (S) is expressed as log N (final numbers of bacteria)/N_0 (initial numbers of bacteria). (a) The matric potential was maintained constant at pF2 (−0.01 MPa); (b) the matric potential was maintained constant at pF4 (−1 MPa); (c) the matric potential was alternating from pF2 to 4 (−0.01 to −1 MPa); (d) bacteria without clay, the matric potential was maintained constant at pF2 (−0.01 MPa) or alternating from pF 2 to 4 (−0.01 to −1 MPa) [127].

vertisols [73]. On the other hand, in oxisols, the kaolinite or halloysite particles are associated into stable micropeds with an interaggregate porosity larger than 2 μm (see Figure 10). In these pores, the bacteria could be subjected to a lack of water, but not to mechanical stresses, which may explain why oxisols are conducive to solanaceous wilt. Recently, irrigation of vertisols in some areas of Guadelupe greatly decreased the effect of the dry season and the disease has developed, giving field confirmation of the experiments of Schmit and Robert [127].

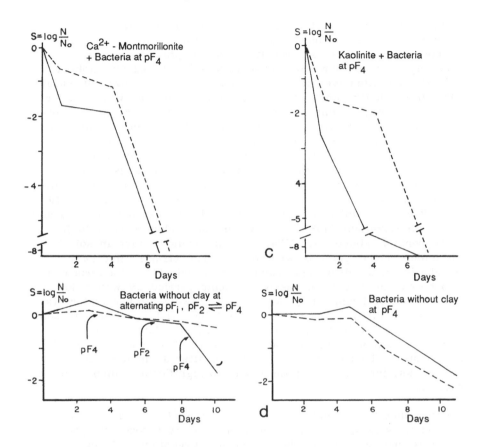

In some other cases of pathogenic microbes, soils were found to be suppressive or conducive, depending on their dominant clay mineralogy [155]. Biotic and abiotic factors are always involved in suppressiveness versus conduciveness. For fusarium wilts, montmorillonitic soils have been shown to be suppressive to the fungi by allowing the development of antagonistic bacteria for banana wilt [155] or for the muskmelon wilt [156]. Similar observations have been made by Stutz et al. [157] for the black root rot of tobacco and its antagonistic bacteria: the ability of the clays to support bacterial growth was vermiculite > montmorillonite > illite. Inasmuch as suppressiveness versus conduciveness occurred in the absence of severe desiccation, factors other than protection by clays against drying have to be considered, among them being mechanical factors. Chemical factors will be discussed in Section III.B.

Mobility of Microbes in Relation to Water Potential. It is generally assumed that there is no limitation to the movement of fungi and actinomycetes in soils as a result of their hyphal system [94]. Scanning electron microscopy observations by Dorioz and Robert [47] showed extensive colonization of clay pores by fungi at -0.01 MPa, although the pores in the various clay minerals were smaller than the average hyphal diameter. For other microorganisms, provided they have genetic capabilities for motility, their mobility depends both on the pore sizes and the volume of water in the pores. Both of these parameters depend on the nature of the mineral constituents and on the water potential. Most microbes require pores larger than their own size and that these pores be filled with water. Amobae and zoospores [115] require pores larger than 10 μm. Bacteria require proportionally smaller pores, between 1.5 and 2 μm, for movement [158]. The water potential necessary to fill these pores is above -0.1 MPa. How do bacteria move in soils? An undetermined point is whether bacterial strains that have flagella in pure culture also have them in soils [1]. Bacteria can also be expected to move passively, being carried by water flow after a rain or irrigation [126,159], by growing hyphae and roots, or by fauna [160]. Without "carriers," bacterial movements in soils would be very limited. Postma et al. [35] inoculated a silty sand and a loam silt with rhizobia and located the bacteria in pores of known size. This distribution persisted after an incubation of several weeks, indicating that there was no significant transport of the bacteria.

Soil Microhabitats in Relation to Water Potential. As discussed in Section II, soils are generally characterized by a complex spectrum of pores that depends on their level of aggregation. Hence, with alternating wet and dry conditions in the field, some pores are saturated and even O_2-limiting, whereas others undergo desiccation. In addition, the pores can be deformed upon desiccation, as seen in vertisols. Microhabitats with contrasting water characteristics exist in soils, as shown by the occurrence in the same soil of species with very different tolerances to desiccation or with very different requirements for O_2 (e.g., nitrification and denitrification frequently occur in the same soil at the same time).

Hattori [2,161] proposed a model for microbial ecology based on two main soil microhabitats: intra-aggregate microhabitats, characterized by small pores with available water under most soil moisture conditions; and interaggregate locations, subject to severe drying. Microbes on the outside of aggregates could easily be washed out of soils by gently shaking the aggregates in water, whereas sonication was required to disperse the microbes in inner positions. Differences in the distribution of microorganisms were shown; most fungi are located on the outside of aggregates, whereas most bacteria occur within aggregates [2]. Gram-negative bacteria, which are very sensitive to desiccation, especially tend to be located inside aggregates [123,162], where they are presumably protected by being imbedded in extracellular polysaccharides and clay minerals [123].

The aggregate model of Hattori [2] was applied to aerobic versus anaerobic habitats and further developed by Klein and Thayer [163]. They differentiated three zones: One was aerobic on the outside of aggregates, another was aerobic with pores filled with both air and water in the outer part of the aggregate, and the last one was anaerobic and situated at the center of the aggregates and within pores filled with water. This model is certainly oversimplified, but it helps to understand organometal transformations (particularly with heavy metals that have various states of oxidation, such as Ag, Hg, Sn, Se, and Pb) [163]. Such a model can also be applied to denitrification: anaerobic microsites within soil aggregates have been invoked to explain the occurrence of denitrification in well-aerated soils. Direct demonstration was possible with O_2 microelectrodes, and anaerobic centers were shown within aggregates, provided that they were large enough (>4 mm) [164]. However, in the presence of very high consumption rates of O_2, as the result of the degradation of organic mater, "hot spots" of denitrification can occur, even when diffusion of O_2 is not significantly reduced [165]. Such a distribution of anaerobic versus aerobic microsites may exist even at a larger scale (e.g., in the prisms of the stable structures of vertisols, which can have considerable variations in the water content from the outside of the prism to the inside) [165].

Several comments can be made about the approach of Hattori [2]. First, the fractionation experiments are considerably affected by the structural stability of the aggregates. The distinction of inner versus outer zones implies that the aggregates are stable, which is probably not true for many soils. As noted by Jocteur Monrozier et al. [59], both inner and outer microniches could be differentiated in the A1 horizon of a vertisol, which was very stable, but only external microniches occurred in the very unstable A1 horizon of an alfisol. The partitioning of the microbes between inner and outer zones is presumably different among soils, depending on the levels of organization and structural stability, and the fractionation method should be adapted to each soil. In addition, several levels of aggregation exist in many soils, with polyaggregates composed of smaller and simpler microaggregates.

It is also necessary to know whether bacteria can enter aggregates that already exist in the soil. For example, in oxisols, the porosity between the micropeds (>10 μm) is accessible to microbes, but the internal porosity, which is smaller than 1 μm is too small (see Figure 10). Similar features exist in the Bt horizon of alfisols [59].

Another approach is to characterize the microhabitats on the basis of the pore size. Kilbertus [150] demonstrated by TEM that most soil bacteria were located in 1- to 2-μm pores. Postma et al. [35] introduced *Rhizobium* into soils maintained at different water potentials. On the basis of different initial moisture contents, soil samples with water-filled pores of different diameters were prepared. It was hypothesized that upon further inoculation, bacteria would locate in pores not previously filled by water. After incubation at −0.01 MPa, the bacteria were washed out. It was observed that the initial contents of rhizobia retained by the soil decreased with a decrease in the initial moisture content of the samples: the less the porosity was initially saturated, the more places were vacant for inoculated bacteria. Retention of cells occurred when the matric potential was higher than pF 2 or 3 (Ψ less than −0.01 or −0.1 MPa). If the poral spectrum is assumed to be unaffected by wetting the sample during the inoculation, which is probably not true, it would correspond to pore sizes smaller than 9 or 0.6 μm. Micrometric pores, which are illustrated in Figures 3 and 15, would then represent protective microniches for introduced bacteria [35,159], and such microniches could be created artificially by incorporation of montmorillonite in the soil [166,167]. However, the protection seems to arise more from a physical protection from predation than from desiccation [168]. Several authors have demonstrated that inner habitats or small pores were not accessible to protozoans grazing on soil bacteria because of the small size of the pores [168−170]. Complex trophic interactions among nematodes, amobae, and bacteria were also related to pore sizes [171].

Thus, the poral spectrum of soils is of major importance for the possible habitats of microbes, as this determines hydric conditions, aeration conditions, trophic conditions, and relations between organisms (predation and competition) [172].

Conclusion

In conclusion, several points can be emphasized. The first is the importance of geometric constraints in soil for the activity of microbes. In particular, the sizes of the pores determine under which conditions of water potential they are filled with water and their accessibility to microbes or to predators. In addition, pores can be deformable, especially those within clay minerals. The second is the complexity of most effects of clays on microbial survival, as it is probable that several factors are involved at the same time, for example, protection against desiccation, availability of iron (see next Section IV.B), and competition or predation relations between soil organisms. Both physical determinations and morphological data seem to be most important in elucidating these aspects of microbial ecology. The micrometric habitat of bacteria, which consists mainly of the interparticle porosity of clays, is of utmost importance. Such a microhabitat is quite complex: for example, when the clays are swollen under water-saturated conditions, the pores are open and allow circulation of bacteria. Under conditions of intermediate water potential the small size of the pores can protect bacteria from their predators. Under low water potentials, the pores can close, resulting in some lethal effects on microbes.

B. Chemical Stresses in Relation to Soil Minerals

Soils constituents are the main source of nutrients for microbes, and microbes obtain these elements (major or trace) through mineral weathering processes in which they participate.

Clay minerals have important chemical effects on microbes, mainly through their cation- or anion-exchange capacity [1]. Other mineral compounds, mostly associated with clay minerals, are of special importance for the microbial life: aluminum, which is toxic in acid conditions; iron, which is necessary for plant and microbes, but the availability of which is very low in certain neutral or alkaline soil conditions.

Effect of Iron on Microorganisms

This subject has received increasing interest in the last 10 years, and some authors have put forward the hypothesis that iron availability can govern competition between pathogenic fungi (e.g., *Fusarium oxysporum*) and certain bacterial species (e.g., *Pseudomonas fluorescens*). Although this subject is controversial and not completely clear, some elements will be discussed, both relative to iron speciation and the effect of iron on the suppressiveness of plant diseases.

Iron Speciation in Soils. Large amounts of total and free iron are present in all kinds of soil, but in aerated soils, this iron is usually in the Fe^{3+} form. Iron is present in silicates, oxides and hydroxides, and in a third form that is of low-range crystallinity (amorphous form) and is, thus, more available. In the latter form, it is also possible to distinguish iron linked to inorganic or organic compounds. Although the content of amorphous iron is less than 1%, this is a significant amount. Iron speciation (Table 4) under these different forms can be deduced from various chemical treatments: treatment according to Mehra and Jackson [173] will give "free iron" (oxides + hydroxides + amorphous). Treatment with NH_4^+ oxalate in the dark yields the amorphous form [174,175].

Table 4   Chemical Data on Iron and Aluminum

a) pk$_a$ Values and Behavior of the Different Elements.

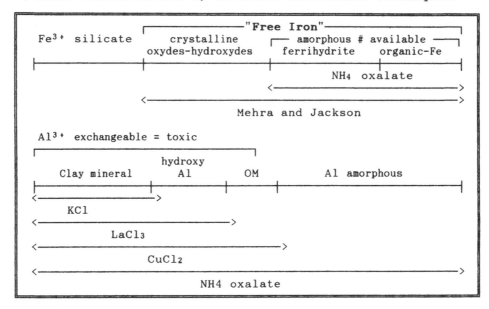

b) Phase Identification by Selective Dissolution Techniques.

c) Stability Constants (in log$_{10}$) of Various Natural and Synthetic Complexes of Iron.

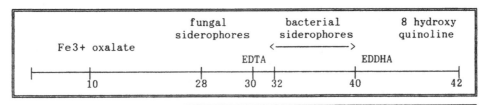

The general behavior of iron in soils is dependent on its chemical characteristics, especially the pKa values of Fe^{3+} and Fe^{2+}, and the conditions of the environmental medium relative to oxidation—reduction and complexation [176]. In terms of the pKa values (see Table 4), Fe^{3+} is mobile only at very low pH values (pH <3); thus, in the normal range of soil pH, the solubility is directly dependent on the very low solubility of the various oxides or hydroxides present, with solubility product (K_S) close to 10^{38}. Amorphous forms or ferrihydrite are relatively the most soluble compounds [176]. Reduction enables the formation of Fe^{2+} (pKa 9.5), which is mobile in the normal range of soil pH. Consequently, if microorganisms and plants are able to reduce Fe^{3+}, they can have an advantage in competition for available iron.

The other means by which iron can be solubilized is to form complexes. If these complexes are "chelates," they can mobilize iron over a wide range of pH. Microorganisms can secrete numerous low-molecular-weight organic acids in a relatively high concentration, the most frequent being oxalic acid from fungi. Even though the stability constant of oxalic acid is relatively low ($K_C < 10^{10}$), it can solubilize amorphous iron, or even other forms of iron, in relatively large quantities. Microorganisms also secrete very strong chelates (siderophores), but in very low concentrations.

Two environmental factors also have to be considered. The first is that even a very strong complex cannot alone release an insoluble element, and protons are generally required for the reaction. In the absence of protons, the complex can form only by precipitation of the organic ligand on the surface of the insoluble iron compound.

The second factor is for soils with relatively high pH values (>7), in which $CaCO_3$ is present. This leads to the specific phenomenon called plant "chlorosis" [177]; even if all the causes are not completely elucidated, it is linked to iron bioavailability. Under these conditions, iron has the same low solubility as at lower pH (from 3 to 7) but the presence of $CaCO_3$ renders inoperable most of the acid or complexing secretions. Under these conditions, one of the best procedures for determining iron availability in plants is extraction with diethylenetriamine pentaacetic acid (DTPA) [178]. The iron content obtained is very often well correlated with the "amorphous" content.

Availability of Iron in Relation to Soil Suppressiveness to Diseases.
Soils suppressive to some of the most important diseases caused by soil-borne plant pathogens have been described from different parts of the world [179]. Among the best known are the soils suppressive to fusarium wilts which can limit the severity of these diseases in banana, carnation, cucumber, cotton, flax, muskmelon, and tomato.

One of the first correlations established between suppressiveness and physicochemical factors concerns the presence of smectitic clays in soils suppressive to fusarium wilt of banana in Central America [155]. More recent examples involve the soils of the Salinas Valley in California and those of Chateaurenard in France, which are naturally suppressive for fusarium wilt. Very often these suppressive soils are compared with conducive soils in which the disease develops. Recently, other soils suppressive of tobacco black root rot by *P. fluorescens* were also described [157].

In the first example (Salinas Valley), the hypothesized mechanism is a competition between *Fusarium* and *Pseudomonas* for iron [180—182]. In the second example (Chateaurenard), the initial hypothesis was an intraspecific competition between the pathogenic and the nonpathogenic strains of *Fusarium*. It was also demonstrated that competition for carbon, related to the activity of the microbial biomass, was also involved in the mechanism of suppressiveness. More recently, however, the presence of montmorillonite [156,183] or vermiculite [157] and iron availability have been suggested [184]. It is worth noting that most suppressive soils have a pH value above 7 and contain a certain amount of $CaCO_3$.

Laboratory experiments showed that it was possible to inhibit the suppressiveness (i.e., enable the fungus to develop in the suppressive soil) if iron was furnished as a complex with a K_C (stability constant) less than 10^{30} (i.e., with ethylenediaminetetraacetic acid; EDTA). Introduction of a ligand, such as ethylenediaminedi[o]hydroxyphenylacetate (EDDHA), which has a K_C close to 10^{40}, enhanced the suppressiveness and converted a conducive soil to a suppressive soil. Fungi and bacteria can mobilize iron by the production of siderophores. Fungal siderophores have a K_C for iron close to 10^{28}, whereas different bacterial siderophores secreted by different species of the genus *Pseudomonas*, with catechol or hydroxamate groups, have a K_C from 10^{32} to more than 10^{40}; they have been called pseudobactin or pyoverdine [185,186].

The main forms of iron that occur in soils and the main K_C of the siderophores or artificial complexes that have been used in various experiments are presented in Table 4, and the structure of pseudobactin is shown in Figure 17 [185]. The ability to release iron is often measured by reference to a very strong complex of Fe (i.e., 8-hydroxy quinoline; $K_C > 10^{42}$).

This competition for complexing iron has been confirmed in a wide range of in vitro experiments, but questions arise related to the natural forms of iron in soils.

Although the iron competition system between bacteria and fungi can be reproduced in the laboratory, it is more complicated in the field as the result of the participation of other constituents such as $CaCO_3$, other kinds of acids, and even of several chemical reactions

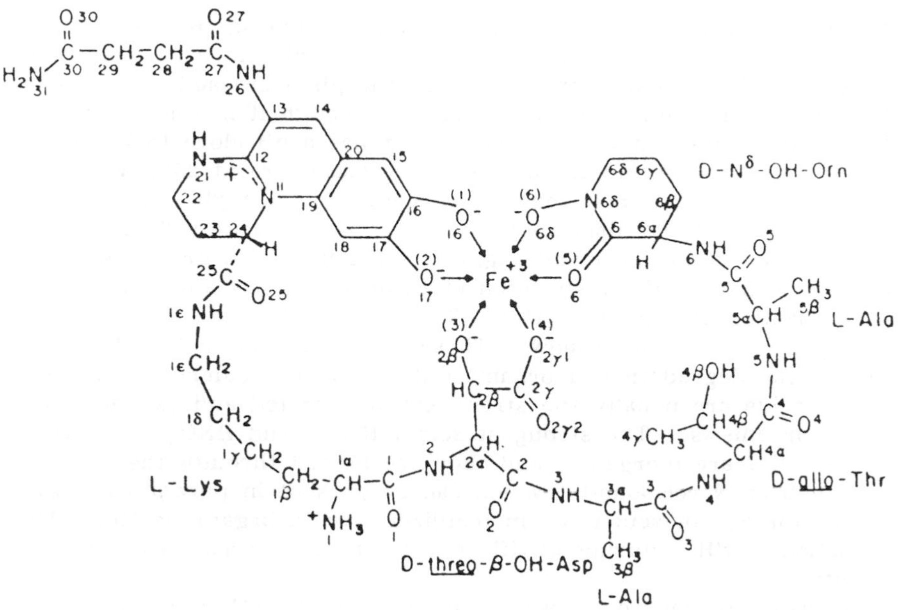

Figure 17 Structure of a siderophore: the pseudobactin. From Ref. 185.

(e.g., dissolution by H^+, iron reduction or complexation). Nevertheless, specific strains of *Pseudomonas* apparently suppress some diseases in soils caused by the subspecies *F. oxysporum* and by *Gaeumannomyces graminis* var. *tritici* [180]. Similarly, introduction into soils of plant growth-promoting rhizobacteria (PGPR), which belong to the genus *Pseudomonas* and are also able to produce siderophores, seems to increase the yield significantly [187]. Even though a better knowledge of siderophore behavior in the presence of $CaCO_3$ and other soil constituents, such as smectites [188] is necessary, such introduction of suppressive bacteria into soil represents a trend for the future.

Effect of Aluminum: A Major Stress for Microorganisms

According to Wright [189], one-fourth of the soils of the world are acid, and soil acidity is increasing for different reasons both in northern and southern parts of continents. So acidity represents a major chemical constraint for either plants or microbes, and all the data converge to focus on the specific role of aluminum in toxicity.

Soil pH and Aluminum Speciation. In the weathering process, a decrease in the pH of soils is normal. For example, in a 30-year field experiment in Versailles on silty soils, the pH decreased from 6.4 to 5. One of the main sources of acidity, although it is weak acidity, is CO_2 dissolution in water. Pure water has a pH close to 7, and the pH of water in equilibrium with the atmosphere at 25°C will not be lower than 5, even if the CO_2 concentration is high. The CO_2 concentration in the atmosphere, which is approximately 0.03%, has been higher in the past and might increase in the future; CO_2 is also derived from the respiration of living organisms and from the mineralization of organic matter. To obtain a pH lower than 5, stronger acids are necessary. These can be organic acids, formed during the degradation of organic matter (low-molecular-weight aliphatic acids are usually the strongest) or secreted and excreted by living organisms. The strongest acids, H_2SO_4 and HNO_3, that exist in soils, are inorganic, and they are formed through the oxidation of S or N compounds, which can be present in parent rock material (for S), in acid rain, in fertilizers, or in organic matter. The oxidation of NH_4^+ during nitrification is one of the main sources of acidity.

In soils, H^+ ions or protons will exchange for other cations such as Ca, Na, or K, present either in primary minerals (dissolution process) or on the exchange complex of clay minerals. When the pH is close to 5, the value of the pKa of Al, this cation can be released and will become the main internal source of acidity. It

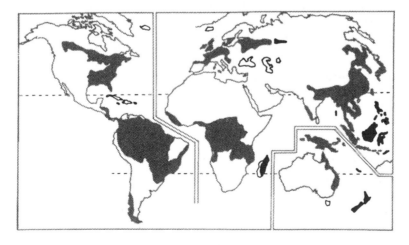

Figure 18   Major areas of naturally acidic soils.   From Ref. 189.

first appears on the exchange complex, as it is derived from aluminosilicates, especially from the octahedral layer of phyllosilicates. The part occupied by Al^{3+} (and H^+) represents the amount of desaturation, which is a part of the cation-exchange capacity. Aluminum is also present in solution as several species (see Table 4), depending on the hydrolytic reactions. Soluble Al, even at low pH values, is more toxic than Al fixed on clays in an exchangeable form, but an equilibrium exists between the different forms of Al. Speciation of Al in solution can be measured either by nuclear magnetic resonance (NMR), which permits the differentiation of monomeric from polymeric species, or by spectrocolorimetric methods using aluminum or Ferron [190,191], and modeling is possible [192].

To determine the speciation of Al on a solid phase, different solutions can be used: (1) Al is displaced from exchangeable positions by KCl solution; this will give exchangeable monomeric Al (or Al^{3+} + H^+, if done by titration), the presence of which is well correlated with soil pH. This solution can also be used to determine other forms of acidity linked to the surface of soil constituents [173,193]. (2) Lanthanum or copper chlorides can exchange Al fixed on hydroxy-Al or organic matter. Stronger extractants, such as sodium citrate [194] are necessary, for the extraction of Al-polymerized interlayers, or ammonium oxalate for amorphous Al in allophanes and imogolite. Tetraborate is used for titration of Al linked to organic matter. Toxic Al corresponds to soluble and exchangeable Al^{3+}, but equilibrium exists with the other forms.

Aluminum Toxicity. Many experiments have shown that living organisms, especially microorganisms, can be grown in acid solutions without serious damage, suggesting that they are not too sensitive to the direct effect of H^+, but rather to the indirect effect of metal toxicities. Even if Mn is more available at low pH, Al toxicity is the most important limitation to plant development in many soils of the world, particularly in forest soils in the northern portions of the continents or in tropical and subtropical soils. The toxicity of Al to plants has been well documented for many years, but the effect of Al on microorganisms is less known, even though it is a major ecological constraint.

Toxicity of Aluminum to Fungi. Most studies on the toxicity of Al to fungi have dealt with either mycorrhizae or fungal plant pathogens, and the experiments sometimes involved both Al and Mn fungitoxicity [195]. In the case of mycorrhizae, studies have been concerned with the ability of fungi to develop in very acid medium (e.g., colonization of soils impacted with acid mine drainage or with acid rain). Thompson and Medve [196] classified the sensitivity to

Mn and Al of different ectomycorrhizal fungi (*Cenococcum grami-forme, Telephon terrestris, Pisolithus trinctorium*): Mn was less toxic than Al, and most of the fungi could grow in solution at Al concentrations between 250 and 500 ppm. According to these au-thors, such mycorrhizae might be used to confer Al tolerance in the reforestation of acid mine soils or areas affected by acid pre-cipitation. Firestone et al. [195] analyzed fungitoxicity in relation to acid rain and confirmed the nontoxicity of Mn. Spore germina-tion was found to be inhibited by relatively high concentrations of Al (>500 ppm), the toxicity being reduced by the addition of a complexant, such as fluorine, to the Al solution.

Ko and Hora [197] showed that fungistasis is related to Al tox-icity in the Hawaiian Islands, where it is widespread. Also in Hawaii, Kobayashi and Ko [198] and Ko and Nishijima [199] showed that soil suppressiveness towards *Phytophthora capsici* and *Rhizoc-tonia solani* seemed to be related to soil pH. They analyzed both the nature of the suppression and the mechanism of lysis of fungal mycelia in soils. Orellana et al. [200] showed that *Verticillium al-loatrum* was almost completely suppressed by 8 ppm in vitro: hy-aline unpigmented mycelia and very few microsclerotia were pres-ent. At pH 4.7 or below, *Whetzelina sclerotiorum* was more tolerant to Al.

Recently, a study was performed in France on the reactivity of soils to *Fusarium solani* or *F. coeruleum*, the principal causes of dry rot in France and South America. Suppressive soils are rela-tively abundant and all have a low pH (<5.5) [201]. Soil suppres-siveness, which is linked to the presence of a higher content of monomeric Al than in conducive soils, can be removed by raising the pH to 5 or above by liming [202]. It was demonstrated that it was also possible to confer suppressiveness by soil solutions or Al^{3+} solutions, but not in the presence of fluorine which complexes with Al. With use of a microprobe, Al was identified on the macroco-nidia that developed in the suppressive soil extract [202]. Re-search conducted in Amazonia [203] demonstrated that naturally acid forest soils are suppressive for different species of *Pythium*, but the soils became conducive under cultivation.

Hence, some species of fungi are very resistant to aluminum and are able to resist up to a 100 ppm of Al, whereas others are very sensitive, and 1—10 ppm in solution is enough to lyse the cells. It seems that aluminum can be a major cause of fungitoxicity and of disease suppressiveness.

Action of Aluminum on Bacteria. Zwarum et al. [204,205] measured the cation-exchange capacity of bacteria and found values from 95 mEq 100 g^{-1} of cells (oven-dried weight) for *Bacillus* to 340 mEq 100 g^{-1} for *Pseudomonas stutzeri*, which is smaller in size. The first organism was inhibited by increasing acidity, but no further detrimental effect was produced by the addition of up to 80 ppm of soluble Al. However, upon saturation with Al, there was a change from gram-positive to gram-negative, and *P. stutzeri* (gram-negative) was more sensitive to Al (10 ppm).

The response of rhizobia to both pH and Al has also been studied. The effects also seem to be complex and differ among strains: slow-growing rhizobia tend to be less acid-sensitive than fast-growing ones, although there are some variations and exceptions [206]. The action of Al reduced the frequency of cell division [207,208]. Acid-tolerant strains of rhizobia seem to produce more exopolysaccharides than do acid-sensitive strains [209]. However, Cunningham and Munns [210] found no correlation between total buffering or chelate ability of Al and production of EPS. Hence, an explanation has to be found, which may be at the level of the charge of the EPS, as such knowledge would be useful for the introduction of selected strains of bacteria, especially of rhizobia, into soils [211].

The toxicity of different species of Al is not yet clear [208]. Work has to be pursued with better control of mono- versus polymeric species, otherwise the exact mechanisms of Al toxicity to bacteria cannot be understood. Another important aspect of bacterial ecology in acid environments that requires clarification is the role of acidity and Al in nitrification. The first range of pH was defined for pure culture of nitrifying bacteria and limits are shown: there is an increase of the rate in nitrification with increases in pH from 4 to 8 [212]. Acidification tends to depress, but not to eliminate, nitrification, which can be because there is less microbial activity at low pH.

There is a general effect of acid rain on the decrease of microbial decomposition rate of organic matter and transformation of N in acid forest soils (ammonification, nitrification, and denitrification) [213, 214]. However, contradictory data exist: several authors [e.g., 215] have recently pointed out that, in forest areas where acid rain is high, nitrification can be very effective even at pH less than 4, and a correlation can be established between the high amount of nitrate in solution and the release and migration of Al. Thus, this nitrification process, which occurs seasonally, is the cause of supplementary soil acidification [216].

In soils, the phenomena are complex, and Boudot et al. [217] demonstrated that Al linked to organic matter can decrease OM mineralization or nitrification, either by chemical (chelate formation) or physical protection of OM. Such phenomena are important for many soils in which OM accumulates, and intervention of Al toxicity can also be suspected.

As with water potential, pH and Al levels are important environmental factors that can effect the antagonism of fungi by bacteria

in soils [218,219]. It should be noted that, as with plants, monomeric Al^{3+} seems to be responsible for the toxicity [220]. For
plants, various mechanisms of the toxicity of Al are hypothesized,
and they can act differently in different species.

1. Coprecipitation with, or adsorption onto phosphate
2. Inhibition of enzymes and growth regulators [221]
3. Fixation on DNA or RNA and blocking their replication [222]

In microorganisms, even if the exact mechanisms of toxicity are
not known, the fixation of Al seems to be external, which emphasizes the role of exopolysaccharides or of exchange capacity of the
organisms on the suppression of toxicity.

Natural chemical stresses, such as acidity and salinity, cannot
be prevented, and solutions need to be found, since these problems
pose major constraints on agriculture, particularly in developing
countries. For acidity, liming is certainly the main remedy, and it
will have the greatest effect on microbial life. Where liming is possible, it will be a remedy both for Al toxicity and some heavy-metal
pollution. However, we have already seen that liming can lead to
some new diseases and problems with iron bioavailability. For both
Al and salinity genetically adapted microbes and plants would certainly be the best solution.

IV. INFLUENCE OF MICROORGANISMS ON THEIR MINERAL ENVIRONMENT

A. Role of Microorganisms in Weathering of Soil Minerals

The overall roles of biological and biochemical factors in the weathering of soil minerals has been recently reviewed [223]. In this chapter, some complementary data on microbial aspects of weathering will
be emphasized.

The general role of microorganisms in weathering, which occurs
through biochemical actions, must be distinguished. Microorganisms are the main cause of oxidation—reduction reactions, which,
for the microbes, are a source of energy. The most general phenomenon is OM mineralization, which results in organic acids, CO_2,
or NH_4^+. These compounds will modify the general weathering conditions by providing an acid pH that can range from 6 to 5 with
CO_2, to less than 5 with organic acids, as well as the formation of
complexes and chelates. Microbes are also the main oxidizers of
NH_4^+, S, and FeS_2, which results in the production of HNO_3 and
H_2SO_4. The presence of such strong mineral acids can explain local pH values of less than 3.

Thus, the weathering of rocks and minerals is increased, making the mobility of metals, such as Al^{3+}, Fe^{3+}, or other transition metals, possible.

In such a general weathering system, which represents a "macrosystem," living microorganisms are one of the four principal factors in soil formation, besides parent rock material, climate, and time [224]. Important soil processes, such as acidification and podzolization, which are located in cold regions of the world (e.g., northern part of continents and mountain areas), are also associated with such biochemical factors related to microorganisms [225,226].

Another more direct role of microorganisms in a weathering microsystem will be illustrated; as in aggregation, the influence of microbes is somewhat limited to the immediate and surrounding environment. Some experimental results illustrate how this microsystem works [227]. In the first experiment, several fungi (*Curvularia lunate*, *Sclerotina sclerotium*, *Sclerotium minor*, and *Aspergillus niger*) were grown on agar that contained glucose, yeast extract, and Ca-saturated vermiculite. After a few days, SEM observations showed both dissolution patterns and abundant precipitates located on the vermiculite or around the fungi. These precipitates, as determined by microprobe and X-ray diffraction, were composed of calcium oxalate (Figure 19). Thus, fungi are able to secrete abundant quantities of oxalic acid and causes Ca to be released. In the second experiment, the same fungi were grown in the presence of crystalline iron phosphates (vivianite). After a few days, observations with SEM showed that dissolution of the vivianite had occurred, and precipitates were again located around the hyphae of the fungi (see Figure 19). Electron microprobe analysis identified the precipitate as a secondary iron phosphate.

These experiments demonstrated that the general microbial microsystem consists of solubilization—precipitation processes that occur on a small scale. Figure 20 shows schematically that these processes include various mechanisms [223]. One of these mechanisms involved in solubilization is cation exchange (for example the Ca of the vermiculite) with the different charged sites of the external part of the microbe [228]. Most often, H^+ is involved, so the exchange reaction is difficult to isolate from acid dissolution mechanisms. Low-molecular-weight organic acids are also excreted by microbes: oxalic acid is frequently excreted by fungi and lichens, but citric and 2-ketogluconic acids are also reported. Under certain conditions, the production of oxalic acid by fungi and mycorrhizae and its precipitation as calcium oxalate can be so abundant that it constitutes a calcium "reservoir" for "calcicol" flora in subsurface horizon developed from acidic rocks (Callot, unpublished data).

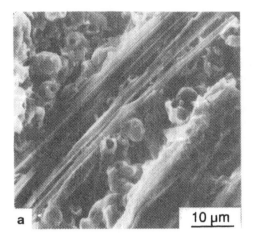 

Figure 19 Microbial weathering of soil minerals: (a) Lichen hyphae invading a crystal mica; (b) precipitation of Ca-oxalate on Ca-vermiculite; (c) precipitation of Ca-oxalate around the hyphae; (d) precipitation of secondary iron phosphate around the hyphae; (e) precipitation of iron on exopolysaccharide around a *Lepothrix* bacteria; (f) precipitation of $CaCO_3$ around a bacteria (the bacterium is at the center). Low-temperature scanning electron microscopy. From (a) [223]; (b, c, d) J. Thompson et al., unpublished; (e) [227]; and (f) [229].

Although microbes can have an important role in dissolution—precipitation of phosphates minerals, such phenomena are also important for carbonates (see Figure 19) [229]. In certain cases, specific bacteria (e.g., *Alcaligenes eutrophus*) can even precipitate heavy metals around them under a carbonate form [230]. In the case of silicates, an increase in solubility generally occurs with acid and complex secretions, and feldspars or micas can be destroyed (see Figure 19). However, the existence of specific microbes has not been proved (e.g., the involvement of *Bacillus siliceus*, quoted by several authors, has not been definitively demonstrated). Specialized microbes do appear to be involved in the transformations of sulfur, iron, and manganese. Sulfur bacteria (*Thiobacillus*) have a major role in mangroves and sediments, and iron and manganese bacteria are of major importance in the dynamics of these elements in soils and sediments. Berthelin [226] has shown the role of bacteria in iron reduction and that an enzymatic mechanism similar to dissimulative nitrate reduction is

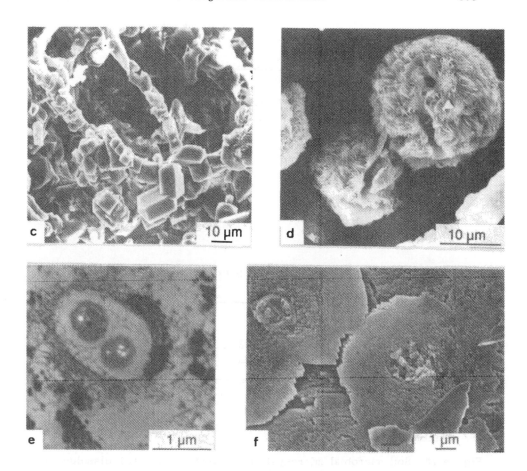

involved. Lefebvre and Rousseau (1991, unpublished data) have
shown that *B. polymyxa* grown at pH 6 is far more effective in dis-
solving goethite and hematite than are concentrated mineral acids
(N/10 or N), even with the addition of a reductive agent. Oxidation of Mn and Fe and their precipitation inside or around bacteria
such as *Gallionella* or *Leptothrix* are frequent phenomena [231].

In several of these examples of the precipitation of elements
around cells, exopolysaccharides seem to have a major role in the
initial steps of readsorption of the element (see Figure 19). Although siderophores are of great importance in iron bioavailability,
they are probably of minor importance, in terms of weathering, because of their low concentration.

In summary, many microbes are able to weather minerals that are
located in their ambient environment, mostly by the secretion of

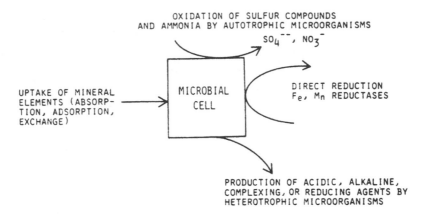

a

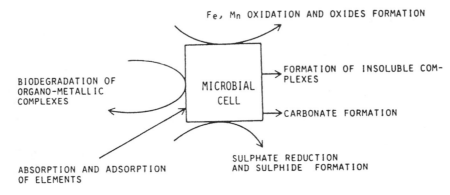

b

Figure 20   Soil microbial microsystems in weathering: (a) dissolution; (b) precipitation.   From Ref. 223.

low-molecular-mass organic acids. In precipitation of the element, polysaccharides are often involved. Very often microbes will derive nutrients, and sometimes energy, from the reactions involved. Such factors should be considered relative to the effects (see Section IV.B) of the majority of microbes on soil organization, which is mainly microaggregation through polysaccharide production.

B.   Microbially Mediated Aggregation in Soils

That microorganisms have a major role in soil aggregation was recognized in the early 1940s by Waksman and co-workers, and studies on this topic have been regularly reviewed [e.g., 232—235]. Evidence for aggregation by microbes came from studies on the incorporation of readily utilizable organic matter in soil [e.g., 231—235] and from studies in which selected microbes were grown in soils or minerals and their effects on aggregation investigated [238,241—244]. Both types of studies showed significant increases in water-stable aggregates. Fungi, rather than bacteria, appear to have the predominant role [232,240,244—246], but efficiency is variable among species [e.g., 237]. Microorganisms contribute to the reorganization and coalescence of mineral particles into new aggregates (i.e., to the molding of aggregates) [247]. They are also involved in the water-stabilization of existing aggregates. In turn, good soil structure enhances biological activity. The role of roots and fauna in soil aggregation will not be considered here, although this separation is quite artificial, as the main concentrations of microbes in soils occur in relation with plant remnants, the rhizosphere, and faunal excretions. The aggregating action of the rhizosphere, earthworms, and termites owes much to microorganisms [47,248—252].

General Mechanisms Involved in Aggregation by Microbes

Several mechanisms have been hypothesized to explain how microorganisms can enhance the formation and stability of soil aggregates. These include the mechanical binding of soil particles by filamentous microorganisms, the adhesion of microbes to soil particles, and the production of aggregating substances (e.g., polysaccharides).

Reorientation of Soil Particles and Mechanical Binding. Changes in the soil or the clay fabric can be observed in the vicinity of fungi or bacteria. When fungi were cultured in clay pastes (at a constant water potential), the clay platelets were observed to be reoriented parallel to the microbial surface and compacted (Figure 21a) [253]. These effects presumably result from the pressures exerted by the fungi during growth, and from local clay shrinkage associated with the adsorption of water by the microbes. Extensive polysaccharide secretion occurred in these experiments (see Figure 21b), and the microorganisms became surrounded by a microenvironment that they constructed [47]. After desiccation and rehydration, cracks originate at the boundary of this microenvironment and tend to separate it from the soil mass (see Figure 21c,d), which represents the first steps in aggregate formation and corresponds to Allison's [247] molding of microaggregates. In the case of bacteria or yeasts, preferential orientation of clay minerals parallel to the cell wall was also observed in soil aggregates [123] or in clay matrices [253]. It is

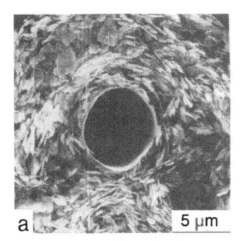 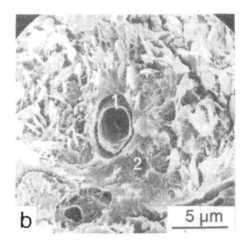

Figure 21 Microbially-mediated fabrics in clay minerals. (a) Parallel orientation of kaolinite platelets around hyphae of *Chaetomium* (−0.01 MPa); (b) polysaccharide secretion in fungal culture in kaolinite (*Chaetomium* sp.) (−0.01 MPa); (c) fissures in rehydrated kaolinite colonized by *Trichoderma* sp. (−0.01 → −100 → −0.01 MPa); (d) fissures around a yeast colony in kaolinite (*Saccharomyces* sp.) (−0.01 → −100 → −0.01 MPa); (e) physical entanglement of clay microaggregate by hyphae of *Chaetomium* (−0.01 → −100 → −0.01 MPa); (f) polysaccharide strands in kaolinite−fungal polysaccharide (scleroglucan) complex (−0.0032 MPa); low-temperature scanning electron microscopy. From Ref. 253 (a−e); Chenu, unpublished (f).

probable that the orientation and compaction of clay minerals in the vicinity of microorganisms results mainly from the shrinkage of the clay during desiccation in soil.

At another level, fungi growing on natural or on artificial aggregates made of clay or sand−clay mixtures [237,248] can be observed to form a network of hyphae all over the aggregate (see Figure 21e) [236,242,245,248,253,254], and Tisdall and Oades [248] found a good correlation between the length of the hyphae of vesicular arbuscular fungi and the water stability of aggregates. The increased stability of aggregates inoculated with fungi was, thus, ascribed to the physical retention of the soil particles by the hyphal network; that is, it was assumed that a network of fungal filaments is sufficient to protect the aggregates from slaking upon sudden wetting or from the dispersion of soil particulates. In addition, hyphae are frequently

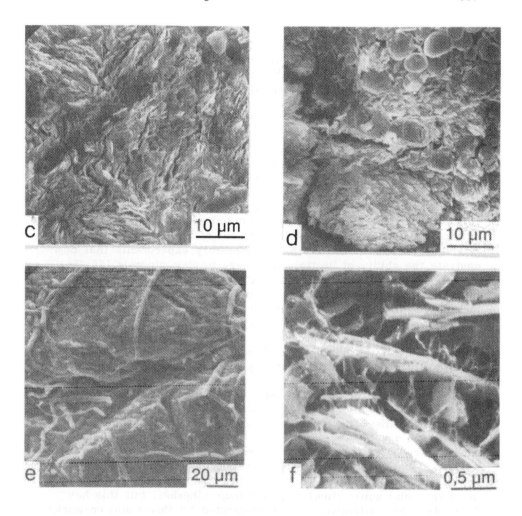

covered with a mucilaginous layer, presumably of polysaccharides [250], as many fungi are known to produce extracellular poly-saccharides to which the clay platelets adhere [243,248,250]. To evaluate both actions, Aspiras et al. grew fungi on soil aggregates and later disrupted the mycelial network by sonication [241]. Most aggregates remained water-stable, demonstrating that binding factors other than the mechanical effect alone were involved, among them some mucilaginous secretion by the fungi. Similar mechanisms presumably occur with other filamentous microorganisms, such as actinomycetes.

Adhesion. The adhesion of microorganisms to mineral particles has often been argued as a possible mechanism in aggregation by microbes [230]. As indicated by Lynch and Bragg [235], three cases happen in soils: adsorbent mineral particles can be larger than microorganisms (e.g., silts and sands), of equal size, or even smaller (e.g., clay minerals, micromicas, very fine quartz particles). In the latter, it is assumed that if several mineral particles adhere to bacteria, the overall association is a microaggregate (i.e., the microbial cell acts as a binding agent) [235]. Actually, clay minerals coating bacteria are a quite common feature in thin sections of soil samples [121–123,257]. Are such associations the result of adhesion processes?

As discussed by Stotzky [1], there is abundant empirical evidence for surface interactions between soil microorganisms and solid surfaces, as microbes are not leached out of soil horizons, but are retained. However, there are few laboratory experiments that demonstrate the ability of bacteria to adhere to clay minerals [66,258, 259]. The mechanisms involved in the adhesion of microorganisms to solid surfaces are physical and nonspecific [1,82,260,261]. They include van der Waals interactions, electrostatic interactions (DLVO theory applied to adhesion) [262], hydrogen-bonding, as well as hydrophobic interactions (thermodynamic approach) [263]. It is often assumed that at the ambient pH of most soils, clays and microorganisms are predominantly negatively charged [82,260]. Therefore, microorganisms must overcome an energy barrier before intimate contact can be achieved, and the electrostatic repulsive forces must be counterbalanced by van der Waals attractive forces. Several mechanisms have been assumed to render this adhesion possible; these are the occurrence of low pH (lower than the pKa of microbial cells) in the vicinity of the cells [260]; the presence of polyvalent cations [82]; and the secretion of polysaccharides that act as binders [264,265]. Surface hydrophobicity (of the microbe or of the solid substratum) may also help adhesion, but this has been given less attention. As demonstrated by Moses and co-workers [266], the adhesion of hydrophilic cells depends very much on electrostatic interactions, but less so for microorganisms with intermediate degrees of hydrophobicity, whereas a hydrophobic microbe, which was also strongly electronegative, adhered to negatively charged surfaces. Physical forces, developed in the drying of soils or in the growth of roots, might also bring the microbes and the clay surfaces close enough to overcome the energy barrier [1], or at least close enough for the exuded polymers to adsorb on the mineral.

Production of Aggregating or Water-Stabilizing Substances. The production of microbial substances is assumed to be a major mechanism by which bacteria contribute to aggregation processes [232,234]. Abundant organic matter, mainly polysaccharides, is produced by microorganisms, as they have an intense secretory activity [117]. Moreover, the microbes themselves have rapid turnovers and hence, the dead cells are then subject to biodegradation, and their constituents are more-or-less quickly released in the soil. Although bacterial and fungal walls have been demonstrated to have high stabilities toward degradation [267–269], other cell constituents are rapidly degraded [269,270], and microbial cells were reported to have half-lives of several weeks [270,271].

A prerequisite for the involvement of organic compounds in aggregation is considered to be their direct interaction with soil minerals, namely, their adsorption on clay minerals. A wide range of microbial substances adsorb on clay minerals, depending on their molecular characteristics as well as on the charge and surface area characteristics of the clay and on the ambient environment. Results and theories on the adsorption of organic polymers on clay minerals have been analyzed by Theng [31,272] and are not discussed here.

Polysaccharides: Binding Agents. Polysaccharides are one of the most efficient fractions of soil organic matter in aggregation processes. A recent review of studies that have demonstrated their importance was provided by Lynch and Bragg [235]. The general trends that emerge from the abundant literature on the topic are the following: Microbial and root-derived polysaccharides are considered to be more active in aggregation than other plant polysaccharides. It is likely that polysaccharide constituents of fungal walls, such as chitin or glucans, are not aggregating agents, as they are essentially nonsoluble and rigid molecules. However, partial biodegradation can change these characteristics. Extracellular polysaccharides (i.e., capsular or slime layer polysaccharides) seem to be the microbial compounds most active in aggregation.

Studies on the mechanisms by which polysaccharides contribute to soil aggregation have been performed with model polysaccharide–mineral (clay) associations. They will be discussed in the following section.

Since polysaccharides are linear macromolecules with long and flexible chains, they were generally hypothesized to adsorb on several mineral particles at the same time and, therefore, bind them [272–274]. This polymer-bridging concept is based on studies of adsorption–flocculation reactions of clays–synthetic polymer in suspension [272,275,276], and on other considerations:

1. Many polysaccharides adsorb on clay minerals, and a general feature is that anionic polysaccharides, which are predominant among EPS, adsorb on clay minerals only at low pH

(<4) or in the presence of di- or trivalent cations. Neutral polysaccharides can be readily adsorbed in high amounts and with high affinity [277–286].

2. Adsorbed polysaccharides can flocculate clay minerals [284, 287,288].

3. Low temperature SEM of clay–polysaccharide complexes allows the polysaccharide strands to be visualized (see Figure 21f) [124,285].

4. Adsorbed microbial polysaccharides reduce the dispersability of clay [285,289].

5. They increase the interparticle cohesion of clay minerals [290]. The tensile strength of a kaolinite increased with the amount of adsorbed polysaccharide. The maximal strength (20 times that of pure kaolinite) was attained when approximately 60% of the surface area was covered by the polysaccharide. Above this threshold, every clay-to-clay contact was presumed to occur through polysaccharide bridges.

The effectiveness of polysaccharides in the bridging process was related to their molecular weight and conformation [272,273,291]. Research on properties and applications of polysaccharides have highlighted other major characteristics of these polymers: In many polysaccharides, intermolecular physical linkages occur (hydrogen bonding, van der Waals forces) that lead to macroscopically viscous solutions and gels [292,293]. It has been observed that polysaccharides can form gels, even when adsorbed on clay minerals [284]. Hence, the cohesion of clay–polysaccharide associations results mainly from an organic network in which the clay particles are embedded (i.e., organomineral gels occur; Figure 22).

The adsorption mechanisms of polysaccharides on clay minerals can be predominantly related to their primary level of structure, but the bridging process depends mainly on their tertiary (shape) and quaternary (intermolecular linkages) structures. Polysaccharides with ordered helical conformations (e.g., xanthan, bacterial alginate, β-1,3-glucans, galactomannans), which form viscous solutions or even gels and have high water-retention properties [292–294], seem to be the most effective in bridging. These polysaccharides are abundant among the extracellular polysaccharides of microbes, as they are major constituents of many capsules and slimes [295–297].

Because polysaccharides that have intermolecular linkages have the highest viscosities, the foregoing hypothesis of organomineral gels might contribute to explaining why the viscosity of the

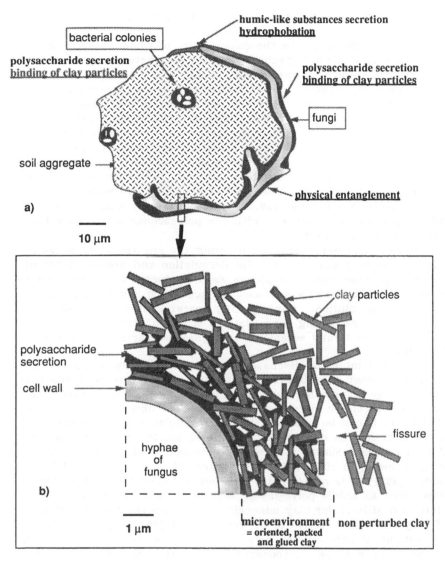

Figure 22 Microbially mediated aggregation: (a) Schematic representation of the binding and stabilization of a soil aggregate by microorganisms; (b) detail of the microenvironment in the vicinity of a fungi. (From C. Chenu and J. H. Dorioz, unpublished.)

polysaccharides in solution was a molecular characteristic that correlated fairly well with the effectiveness of these polysaccharides in aggregation [298,332]. In the experiments of Channey and Swift [299], bacterial alginate probably formed gels between the soil microaggregates and then established "bridges" on a larger scale to form bigger aggregates. The effectiveness of Ca in microbially mediated aggregation [300] could also be related to the ability of Ca to gelify many anionic polysaccharides by cross-linking [292].

Several points need further discussion to demonstrate how all this works in soil. First, do microbial strains that are known to produce polysaccharides in culture really produce them in the soils, where starvation conditions tend to prevail [1]? Electron microscopy of soils reveal that extracellular polysaccharides are widely present around bacteria and fungi [48,123,301], and many analyses of soil polysaccharide are consistent with the polysaccharides of microbial slimes or capsules [302].

Second, most experiments concerning the adsorption of polysaccharides on clay minerals and the flocculation and dispersability of the clay—polysaccharide complexes were conducted in dilute suspensions, in which the clay surfaces are accessible and the polysaccharide is in solution. Consequently, careful attention is necessary when extrapolating these data to soils, in which the mineral surfaces are not fully accessible to high-molecular-mass polymers [303,304].

A third point to be discussed is whether the polysaccharides are in solution when they interact with clays. It is more probable that extracellular polysaccharides are exuded by soil microorganisms in the form of gels or viscous slimes. From the observations of Foster [122,301], it can be deduced that extracellular polysaccharides of bacteria remain as discrete zones several microns thick around microbial bodies. Similar features were recorded with yeasts and fungi growing in clays [47] (see Figure 21b). As discussed by Oades [305], it is questionable whether extracellular polysaccharides occur as mobile polymers in soil, because of their physical state and affinity for clay minerals. Furthermore, the diffusion of polymers in soil seems to be very limited [303,304]. The concentrations of extracellular polysaccharides are probably very high in the vicinity of soil microorganisms, and it is possible that clay—polysaccharide interactions result from the adsorption of clays on polysaccharide gels and the embedment of clays in polysaccharide gels, as well as from the adsorption of polysaccharides on clay minerals.

Polysaccharides thus act as binding agents that enhance the resistance of aggregates to both slaking and clay dispersion. They are thus involved in aggregate formation as well as in aggregate stabilization.

Lipids and Humic-like Substances: Water-Stabilizing Substances. Another important class of microbial organic matter includes the lipids and humic-like substances. Their possible involvement in aggregation has received considerably less attention than have polysaccharides. Microbial lipids mainly consist of aliphatic hydrocarbons, ketones, wax esters, and very complex high-molecular-weight waxes [306]. Although bacteria, fungi, and algae generally contain less than 10% lipids [306], microbial lipids contribute to the stability of soil aggregates [307,308; G. Coulibali, doctoral, thesis Université de Poitiers, 1984] and are strongly associated with soil clay mineral [309]. Among the humic-like substances, the fungal melanins are of particular importance. These are intra- or extracellular dark-colored polymers that resemble humic acids in many characteristics (e.g., in elemental composition, functional group content, exchange acidity) [310—314]. Several authors have established, by using polar or nonpolar solvents, that the water-stabilizing action of various fungi is caused by such compounds [241,315,316].

Bond [245] and Monnier [236] demonstrated that when fungi were grown on soil aggregates, these aggregates became hydrophobic and water-stable. Microbial lipids and humic-like substances were presumably responsible for this. When lipids [317] or humic substances [318,319] are adsorbed onto clay minerals, water contact angles (θ) of more than 90° are recorded, whereas clean silicate surfaces are highly hydrophilic ($\theta \approx 0°$). The clay surfaces become hydrophobic at low organic contents ($\approx 1\%$); i.e., at low coverage of the clay surface area [319]. Hence, humic-like substances can be expected to be quite effective in stabilizing the soil aggregates. Another important point is that the hydrophobic properties of organic compounds are expressed upon desiccation [315]. This has been interpreted as conformational rearrangements during desiccation, leading to the exposure of hydrophobic moieties on the molecules [320]. There are several other kinds of organic compounds (e.g., proteins) [Jouany, unpublished results] and polysaccharides [Chenu and Jouany, in preparation] that can induce some water repellency upon desiccation. Water repellency of soil aggregates reduces the wetting rate and, thus, the slaking of aggregates [236]; lipids and humic-like substances are, therefore, involved in the stabilization of existing aggregates.

To conclude, soil microorganisms can contribute to the formation and stabilization of soil aggregates by several mechanisms. The specific effectiveness of fungi in aggregation processes is the result of several mechanisms that are summarized in Figure 22: physical entanglement is associated with the secretion of binding substances (many fungi produce β-1,3-glucans that are very effective binders). In addition, many fungi also produce humic-like substances that can render the aggregates hydrophobic. Physical entanglement and polysaccharide secretion are also involved with actinomycetes [242]. Bacteria act mainly through the secretion of aggregating substances, such as polysaccharides, or to a lesser extent, water-stabilizing ones, such as hydrophobic compounds. The role of bacterial adhesion in soil aggregation is yet to be confirmed.

In many cases, when aggregates stabilized by organic matter are allowed to dry or to settle for some time, an increase in water stability is recorded. This "ageing" phenomenon has recently been emphasized [250,321—323] and L. Habbib, doctoral thesis, Université de Nancy]. Thixotropy has been proposed as a mechanism [317]. Other factors that have been less considered in the literature might also be involved and may lead to increases in water stability (e.g., enhanced repellency of organics after desiccation and increases in organic versus mineral bonds as the constituents come closer together on desiccation).

Levels of Microbial Aggregation

Bacterial Microaggregates. In thin sections of soil, bacteria or bacterial colonies are frequently surrounded by polysaccharides [48, 301] and by oriented clay particles [121—123,263,324]. These features, described as microaggregates [305], were discussed previously in this chapter (Section III.A). Indirect evidence of such microaggregates has also been given: When small-sized fractions are separated from soils by physical dispersion methods, ultrasonic-resistant microaggregates remain in the <20-μm fractions [6,7,59]. Labile organic matter that is primarily aliphatic is characteristic of this-sized fractions [6,325] and coexists with highly condensed aromatic moieties [59,325], which suggests that both microbial substances and humic-like components are involved in the stability of these microaggregates. The TEM studies have shown that ultrasonic-resistant microaggregates mainly consist of bacterial microaggregates and, in the coarse clay fraction of a loessic cultivated soil, 90% of the enumerated bacteria were coated with clay minerals [Chenu and Balesdent, unpublished results]. Jocteur Monrozier et al. [59] demonstrated that 40—60% of the microbial biomass may be associated with the microaggregates of 2—20 μm in size, depending on the abundance and nature of the clay minerals. Even if bacteria make up only a few percent of the total organic carbon in most soils (0.1—5%), such microaggregates have a great significance because: (1) they persist after the death of microbes; (2) they control microbial efficiency of substrate oxidation [326]; (3) they limit the numbers of microbes enumerated [Richaume, personal communication]; and (4) successive microbial populations contribute to build up a stable pool of microaggregates rich in organic materials.

Fungi and Actinomycetes. Fungi are preferentially located outside of soil aggregates, as demonstrated by direct observation [236,245] or by washing—sonications of aggregates [2]. Hence, fungi stabilize larger aggregates than do bacteria (e.g., aggregates of several tens of microns in diameter) [47,58,327].

The size and shape of microorganisms and their secretion, if any, of aggregating substances determines the scale at which microorganisms act. Fungi and bacteria, through their secretion of binding substances, can generate aggregates in their micro-environment from a few microns for bacteria to several tens of microns for fungi [253]. Microbially mediated aggregation results overall in large stable microaggregates (>200 μm) [60,239,300,328, 329], which are composed of several hierarchical levels of particle aggregation, each level being the result of a major binding agent [58].

Persistence of Microbial Aggregation

Microbial aggregation is relatively transitory and decreases over several weeks [236,240,242,330,331], primarily as the result of the short life of the microbes themselves and because of the lability of their organic constituents, especially of polysaccharides [234,243, 270,332—334]. This lability has frequently been invoked to lessen the importance of microbial aggregation. It is, indeed, a serious handicap in the use of polysaccharides as soil conditioners. However, under natural conditions, microbial generations succeed each other and maintain an average level of aggregation, with seasonal variations [281].

Several factors related to soil minerals are likely to enhance the persistence of microbial compounds (e.g., metal complexation of polysaccharides [234,334] and adsorption of humic substances [332]), which are considered to be chemical mechanisms of protection. Physical mechanisms of protection (*physical* means that the protection can be suppressed by physical dispersion agents) have also been described: adsorption of the labile compounds on clay minerals [335—337] or on amorphous Al—Fe compounds [217] and occlusion within the clay porosity [303]. For example, the clay-coated bacteria or amorphous organic matter that are observed in ultrathin sections of soils are presumably protected from microbial degradation (see Figure 15) [48,122,333]; and many studies concerning organic matter turnover emphasize the physical protection of organic matter by entrapment in pores that are inaccessible to microbes [339—341]. A close relationship between the turnover of organic matter and stable aggregation can be postulated, as once aggregates are disrupted, the enclosed organic matter is available for degradation.

Consequences for Microbial Development

By the secretion of organic polymers and by mechanical and physical actions, microorganisms change the organization and physical characteristics of the media in which they live. However, little information is available on the influence of microbes on physical properties of soil, other than aggregation. Microbial exocellular polymers can increase the water-retention properties of minerals, especially the retention of readily available water [124] (see Section II.A). Microbes themselves can create porosities: in vertisols, tubular pores of several microns in diameter are voids left by dead fungi [Cabidoche, 1990 personal communication].

In a stable aggregated structure, at least two kinds of pore spaces occur: intra-aggregate pores provide storage for water and interaggregate pores ensure the circulation of air and water. These porosities constitute different ecological habitats [2], as previously discussed in this chapter (Section III.A). The amount and stability of the protected intra-aggregate microniches are undoubtedly dependent on some stable aggregation (e.g., it was demonstrated that both microenvironments occurred in a vertisol A1 horizon, whereas external microniches were dominant in an unstable and poorly aggregated alfisol A1 horizon [59]).

Management of Microbial Aggregation

What are the possible applications of microbially mediated aggregation for the improvement or preservation of soil structure and fertility? The aggregating effect of microbes can be promoted by classic or more recent agricultural practices, such as the growing of plants that encourage rhizosphere activity, the return of straw and other plant residues to the soil, the addition of animal or industrial organic manures, and the conditioning of the soil with polysaccharides or polysaccharide-producing microorganisms. The conditioning of agricultural soils with polysaccharides directly sprayed on the surface has been considered since the 1950s [342], but this was not practical as the result of the high costs and low persistence of the conditioners. Nevertheless, there is renewed interest in this subject [343—346]. However, application of polysaccharide solutions to soils is likely to be restricted to particular situations, such as irrigated soils [346—348] or slope stabilization [349].

More exciting fields of research involve the inoculation and management in soils of microbial populations to promote their aggregating activity. Polysaccharide-producing unicellular algae (e.g., *Chlamydomonas mexicana* and *C. sajao*) sprayed on soils in the temperate regions grew and improved the stability of the soil aggregates [350—352]. However, only surface stabilization was achieved: increases in aggregate stability and in polysaccharide contents occurred mainly in the top 2—3 mm of the soil [350,351,353]. There is a potential for soil conditioning with algae, although it appears to be restricted to irrigated farmlands, as algae are very sensitive to water stress. Species of *Chlamydomonas* reproduce well on soil between water potentials of 0.0 and −0.1 MPa [354]. However, the autotrophy of algae is a great advantage, as they do not need supplemental carbon, nor do they compete with indigenous heterotrophic microbes.

Another field of applied research is the management of bacterial populations to promote their production of polysaccharides and their aggregating activity. It is hypothesized that soil carbohydrate levels and aggregation can be controlled by short-term management practices that control N supply and the C/N ratio [Roberson, E. B., S. Sarig, and M. K. Firestone, 1989. Management of polysaccharide mediated aggregation in agricultural soils. Agron. Abst. p. 225]. The C/N ratio and water stress have been shown to control polysaccharide production by soil a pseudomonad [132].

Conclusion

Extracellular polysaccharides are emphasized as constituents of major importance in microbial functions in soils, as they form the interface between microbes and the soil constituents. Extracellular polysaccharides are involved in aggregation of soil particles and in the resistance of microbes to wetting and drying cycles and to high osmotic stress. They may also be involved, through their chelating properties, in the resistance to the toxicity of Al and heavy metal and in metal accumulation at the cell surface (e.g., Ca, Fe, Mn). Many of these properties are concerned with the various levels of structure of polysaccharides, from the primary structure (functional groups) to the tertiary and quaternary levels.

V. GENERAL CONCLUSION

Soil is a place of interactions between microorganisms and soil constituents and where microbial events are localized in microsystems. Physical soil factors are at least as important as chemical soil factors for microbial life. The presence of pores, their size and filling by water and air, determines mobility, competition between microorganisms, and type of dominant microbial function (e.g., aerobic versus anaerobic). Here, microhabitats derived from associations with clay particles are of great importance for soil microorganisms, especially bacteria, in soil.

Knowledge of soil organization and related physical properties allows some general predictions of microbial function. Among the

different levels of soil organization, the aggregate level represents
the main microsystem. Even if reality is more complex than a simple distinction between inner and outer positions relative to stable
aggregates, it is a convenient way to understand the distribution
of microorganisms and their activity.

Different methods exist to determine and study soil aggregation:
direct observation, fractionation, or an indirect approach through
volumetric moisture curves. These methods can also be used for
microorganisms, and good tracers exist, such as ATP or C and N
isotopes. Ideally, such methods should be used from the macro-
scale (macroaggregate) to the microscale, which represents the interface between clay, organic matter, or biopolymers (polysaccharides), and microbes.

Concerning chemical constraints, acidity, which is mainly related
to aluminum toxicity, was discussed as an example. Acidity is the
main chemical stress in many soils of the world, and its importance
is increasing in the remaining parts because of acid rains. However, in chemical stress, heavy metals as well as pesticides also
constitute major constraints. It is important to study further the
effect of all these factors on microbial life in soil and the remedies
that can be applied (e.g., liming, organic matter management, genetic adaptation of microbes and plants) especially for Al tolerance.

Considering soils as places of interactions permits making some
general statements. If environmental factors of soil are important
for microbial life, microbial life is, in return, a major factor in the
soil environment. Microbes are able to transform the physical properties of their microenvironment through aggregation and the chemical properties through weathering (i.e., microbes modify both the
structure and the geochemical cycles in soil). Although nutrients
were not discussed in this chapter, it is important to consider that
microbes can fix elements (N) or weather minerals to extract insoluble elements (P, K, Fe) if agriculture is to become more biological.

All these aspects need to be considered if new orientations have
to be devised for the future. The first way could be to enhance
the functions of microbes through in situ management of microbial
functions. Introduction of microbes into soils is increasing, primarily for rhizobia and mycorrhizae and for biological control of
plant diseases. However, inoculation may also be a way to enhance
plant growth through plant growth-promoting bacteria (PGR) and
to improve bioremediation of excess of nitrite, pesticides, and heavy
metals in soil. These aspects represent new trends of research for
the future and point out the necessity of better understanding the
microbial ecology of soils.

ACKNOWLEDGMENTS

The authors express their gratitude to Dr. G. Stotzky whose contribution greatly improved the manuscript. They thank also Drs. Jocteur Monrozier (Lyon), Alabouvette (Dijon), Schmit (Versailles) for their advice and T. Dimey for word processing of the manuscript, A. M. Jaunet for the figures, and J. A. Marie for reviewing the English.

REFERENCES

1. Stotzky, G. 1986. Influence of soil minerals on metabolic processes, growth, adhesion, and ecology of microbes and viruses, p. 305–428. *In* P. M. Huang and M. Schnitzer (eds.), Interactions of soil minerals with organics and microbes. SSSA Special Publication No. 17, Soil Science Society of America, Madison, Wisconsin.

2. Hattori, T. 1973. Microbial life in the soil. Marcel Dekker, New York.

3. Robert, M., M. Hardy, and F. Elsass. 1991. Crystallochemistry, properties and organization of soil clays derived from major sedimentary rocks in France. Clay Miner. (in press).

4. Righi, D., and P. Jadault. 1988. Improving soil clay mineral studies by high-gradient magnetic separation. Clay Miner. 23: 225–232.

5. Feller, C. 1979. Une méthode de fractionnement granulométrique de la matière organique des sols. Application aux sols tropicaux à textures grossières, très pauvres en humus. Cah. ORSTOM Ser. Pédol. 17:339–346.

6. Turchenek, L. W., and J. M. Oades. 1979. Fractionation of organo-mineral complexes by sedimentation and density techniques. Geoderma 21:311–343.

7. Balesdent, J., J. P. Petraud, and C. Feller. 1991. Effets des ultrasons sur la distribution granulométrique des matières organiques de sols. Sci. Sol (in press).

8. McDonald, R. M. 1986. Sampling soil microfloras: optimization of density gradient in Percoll to separate microorganisms from soil suspensions. Soil Biol. Biochem. 18:407–410.

9. Feller, C., G. Burtin, B. Gerard, and J. Balesdent. 1991. Utilisation des résines sodiques et des ultrasons sur le fractionnement granulométrique de la matière organique des sols. Intérêt et limite. Sci. Sol (in press).

10. Ramsay, A. J. 1984. Extraction of bacteria from soil: efficiency of shaking or ultrasonication as indicated by direct counts and autoradiography. Soil Biol. Biochem. 16:475–481.

11. Bailey, S. W., G. W. Brindley, W. D. Jones, R. T. Martin, and M. Ross. 1971. Summary of national and international recommendations on clay mineral nomenclature. Clays Clay Miner. 19:129–132.

12. Tessier, D. 1984. Etude expérimentale de l'organisation des matériaux argileux. Hydration, gonflement et structuration au cours de la dessiccation et de la réhumectation. Thèse de Doctorat d'Etat. Univ. Paris VII, publication INRA, 361 pp.

13. Tessier, D. 1991. Behavior and microstructure of clay minerals, p. 387–415. *In* M. F. De Boodt, M. H. B. Hayes, and A. Herbillon (eds.), Soil colloids and their associations in aggregates. NATO ASI Series, Serie B:Physics Vol. 215.

14. Tessier, D., and A.-M. Jaunet. 1987. Some applications of S.E.M.-T.E.M. to clay and soil microstructure research. *In* Proc. 24th Annual Microscopy Colloquium, 1987, Ames, Iowa.

15. Tessier, D., and G. Pédro. 1987. Mineralogical characterization of 2:1 clays in soils: importance of the clay texture, p. 78–84. *In* S. G. Schultz, H. van Olphen, and F. A. Mumpton (eds.), Proc. Int. Clay Conf. Denver, 1985. The Clay Minerals Society, Bloomington.

16. Nadeau, P. H., M. J. Wilson, W. J. McHardy, and J. M. Tait. 1984. Interstratified clays as fundamental particles. Science 225:923–925.

17. Quirk, J. P., and L. A. G. Aylmore. 1971. Domains and quasi-crystalline regions in clay systems. Soil Sci. Soc. Am. Proc. 35:652–654.

18. Ben Rhaiem, H., C. H. Pons, and D. Tessier. 1987. Factors affecting the microstructure of smectites, p. 292–297. *In* L. G. Schultz, H. Van Olphen, and F. Mumpton (eds.), Proc. of Int. Clay Conf. Denver, 1985.

19. Dixon, J. B. 1982. Mineralogy of vertisols, p. 48–59. *In* Vertisols and rice soils of the tropics. Symposia papers II, 12th ICSS New Delhi, India.

20. Wilson, M. J. 1987. Soil smectite and related interstratified minerals: recent developments, p. 167–173. *In* L. C. Schultz, H. van Olphen, and F. A. Mumpton (eds.), Proc. Int. Clay Conf. Denver, The Clay Minerals Society, Bloomington.

21. Delvaux, B., A. J. Herbillon, L. Vielvoye, and M. M. Mestdagh. 1990. Surface properties and clay mineralogy of hydrated halloysitic soil clays. II. Evidence for the presence of halloysite/smectite (H/Sm) mixed layer clays. Clay Miner. 25:141–160.

22. Jones, R. C., and G. Uehara. 1973. Amorphous coatings on mineral surfaces. Soil Sci. Soc. Am. Proc. 37:792–798.

23. Saleh, A. M., and A. A. Jones. 1984. The crystallinity and surface characteristics of synthetic ferrihydrite and its relationship to kaolinite surfaces. Clay Miner. 19:745–755.

24. Robert, M., G. Veneau, and M. Abreu. 1987. Etudes microscopiques d'associations aluminium-argiles ou fer-argiles, p. 467-474. *In* N. Fedoroff, L.-M. Bresson, and M.-A. Courty (eds.), Soil micromorphology. Int. Working Meeting on Soil Micromorphology, Paris, 1985.

25. Van Raij, B., and H. Peech. 1972. Electrochemical properties of some oxisols and alfisols of the tropics. Soil Sci. Soc. Am. Proc. 36:587–593.

26. Robert, M. 1986. Some general aspects of K dynamics and new trends in soil mineralogy, p. 1121–1132. *In* Trans. 13th Cong. ISSS, Hambourg, Vol. 6.

27. Rengasamy, P., and J. M. Oades. 1977. Interaction of monomeric and polymeric species of metal ions with clay surfaces. II Changes in surface properties after additions of FeIII. Aust. J. Soil Res. 15:235–242.

28. Robert, M., and M. Tercé. 1989. Role of clays and coatings on chemical properties, p. 57–71. *In* B. Bar Yosef, N. J. Barrow, and J. Goldschmid (eds.), Inorganic contaminants in the vadose zone. Ecological studies, vol. 74, Springer Verlag, New York.

29. Bruand, A., D. Tessier, and D. Baize. 1988. Contribution à l'étude des propriétés de rétention en eau des sols argileux: importance de la prise en compte de l'organisation de la phase argileuse. C.R. Acad. Sci [Paris], 307:1937–1941.

30. Mortland, ιM. M. 1970. Clay–organic complexes and interactions. Adv. Agron. 22:75–117.

31. Theng, B. K. G. 1982. Clay-polymer interactions: summary and perspectives. Clays Clay Miner. 30:1–10.

32. Berkeley, R. C. W., J. M. Lynch, J. Melling, P. R. Rutter, and B. Vincent. 1980. Microbial adhesion to surfaces. Ellis Horwood, Chichester.

33. International Soil Science Society. 1976. Soil physics terminology. ISSS Bull. 48:16–22.

34. Hattori, T. 1967. Microorganisms and soil aggregates as their microhabitat. Bull. Inst. Agric. Res. Tohoku Univ. 159–193.

35. Postma, J., J. A. Van Veen, and S. Walter. 1989. Influence of different initial moisture on the distribution and population dynamics of introduced *Rhizobium leguminosarum* biovar *trifolii*. Soil Biol. Biochem. 21:437–442.

36. Harris, R. F., W. R. Gardner, A. A. Adebayo, and L. E. Sommers. 1970. Agar dish isopiestic equilibration method for controlling the water potential of solid substrates. Appl. Microbiol. 19:536–537.

37. Mexal, J., J. T. Fisher, J. Osteryoung, and C. P. P. Reid. 1975. Oxygen availability in polyethylene glycol solutions

and its implications in plant—water relations. Plant Physiol.
55:20—24.

38. Richards, L. A. 1947. Pressure membrane apparatus—construction and use. Soil Sci. 51:377—386.

39. Papendick, R. I., and G. S. Campbell. 1981. Theory and
measurement of water potential, p. 1—22. *In* Water potential
relations in soil microbiology. SSSA Special Publication No.
9, Soil Science Society of America, Madison, Wisconsin.

40. Tessier, D., and J. Berrier. 1979. Utilisation de la micro-
scopie électronique à balayage dans l'étude des sols. Observation de sols soumis à différents pF. Sci. Sol 1:67—82.

41. Monnier, G., P. Stengel, and J.-C. Fies. 1973. Une méthode
de mesure de la densité apparente de petits agglomérats ter-
reux. Application à l'analyse des systèmes de porosité du sol.
Ann. Agron. 24:533—545.

42. Stevenson, I. L., and M. Schnitzer. 1982. Transmission electron microscopy of extracted fulvic and humic acids. Soil Sci.
133:179—185.

43. Oades, J. M. 1989. An introduction to organic matter in min-
eral soils, p. 89—160. *In* J. B. Dixon and S. B. Weed (eds.),
Minerals in soil environments, 2nd ed. Soil Science Society of
America, Madison, Wisconsin.

44. Martin, J. P., W. P. Martin, J. B. Page, W. A. Raney, and
J. D. De Ment. 1955. Soil aggregation. Adv. Agron. 7:
1—37.

45. Callot, G., and B. Jaillard. 1987. Apports de la loupe binocu-
laire à l'étude des interfaces sol/racine et sol/champignon, p.
73—80. *In* N. Fedoroff, L.-M. Bresson, and M. A. Courty
(eds.), Soil micromorphology. Proc. 7th Int. Working Meeting
on Soil Micromorphology, Paris, 1985.

46. Campbell, R., and R. Porter. 1982. Low temperature and
scanning electron microscopy of microorganisms in soil. Soil
Biol. Biochem. 14:241—245.

47. Dorioz, J.-M., and M. Robert. 1987. Aspects microscopiques
des relations entre microorganismes ou végétaux et les argiles.
Conséquence sur la microorganisation et la microstructuration
des sols, p. 353—361. *In* N. Fedoroff, L.-M. Bresson, and
M.-A. Courty (ed.), Soil micromorphology. Proc. 7th Int.
Working Meeting on Soil Micromorphology, Paris, 1985.

48. Foster, R. C. 1981. Localization of organic materials in situ
in ultrathin sections of natural soil fabrics using cytochemical
techniques, p. 309—319. *In* E. B. A. Bisdom (ed.), Int.
Working group on Submicroscopy, Wageningen, The Netherlands.

49. Altemüller, H. J., and B. van Vliet-Lanoe. 1990. Soil thin
sections fluorescence miscroscopy. *In* L. A. Douglas (ed.),

Soil micromorphology: a basic and applied science. Dev. Soil Sci. 19:565—580.

50. Postma, J., and H. J. Altemüller. 1990. Bacteria in thin soil sections stained with the fluorescent brightener calcofluor white M2R. Soil Biol. Biochem. 22:89—96.

51. Jongerius, A., D. Schoonderbeek, A. Jager, and S. T. Kowalinski. 1972. Electrooptical soil porosity investigation by means of Quantimet—B equipment. Geoderma 7:177—198.

52. Murphy, C. P., P. Bullock, and R. H. Turner. 1977. The measurement and characterization of voids in soil thin sections by image analysis. I. Principles and techniques. J. Soil Sci. 28:498—508.

53. Bisdom, E. B. A. 1981. Submicroscopy of soils and weathered rocks. *In* E. B. A. Bisdom (ed.), Pudoc. Wageningen, The Netherlands.

54. Darbyshire, J. F., L. Robertson, and L. A. Mackie. 1985. A comparison of two methods of estimating the soil pore network available to protozoa. Soil Biol. Biochem. 17:619—624.

55. Scher, F. M., and R. Baker. 1983. Fluorescent microscopic technique for viewing fungi in soil and its application to studies of a *Fusarium* suppressive soil. Soil Biol. Biochem. 15:715—718.

56. Chauvel, A. 1977. Recherches sur la transformation des sols ferrallitiques de Casamance (Sénégal). Trav. Doc. ORSTOM 62:532 p.

57. Cambier, P., and R. Prost. 1981. Etude des associations argile-oxyde: organisation des constituants d'un matériau ferrallitique. Agronomie 1:713—722.

58. Tisdall, J. M., and J. M. Oades. 1982. Organic matter and water stable aggregates. J. Soil Sci. 33:141—163.

59. Jocteur Monrozier, L., J. N. Ladd, R. W. Fitzpatrick, R. W. Foster, and M. Raupach. 1991. Physical properties, mineral and organic components and microbial biomass content of size fraction in soils of contrasting aggregation. Geoderma (in press).

60. Elliott, E. T. 1986. Aggregate structure and carbon, nitrogen and phosphorus in native and cultivated soils. Soil Sci. Soc. Am. J. 50:627—633.

61. Yoder, R. E. 1936. A direct method of aggregate analysis of soils and a study of the physical nature of erosion losses. J. Am. Soc. Agron. 28:337—351.

62. Emerson, W. W. 1954. The determination of the stability of soil crumbs. J. Soil Sci. 5:233—250.

63. Emerson, W. W. 1967. A classification of soil aggregates based on their coherence in water. Aust. J. Soil Res. 5:47—57.

64. Gregorich, E. G., R. G. Kachanoski, and R. P. Voroney. 1988. Ultrasonic dispersion of aggregates: distribution of organic matter in size fractions. Can. J. Soil Sci. 68:395–403.

65. Ahmed, M., and J. M. Oades. 1984. Distribution of organic matter and adenosine triphosphate after fractionation of soil by physical procedures. Soil Biol. Biochem. 16:465–470.

66. Ozawa, T., and M. Yamaguchi. 1986. Fractionation and estimation of particle-attached and unattached *Bradyrhizobium japonicum* strains in soils. Appl. Env. Microbiol. 52:911–914.

67. Chretien, J., and D. Tessier. 1988. Influence du squelette sur les propriétés physiques des sols: hydratation, gonflement et aération. Sci. Sol. 26:255–268.

68. Bruand, A., and R. Prost. 1987. Effect of water content on the fabric of a soil material: an experimental approach. J. Soil Sci. 38:461–472.

69. Linn, D. M., and J. W. Doran. 1984. Aerobic and anaerobic microbial populations in no-till and plowed soils. Soil Sci. Soc. Am. J. 48:794–799.

70. Linn, D. M., and J. W. Doran. 1984. Effect of water-filled pore space on carbon dioxide and nitrous oxide production in tilled and non-tilled soils. Soil Sci. Soc. Am. J. 48:1267–1272.

71. Doran, J. W., L. N. Mielke, and J. F. Power. 1990. Microbial activity as regulated by soil water-filled pore space, p. 94–99. Trans. 14 Int. Congr. of Soil Science, Kyoto, Japan, vol. 3.

72. Selino, D., J. Proth, S. Bruckert, and G. Kilbertus. 1978. Sols acides structurés en agrégats: analyse d'un mode d'action d'origine biologique, p. 209–225. *In* 103 Congrès National des Sociétés Savantes, Nancy, 1978, Sciences, fasc. IV.

73. Schmit, J., P. Prior, H. Quiquampoix, and M. Robert. 1990. Studies on survival and localization of *Pseudomonas solanacearum* in clays extracted from vertisols, p. 1001–1009. *In* Z. Klement (ed.), Plant pathogenic bacteria. Proc. of 7th Int. Conf. on plant pathogenic bacteria, Budapest. Akademiai Kaido.

74. Camacho, E., M. Robert, and A.-M. Jaunet. 1990. Mineralogy and structural organization of a red to yellow soil sequence in Cuba—relationships with soil properties. *In* L. A. Douglas (ed.), Soil micromorphology. Int. Working Meeting on soil micromorphology. San Antonio, Texas, 1988. Dev. Soil Sci. 19:183–190.

75. Rosello, V. 1984. Les sols bruns des hauts (Ile de la Réunion). Thèse Univ. Paris VII, 200 p.

76. Huang, P. M. 1988. Ionic factors affecting aluminum transformation and the impact on soil and environmental sciences. Adv. Soil Sci. 8:1–78.

77. Marshall, K. C. 1964. Survival of root nodule bacteria in dry soils exposed to high temperature. Aust. J. Agric. Res. 15:273–281.

78. Marshall, K. C. 1975. Clay mineralogy in relation to survival of soil bacteria. Annu. Rev. Phytopathol. 13:357–373.

79. Stotzky, G. 1966. Influence of clay minerals on microorganisms: II. Effect of various clay species, homoionic clays, and other particles on bacteria. Can. J. Microbiol. 12:831–848.

80. Stotzky, G. Influence of clay minerals on microorganisms: III. Effect of particle size, cation exchange capacity, and surface area on bacteria. Can. J. Microbiol. 12:1235–1246.

81. Stotzky, G., and L. T. Rem. 1966. Influence of clay minerals on microorganisms. I. Can. J. Microbiol. 12:547–563.

82. Stotzky, G. 1980. Surface interactions between clay minerals and microbes, viruses and soluble organics, and the probable importance of these interactions to the ecology of microbes in soil, p. 231–249. *In* R. C. W. Berkeley, J. M. Lynch, J. Melling, P. R. Rutter, and B. Vincent (eds.), Microbial adhesion to surfaces. Ellis Horwood, Chichester.

83. Griffin, D. M. 1981. Water and microbial stress. Adv. Microb. Ecol. 5:91–136.

84. Griffin, D. M. 1981. Water potential as a selective factor in microbial ecology of soils, p. 141–151. *In* J. F. Parr, W. R. Gardner, and L. F. Elliott (eds.), Water potential relations in soil microbiology. SSSA Special Publication No. 9, Soil Science Society of America, Madison, Wisconsin.

85. Rhoades, J. D. 1972. Quality of water for irrigation. Soil Sci. 113:277–284.

86. Adebayo, A. A., and R. F. Harris. 1971. Fungal growth responses to osmotic as compared to matric water potential. Soil Sci. Soc. Am. Proc. 35:465–469.

87. Cook, R. J., R. I. Papendrick, and D. M. Griffin. 1972. Growth of two root-rot fungi as affected by osmotic and matric water potentials. Soil Sci. Soc. Am. Proc. 36:78–82.

88. Brown, A. D. 1976. Microbial water stress. Bacteriol. Rev. 40:803–846.

89. Seifert, T. J. 1964. Influence of the size of soil structural aggregates on the degree of nitrification. II. The role of aeration. Folia Microbiol. 9:363–377.

90. Williams, S. T., M. Shameemullah, E. T. Watson, and C. I. Mayfield. 1972. Studies on the ecology of actinomycetes in soil. VI. The influence of moisture tension and growth and survival. Soil Biol. Biochem. 4:215–225.

91. Miller, R. D., and D. D. Johnson. 1964. Effect of soil moisture tension on carbon dioxide evolution, nitrification, and nitrogen mineralization. Soil Sci. Soc. Am. Proc. 28:644–647.

92. Stanford, G., and E. Epstein. 1974. Nitrogen mineraliza-
    tion—water relations in soils. Soil Sci. Soc. Am. Proc. 38:
    103—107.

93. Sommers, L. E., C. M. Gilmour, R. E. Wildung, and S. M.
    Beck. 1981. The effect of water potential on decomposition
    processes in soils, p. 97—117. *In* Water potential relations in
    soil microbiology. SSSA Special Publication No. 9, Soil Sci-
    ence Society of America, Madison, Wisconsin.

94. Dommergues, Y. R. 1962. Contribution à la dynamique mi-
    crobienne des sols en zone semi-aride et en zone tropicale
    sèche. Ann. Agron. 4:265—324; 5:391—468.

95. Griffin, D. M. 1969. Soil water and the ecology of fungi.
    Annu. Rev. Phytopathol. 7:289—310.

96. Baker, K. F., and R. J. Cook. 1974. Biological control of
    plant pathogens. Freeman, San Francisco.

97. Harris, R. F. 1981. Effect of water potential on microbial
    growth and activity, p. 23—95. *In* J. F. Parr, W. R. Gard-
    ner, and L. F. Elliott (eds.), Water potential relations in
    soil microbiology. Soil Science Society of America, Madison,
    Wisconsin.

98. Measures, J. C. 1975. Role of amino acids in osmoregulation
    of non-halophilic bacteria. Nature 257:398—400.

99. Csonka, L. N. 1989. Physiological and genetic responses to
    osmotic stress. Microbiol. Rev. 53:121—147.

100. Le Rudulier, D., A. R. Strom, A. M. Dandekar, L. T. Smith,
     and R. C. Valentine. 1984. Molecular biology of osmoregula-
     tion. Science 224:1064—1068.

101. Hua, S. T., V. Y. Tsai, G. M. Lichens, and A. T. Noma.
     1982. Accumulation of amino acids in *Rhizobium* sp. strain
     WR1001 in response to sodium chloride salinity. Appl. Env.
     Microbiol. 43:135—140.

102. Tempest, D. W., and J. L. Meers. 1970. Influence of en-
     vironment on the content and composition of free amino acids
     pools. J. Gen. Microbiol. 64:171—185.

103. Killham, K., and M. Firestone. 1984. Salt stress control of
     intracellular solutes in streptomycetes indigenous to saline
     soils. Appl. Env. Microbiol. 47:301—306.

104. Brown, E. J. 1978. Compatible solutes and extreme water
     stress in eucaryotic microorganisms. Adv. Microb. Physiol.
     17:181—242.

105. Smith, D. C. 1979. Is a lichen a good model of biological
     interactions in nutrient-limited environments, p. 291—303.
     In M. Shilo (ed.), Strategies of microbial life in extreme en-
     vironments. Verlag Chemie, Berlin.

106. Luard, E. J. 1982. Accumulation of intracellular solutes by
     two filamentous fungi in response to growth at low steady
     state osmotic potential. J. Gen. Microbial. 128:2563—2574.

107. Luard, E. J. 1982. Effect of osmotic shock on some intracellular solutes in two filamentous fungi. J. Gen. Microbiol. 128:2575–2581.

108. Ernst, A., T. W. Chen, and P. Böger. 1987. Carbohydrate formation in rewetted terrestrial cyanobacteria. Oecologia [Berlin] 72:574–576.

109. Reed, R. H., and W. D. P. Stewart. 1983. Physiological response of *Rivularia atra* to salinity: osmotic adjustment in hypolsaline media. New Phytol. 95:595–603.

110. Blumwald, E., and E. Tel-Or. 1982. Osmoregulation and cell composition in cell adaptation of *Nostoc muscorum*. Arch. Microbiol. 132:168–172.

111. McAneney, K. J., R. F. Harris, and W. R. Gardner. 1982. Bacterial water relations using polyethylene glycol 4000. Soil Sci. Soc. Am. J. 46:542–547.

112. Busse, M. D., and P. J. Bottomley. 1989. Growth and nodulation responses of *Rhizobium meliloti* to water stress induced by permeating and nonpermeating solutes. Appl. Env. Microbiol. 55:2431–2436.

113. Kieft, T. L., E. Soroker, and M. K. Firestone. 1987. Microbial biomass response to rapid increase in water potential when dry soil is wetted. Soil Biol. Biochem. 19:119–126.

114. Salema, M. P., C. A. Parker, D. K. Kigby, D. L. Chatel, and T. M. Armitage. 1982. Rupture of nodule bacteria on drying and rehydration. Soil Biol. Biochem. 14:15–22.

115. Cook, R. J., and J. M. Duniway. 1981. Water relations in the life cycles of soil-borne plant pathogens. *In* J. F. Parr, W. R. Gardner, and L. F. Elliott (eds.), Water potentials relations in soil microbiology. SSSA Special Publication No. 9, Soil Science Society of America, Madison, Wisconsin.

116. Duniway, J. M. 1979. Water relations of water molds. Annu. Rev. Phytopathol. 17:431–460.

117. Hepper, C. 1975. Extracellular polysaccharides of soil bacteria. *In* N. Walker (ed.), Soil microbiology. John Wiley & Sons, New York.

118. Dudman, W. F. 1977. The role of surface polysaccharides in natural environments, p. 357–414. *In* I. Sutherland (ed.), Surface carbohydrates of the procaryotic cell. Academic Press, New York.

119. Wilkinson, J. F. 1958. The extracellular polysaccharides of bacteria. Bacteriol. Rev. 22:46–73.

120. Bayer, M. E. 1967. Response of cell walls of *Escherichia coli* to a sudden reduction of the environmental osmotic pressure. J. Bacteriol. XX:1104–1112.

121. Kilbertus, G., J. Proth, and F. Mangenot. 1977. Sur la répartition et la survivance des microorganismes du sol. Bull. Soc. Acad. Lorr. Sci. 16:93–104.

122. Foster, R. C. 1988. Microenvironments of soil microorgan-
     isms. Biol. Fertil. Soils 6:189–203.
123. Kilbertus, G., J. Proth, and B. Verdier. 1979. Effets de
     la dessiccation sur les bactéries gram-négatives d'un sol.
     Soil Biol. Biochem. 11:109–114.
124. Chenu, C. 1989. Influence of a fungal polysaccharide,
     scleroglucan, on clay microstructures. Soil Biol. Biochem.
     21:299–305.
125. Pena-Cabriales, J. J., and M. Alexander. 1979. Survival
     of *Rhizobium* in soils undergoing drying. Soil Sci. Soc. Am.
     J. 43:962–966.
126. Bitton, G., Y. Henis, and N. Lahav. 1976. Influence of
     clay minerals, humic acid and bacterial polysaccharide on the
     survival of *Klebsiella aerogenes* exposed to drying and heat-
     ing in soils. Plant Soil 45:65–74.
127. Schmit, J., and M. Robert. 1984. Action des argiles sur la
     survie d'une bactérie phytopathogene *Pseudomonas solanacear-
     um* E. F. S. C. R. Acad. Sci. [Paris] 299:733–738.
128. Bushby, H. V. A., and K. C. Marshall. 1977. Some fac-
     tors affecting the survival of root-nodule bacteria on desic-
     cation. Soil Biol. Biochem. 9:143–147.
129. Osa-Afiana, L. O., and M. Alexander. 1982. Differences
     among cowpea rhizobia in tolerance to high temperatures and
     desiccation in soil. Appl. Environ. Microbiol. 53:435–439.
130. Chao, W. L. 1983. Survival of *Rhizobium* in soils undergoing
     drying and the use of these soils as inoculant carriers. PhD
     thesis, Cornell Univ., Ithaca, NY Diss. Abstr. Int. 43–3852B.
131. Hartel, P. G., and M. Alexander. 1986. Role of extracellu-
     lar polysaccharide production and clays in the desiccation tol-
     erance of cowpea *Bradyrhizobia*. Soil Sci. Soc. Am. J. 50:
     1193–1198.
132. Roberson, E. B., and M. K. Firestone. 1990. Environmental
     control of exopolysaccharide production by soil bacteria. Amer-
     ican Society for Microbiology Meetings, Anaheim, May 1990.
133. Mugnier, J., and G. Jung. 1985. Survival of bacteria and
     fungi in relation to water activity and the solvent properties
     of water in biopolymer gels. Appl. Environ. Microbiol. 50:
     108–114.
134. Bashan, Y. 1986. Alginate beads as synthetic inoculant
     carrier for slow release of bacteria that affect plant growth.
     Appl. Environ. Microbiol. 51:1089–1098.
135. Lewis, J. A., and G. C. Papavizas. 1987. Application of
     Trichoderma and *Gliocladium* in alginate pellets for control
     Rhizoctonia damping-off. Plant Pathol. 36:438–446.
136. Kloepper, R. J., and M. N. Schroth. 1981. Development of
     a formulation of rhizobacteria for inoculation of potato seed
     pieces. Phytopathology 71:590–592.

137. Bauer, L. D., W. H. Gardner, and W. R. Gardner. 1972. Soil physics. John Wiley & Sons, New York.

138. Letey, J., W. A. Jury, A. Hadas, and N. Valoras. 1980. Gas diffusion as a factor in laboratory incubation studies on denitficiation. J. Environ. Qual. 9:223–226.

139. Yashida, T. 1975. Microbial metabolism of flooded soils. Soil Biochem. 3:83–122.

140. Aulakh, M. S., D. A. Rennie, and E. A. Paul. 1982. Gaseous nitrogen losses from cropped and summer-fallowed soils. Can. J. Soil Sci. 62:187–196.

141. Grundmann, G. L., and D. E. Rolston. 1987. A water function approximation to degree of anaerobiosis associated with denitrification. Soil Sci. 144:437–441.

142. Osa-Afiana, L. O., and M. Alexnader. 1982. Clays and the survival of *Rhizobium* in soil during desiccation. Soil Sci. Soc. Am. J. 46:285–288.

143. Al-Rashidi, R. K. Loynachan, and L. R. Frederick. 1982. Desiccation tolerance of four strains of *Rhizobium japonicum*. Soil Biol. Biochem. 14:489–493.

144. Jansen van Rensburg, H., and B. W. Stridjom. 1980. Survival of fast and slow-growing *Rhizobium* spp. under conditions of relatively mild desiccation. Soil Biol. Biochem. 12:353–356.

145. Hartel, P. G., and M. Alexander 1984. Temperature and desiccation tolerance of cowpea rhizobia. Can. J. Microbiol. 30:820–823.

146. Chao, W. L., and M. Alexander. 1984. Mineral soils as carriers for *Rhizobium* inoculants. Appl. Env. Microbiol. 47:94–97.

147. Dupler, M., and R. Baker. 1984. Survival of *Pseudomonas putida*, a biological control agent, in soil. Phytopathology 74:195–200.

148. Zechman, J. M., and L. E. Casida. 1982. Death of *Pseudomonas aeruginosa* in soil. Can. J. Microbiol. 28:788–794.

149. Bushby, H. V. A., and K. C. Marshall. 1977. Water status of rhizobia in relation to their susceptibility to desiccation and to their protection by montmorillonite. J. Gen. Microbiol. 99:19–27.

150. Kilbertus, G. 1980. Etude des microhabitats contenus dans les agrégats du sol relation avec la biomasse bactérienne et la taille des protocaryotes présents. Rev. Ecol. Biol. Sol 17:543–557.

151. Ganry, F., H. G. Diem, J. Wey, and Y. R. Dommergues. 1985. Inoculation with *Glomus mosseae* improves N_2 fixation by field-grown soybeans. Biol. Fertil. Soils 1:15–24.

152. Fargues, J., O. Reisinger, P. H. Robert, and C. Aubart. 1983. Biodegradation of entomopathogenic hyphomycetes:

influence of clay coating on *Beauveria bassiana* blastospore survival in soil. J. Invertebr. Pathol. 41:131–142.

153. Fravel, D. R., J. J. Marois, R. D. Lumdsen, and W. J. Connick, 1985. Encapsulation potential biocontrol agents in an alginate–clay matrix. Phytopathology 75:774–777.

154. Béreau, M., and C. M. Messian. 1975. Réceptivités comparées des sols à l'infestation par *Pseudomonas solanacearum*. Ann. Phytopathol. 7:191–193.

155. Stotzky, G., and R. T. Martin. 1963. Soil mineralogy in relation to the spread of fusarium wilt of banana in Central America. Plant Soil 18:317–337.

156. Alabouvette, C., F. Rouxel, and J. Louvet. 1979. Characteristics of fusarium wilt suppressive soils and prospect for their utilisation in biological control, p. 165–182. *In* B. Schippers and W. Gams (eds.), Soil borne plant pathogens. Academic Press, London.

157. Stutz, E., G. Kahr, and G. Defago. 1989. Clays involved in suppression of tobacco black root rot by a strain of *Pseudomonas fluorescens*. Soil Biol. Biochem. 21:361–366.

158. Wong, P. T. W., and D. M. Griffin. 1976. Bacterial movement at high matric potentials. I. In artificial and natural soils. Soil Biol. Biochem. 8:215–218.

159. Van Elsas, J. D., A. F. Dijkstra, J. M. Govaert, and J. A. Van Veen. 1986. Survival of *Pseudomonas fluorescens* and *Bacillus subtilis* introduced into two soils of different texture in field microplots. FEMS Microbiol. Ecol. 38:151–160.

160. Wong, P. T. W., and D. M. Griffin. 1976. Bacterial movement at high matric potentials. II. In fungal colonies. Soil Biol. Biochem. 8:219–223.

161. Hattori, T. 1988. Soil aggregates as microhabitats for microorganisms. Rep. Inst. Agric. Res. Tohoku Univ. 37:23–26.

162. Hattori, T., and R. Hattori. 1976. The physical environment in soil microbiology: an attempt to extend principles of microbiology to soil microorganisms. Crit. Rev. Microb. 4:423–461.

163. Klein, A. D., and J. S. Thayer. 1990. Interactions between soil microbial communities and organometallic compounds, p. 431–481. *In* J. M. Bollag and G. Stotzky (eds.), Soil Biochemistry, Vol. 6. Marcel Dekker, New York.

164. Sexstone, A. J., N. P. Reusbech, T. N. Parkin, and J. M. Tiedje. 1985. Direct measurement of oxygen profiles and denitrification rates in soil aggregates. Soil Sci. Soc. Am. J. 49:645–651.

165. Jaillard, B., and Y. M. Cabidoche. 1984. Etude de la dynamique de l'eau dans un sol argileux gonflant: dynamique hydrique. Sci. Sol 3:239–251.

166. Heijnen, C. E., and J. A. Van Veen. 1991. A determination of protective microhabitats for bacteria introduced to soils. FEMS Microbiol. Ecol. (in press).

167. Heijnen, C. E., J. Postma, and V. A. Van Veen. 1990. The significance of artificially formed and originally present protective microniches for the survival of introduced bacteria in soil. Proc. Int. Soil Science Conference, Kyoto, August 1990. III, 88–93.

168. Heijnen, C. E., J. D. van Elsas, P. J. Kuikman, and J. A. van Veen. 1988. Dynamics of *Rhizobium leguminosarum* biovar *trifolii* introduced into soil; the effect of bentonite clay on predation by protozoa. Soil Biol. Biochem. 20:483–488.

169. Vargas, R., and T. Hattori. 1986. Protozoan predation of bacterial cells in soil aggregates. FEMS Microbiol. Ecol. 38:233–242.

170. Vargas, R., and T. Hattori, 1990. The distribution of protozoa among soil aggregates. FEMS Microbiol. Ecol. 74:73–78.

171. Elliott, E, T., R. V. Anderson, D. C. Coleman, and C. V. Cole. 1980. Habitable pore space and microbial trophic interactions. Oikos 35:327–335.

172. Coûteaux, M. M., G. Faurie, L. Palka, and C. Steinberg. 1988. La relation prédateur proie (protozoaires-bactéries) dans les sols: rôle dans la régulation des populations et conséquences sur les cycles du carbone et de l'azote. Rev. Ecol. Biol. Sol 25:1–31.

173. Mehra, O. P., and M. L. Jackson. 1960. Iron oxide removal from soils and clays by a dithionite-citrate system buffered with sodium bicarbonate. Clays Clay Miner. 7:317–327.

174. Schwertmann, U. 1964. Differenzierung des Eisenoxides des Bodens durch extraction mit Ammonium oxalate lösung. Z. Pflanzenernahr 105:194–202.

175. McKeague, J. A., and J. H. Day. 1966. Dithionite and oxalate extractable Fe and Al as acids in differentiating various classes of soils. Can. J. Soil Sci. 46:13–22.

176. Lindsay, W. B. 1988. Solubility and redox equilibria of iron compounds in soils. *In* J. W. Stucki, B. A. Goodman, and U. Schwertmann (eds.), Iron in soil and clay minerals. NATO ASI Ser. 217:37–62.

177. Loeppert, R. H. 1988. Chemistry of iron in calcerous systems. *In* J. W. Stucki, B. A. Goodman, and U. Schwertmann (eds.), Iron in soil and clay minerals. NATO ASI Ser. 217:689–714.

178. Geiger, S. C., and R. H. Loeppert. 1986. Correlation of DTPA extractable Fe with indigenous properties of selected calcareous soils. J. Plant Nutr. 9:229–240.

179. Alabouvette, C. 1990. Biological control of fusarium wilt pathogens in suppressive soils, p. 27–43. *In* D. Hornley (ed.), Biological control of soil borne plant pathogens. CAB International.

180. Kloeper, J. W., J. Leong, M. Teintze, and M. N. Schnoth. 1980. Enhanced plant growth by siderophores produced by plant growth-promoting rhizobacteria. Nature 286:885–886.

181. Scher, F. M. 1986. Biological control of fusarium wilts by *Pseudomonas putida* and its enhancement by EDDHA. *In* T. R. Swinburne (ed.), Iron, siderophores, plant diseases. NATO ASI Ser. A, 117:109–117.

182. Scher, F. M., M. Dupler, and R. Baker. 1984. Effect of synthetic iron chelates on population densities of *Fusarium oxysporum* and the biological control agent *Pseudomonas putida* in soil. Can. J. Microbiol. 30:1271–1275.

183. Alabouvette, A. 1986. Fusarium wilt suppressive soils from the Châteaurenard region: review of a 10 year study. Agronomie 6:273–284.

184. Lemanceau, P. 1988. Réceptivité des sols aux fusarioses vasculaires. Etude critique des théories proposées. Thèse Univ. Claude Bernard, Lyon I, 99 pp.

185. Tientze, M., M. B. Hossain, C. L. Barnes, J. Leong, and D. van der Helm. 1981. Structure of ferric pseudobactin, a siderophore from a plant growth promoting *Pseudomonas*. Biochemistry 20:6446–6457.

186. Meyer, J. M., F. Halle, D. Hohnadel, P. Lemanceau, and H. Ratefidarivelo. 1987. Siderophores of *Pseudomonas*—biological properties, p. 198–205. *In* G. Winkelmann, D. van der Helm, and J. B. Nielands (eds.), Iron transport in microbes, plants and animals. VCH, Weinheim.

187. Schroth, M. N., and J. G. Hancock, 1982. Disease-suppressive soil and root-colonizing bacteria. Science 216:1376–1381.

188. Lavie, S., and G. Stotzky. 1986. Interactions between clay minerals and siderophores affect the respiration of *Histoplasma capsulatum*. Appl. Environ. Microbiol. 51:74–79.

189. Wright, R. J. 1989. Soil aluminum toxicity and plant growth. Commun. Soil Sci. Plant Anal. 20:1479–1497.

190. Parker, D. R., L. W. Zelazny, and T. B. Kinraide. 1988. Comparison of three spectrophotometric methods for differentiating mono and polynuclear hydroxy-aluminum complexes. Soil Sci. Soc. Am. J. 52:67–75.

191. Parker, D. R., T. B. Kinraide, and L. W. Zelazny. 1988. Aluminum speciation and phytotoxicity in dilute hydroxy-aluminum solutions. Soil Sci. Soc. Am. J. 52:438–444.

192. Driscoll, C. T. 1989. The chemistry of aluminum in sur-
     face waters, p. 241–247. *In* G. Sposito (ed.), The en-
     vironmental chemistry of aluminum. CRC Press, Boca Raton,
     Florida.

193. Rouiller, J., B. Guillet, and S. Bruckert. 1980. Cations
     acides échangeables et acidités de surface. Sci. Sol 2:
     161–175.

194. Tamura, T. 1957. Identification of the 14 A clay mineral
     component. Am. Miner. 42:107–110.

195. Firestone, M. K., K. Killham, and J. G. McColl. 1983. Fun-
     gal toxicity of mobilized soil aluminum and manganese. Appl.
     Environ. Microbiol. 46:758–761.

196. Thompson, G. W., and R. J. Medve. 1984. Effects of alu-
     minum and manganese on the growth of ectomycorrhizal fungi.
     Appl. Environ. Microbiol. 48:556–560.

197. Ko, W. H., and F. K. Hora. 1971. Fungitoxicity in certain
     Hawaiian soils. Soil Sci. 112:276–279.

198. Kobayashi, N., and W. H. Ko. 1985. Nature of suppression
     of *Rhizoctonia solani* in Hawaiian soils. Trans. Br. Mycol.
     Soc. 84:691–694.

199. Ko, W. H., and K. A. Nishijima. 1985. Nature of suppres-
     sion of *Phytophthora capsici* in a Hawaiian soil. Phytopath-
     ology 75:683–685.

200. Orellana, R. G., C. D. Foy, and A. L. Fleming. 1975. Ef-
     fect of soluble aluminum on growth and pathogenicity of *Ver-
     ticillium albo-atrum* and *Whetzelinia sclertiorum* from sunflower.
     Phytopathology 65:202–205.

201. Tivoli, B., R. Corbière, and E. Lemarchand. 1989. Rela-
     tion entre le pH des sols et leur niveau de réceptivité à *Fu-
     sarium solani* var. *coeruleum* et *F. roseum* var. *sambucinum*
     agents de la pourriture sèche des tubercules de pomme de
     terre. Agronom. 10:63–68.

202. Ridao, A. C. 1990. La réceptivité des sols aux fusarioses
     de la pomme de terre: mécanismes de résistance à *Fusarium
     solani* var. *coeruleum*. Thèse Univ. Rennes I.

203. Lourd, M., and D. Bouhot. 1987. Recherche et caractérisa-
     tion de sols résistants aux *Pythium* spp. en Amazonie brésil-
     ienne. Bull. OEPP/EPPO 17:569–575.

204. Zwarun, A. A., G. W. Bloomfield, and G. W. Thomas. 1971.
     Effect of soluble and exchangeable aluminum on a soil bacil-
     lus. Soil Sci. Am. Proc. 35:460–463.

205. Zwarun, A. A., and G. W. Thomas. 1973. Effect of soluble
     and exchangeable aluminum on *Pseudomonas stuzeri*. Soil Sci.
     Soc. Am. Proc. 37:386–387.

206. Munns, D. N. 1984. Acid soil tolerance in legumes and rhi-
     zobia. Adv. Plant Nutr. 63–91.

207. Munns, D. N., and H. H. Keyser. 1981. Response of *Rhizobium* strains to acid and Al stress. Soil Biol. Biochem. 13: 115–118.

208. Wood, M., and J. E. Cooper. 1988. Acidity, aluminium and multiplication of *Rhizobium trifolii*: possible mechanisms of aluminum toxicity. Soil Biol. Biochem. 29:95–99.

209. Cunningham, S. D., and D. N. Munns. 1984. Effect of rhizobial extracellular polysaccharide on pH and aluminum activity. Soil Sci. Soc. Am. J. 48:1276–1280.

210. Cunningham, S. D., and D. N. Munns. 1984. The correlation between extracellular polysaccharide production and acid tolerance in *Rhizobium*. Soil Sci. Soc. Am. J. 48:1273–1276.

211. Crozat, Y., J.-C. Cleyet-Marel, J.-J. Giraud, and M. Obaton. 1982. Survival rates of *Rhizobium japonicum* populations introduced into different soils. Soil Biol. Biochem. 14:401–427.

212. Fu, M. H., X. C. Xu, and M. A. Tabatabai. 1987. Effect of pH on nitrogen mineralization in crop-residue treated soils. Biol. Fertil. Soils 5:115–119.

213. Francis, A. J. 1982. Effects of acidic precipitation and acidity on soil microbial processes. Water Air Soil Pollut. 18:375–394.

214. Tabatabai, M. A. 1985. Effect of acid rain on soils. Crit. Rev. Environ. Control 15:65–110.

215. Berthelin, J., M. Bonne, G. Belgy, and F.-X. Wedrago. 1985. Major role of nitrification in the weathering of minerals of brown acid forest soils. Geomicrobiol. J. 4:175–190.

216. Simon-Sylvestre, G., M. Robert, G. Veneau, and A. Beaumont. 1990. Experimental nitrification related to acidification and silicate weathering. *In* J. Berthelin (ed.), Diversity of environmental biochemistry. Elsevier. p. 371–378.

217. Boudot, J.-P., A. Bel Hadj Brahim, R. Steimen, and F. Seigle Murandi. 1989. Biodegradation of synthetic organo-metallic complexes of iron and aluminium with selected metal to carbon ratios. Soil Biol. Biochem. 21:961–966.

218. Rosenzweig, W. D., and G. Stotzky. 1979. Influence of environmental factors on antagonism of fungi by bacteria in soils: clay minerals and pH. Appl. Environ. Microbiol. 38:1120–1126.

219. Cook, R. J., and M. F. Baker. 1983. The nature and practice of biological control of plant pathogen. American Phytopathology Society, St. Paul, Minnesota, 539 p.

220. Kinraide, T. B., and D. R. Parker. 1989. Assessing the phytotoxicity of mononuclear hydroxy-aluminum. Plant Cell Environ. 12:479–487.

221. Viala, R. E., J. F. Morrison, and W. W. Cleland. 1980. Interaction of metal (III)-adenosine 5'-triphosphate complexes with yeast hexokinase. Biochemistry 19:3131–3137.

222. Matsumoto, H., and S. Morimura. 1980. Repressed template activity of chromatin of pea roots treated by aluminium. Plant Cell. Physiol. 21:951–959.

223. Robert, M., and J. Berthelin. 1986. Role of biological and biochemical factors in soil mineral weathering, p. 453–496. *In* P. M. Huang and M. Schnitzer (eds.), Interactions of soil minerals with natural organics and microbes. SSSA Special Publication No. 17, Soil Science Society of America, Madison, Wisconsin.

224. Jenny, H. 1941. Factors of soil formation. McGraw-Hill, New York.

225. Pédro, G., and G. Sieffermann. 1979. Weathering of rocks and formation of soils. Review in modern problems of geochemistry. *In* F. Siegen (ed.), UNESCO 39–55.

226. Berthelin, J. 1988. Microbial weathering processes in natural environments, p. 33–59. *In* A. Lerman and M. Meybeck (eds.), Geochemical cycles. Klune Academic Publishers.

227. Thompson, J., M. Robert, and J. Berrier. 1988. Fungal activity in dissolution and precipitation of minerals. Int. Working Meeting on Soil Micromorphology. San Antonio, Texas, July 1988.

228. Beverigde, T. J., and R. G. E. Murray. 1976. Uptake and retention of metals by cell walls of *Bacillus subtilis*. J. Bacteriol. 127:1502–1518.

229. Jaillard, B. 1987. Les structures rhizomorphes calcaires: modèle de réorganisation des minéraux du sol par les racines. Pub. INRA Montpellier, France, 219 p.

230. Diels, L. 1989. Accumulation and precipitation of Cd and Zn ions by *Alcaligenes eutrophus* strains. Biohydrometallurgy XX:369–377.

231. Houot, S., and J. Berthelin. 1987. Dynamique du fer et de la formation du colmatage ferrique des drains dans des sols hydromorphes à amphigley (Aeric haplaquepts), p. 345–352. *In* N. Fedoroff, L. M. Bresson, and M. A. Courty (eds.), Micromorphologie des sols. AFES, Paris.

232. Griffiths, E. 1965. Microorganisms and soil structure. Biol. Rev. 40:129–142.

233. Harris, R. F., G. Chesters, and O. N. Allen. 1966. Dynamics of soil aggregation. Adv. Agron. 18:107–168.

234. Martin, J. P. 1971. Decomposition and binding action of polysaccharides in soil. Soil Biol. Biochem. 3:33–41.

235. Lynch, J. M., and E. Bragg. 1985. Microorganisms and soil aggregate stability. Adv. Soil Sci. 2:134–170.

236. Monnier, G. 1965. Action des matières organiques sur la stabilité structurale des sols. Thèse Fac. des Sciences, Paris.

237. Harris, R. F., O. N. Allen, G. Chesters, and O. J. Attoe. 1963. Evaluation of microbial activity in soil aggregate stabilization and degradation by the use of artificial aggregates. Soil Sci. Soc. Am. Proc. 27:542–545.

238. Griffiths, E., and D. Jones. 1965. Microbiological aspects of soil structure. I. Relationships between organic amendments, microbial colonization and changes in aggregate stability. Plant Soil 23:17–33.

239. Nussbaumer, E., A. Guckert, and F. Jacquin. 1970. Nature et répartition de la matière organique cimentant les agrégats d'un sol après incubation en présence de glucose radioactif. C. R. Acad. Sci. 270:3235–3238.

240. Metzger, L., D. Levanon, and U. Mingelgrin. 1987. The effect of sewage sludge on soil structural stability: microbiological aspects. Soil Sci. Soc. Am. J. 51:346–351.

241. Aspiras, R. B., O. N. Allen, G. Chesters, and R. F. Harris. 1971. Chemical and physical stability of microbially stabilized aggregates. Soil Sci. Soc. Am. Proc. 35:283–286.

242. Aspiras, R. B., O. N. Allen, R. F. Harris, and G. Chesters. 1971. Aggregate stabilization by filamentous microorganisms. Soil Sci. 112:282–284.

243. Martin, J. P., J. O. Erwin, and R. A. Shepherd. 1959. Decomposition and aggregating effect of fungus cell material on soil structure. Soil Sci. Soc. Am. Proc. 23:217–220.

244. Harris, R. F., G. Chesters, O. N. Allen, and O. J. Attoe. 1964. Mechanism involved in soil aggregate stabilization by soil fungi and bacteria. Soil Sci. Soc. Am. Proc. 28:529–532.

245. Bond, R. D., and J. R. Harris. 1964. The influence of the microflora on physical properties of soil. I. Effects associated with filamentous algae and fungi. Aust. J. Soil Res. 2:111–122.

246. Molope, M., B. Grieve, and E. R. Page. 1987. Contributions by fungi and bacteria to aggregate stability of cultivated soils. J. Soil Sci. 38:71–77.

247. Allison, F. E. 1968. Soil aggregation. Some facts and fallacies as seen by a microbiologist. Soil Sci. 2:136–143.

248. Tisdall, J. M., and J. M. Oades. 1979. Stabilization of soil aggregates by the root system of ryegrass. Aust. J. Soil Res. 17:429–441.

249. Shipitalo, M. J., and R. Protz. 1989. Chemistry and micromorphology of aggregation in earthworm casts. Geoderma 45:357–374.

250. Marinissen, J. C. Y., and A. R. Dexter. 1990. Mechanisms of stabilization of earthworm casts and artificial casts. Biol. Fertil. Soils 9:163–167.

251. Garnier-Sillam, E., F. Toutain, G. Villemin, and F. Renoux. 1987. Contribution à l'étude du rôle des termites dans l'humification des sols forestiers tropicaux, pp. 331–335. *In* N. Fedoroff, L. M. Bresson, and M. A. Courty (eds.), Soil micromorphology, Proc. 7th Int. Working Meeting on Soil Micromorphology, Paris, 1985.

252. Garnier-Sillam, E., F. Toutain, G. Villemin, and F. Renoux. 1989. Etudes préliminaires des meules originales du termite xylophage *Sphaerotermes sphaerothorax* (Stosdedt). Insectes Soc. 36:293–312.

253. Dorioz, J.-M., and M. Robert. 1982. Etude expérimentale de l'interaction entre champignons et argiles: conséquences sur la microstructuration des sols. C. R. Acad. Sci. 295: 511–516.

254. Metzger, L., and M. Robert. 1985. A scanning electron microscopy study of the interaction between sludge organic components and clay particles. Geoderma 36:159–167.

255. Clough, K. S., and J. C. Sutton. 1978. Direct observation of fungal aggregates in sand dune soil. Can. J. Microbiol. 24:333–335.

256. Campbell, R. 1983. Ultrastructural studies of *Ganeumannomyces graminis* in the water films on wheat roots and the effect of clay on the interaction between this fungus and antagonistic bacteria. Can. J. Microbiol. 29:39–45.

257. Foster, R. C. 1978. Ultramicromorphology of South Australian soils. *In* W. W. Emerson, R. D. Bond, and A. R. Dexter (eds.), Modification of soil structure. John Wiley & Sons, New York, 438 p.

258. Santoro, T., and G. Stotzky. 1968. Sorption between microorganisms and clay minerals as determined by the electrical sensing zone particle analyser. Can. J. Microbiol. 14: 299–307.

259. Marshall, K. C. 1968. Interaction between colloidal montmorillonite and cells of *Rhizobium* species with different ionogenic surfaces. Biochim. Biophys. Acta 156:179–186.

260. Marshall, K. C. 1980. Bacterial adhesion in natural environments, p. 187–196. *In* R. C. W. Berkeley, J. M. Lynch, T. Melling, P. R. Rutter, and B. Vincent (eds.), Microbial adhesion to surfaces. Ellis Horwood, Chichester.

261. Fletcher, M., N. J. Latham, J. M. Lynch, and P. R. Rutter. 1980. The characteristics of interfaces and their role in microbial attachment, p. 67–78. *In* R. C. W. Berkeley, J. M. Lynch, J. Melling, and B. Vincent (eds.), Microbial adhesion to surfaces. Ellis Horwood, Chichester.

262. Rutter, P. R., and B. Vincent. 1980. The adhesion of microorganisms to surfaces: physicochemical aspects, p. 79–91.

In R. C. W. Berkeley, J. M. Lynch, J. Melling, P. R. Rutter, and B. Vincent (eds.), Microbial adhesion to surfaces. Ellis Horwood, Chichester.

263. Absolom, D. R., F. V. Lamberti, Z. Policova, W. Zingg, C. J. Oss, and A. W. Neuman. 1983. Surface thermodynamics of bacterial adhesion. Appl. Environ. Microbiol. 46: 90–97.

264. Sutherland, I. W. 1980. Polysaccharides in the adhesion of marine and freshwater bacteria, p. 329–338. *In* R. C. W. Berkeley, J. M. Lynch, J. Melling, P. R. Rutter, and B. Vincent (eds.), Microbial adhesion to surfaces. Ellis Horwood, Chichester.

265. Hermesse, M. P., C. Dereppe, Y. Bartolome, and P. G. Rouhet. 1988. Immobilization of *Acetobacter aceti* by adhesion. Can. J. Microbiol. 34:638–644.

266. Moses, N., D. E. Amory, A. J. Leonard, and P. G. Rouxhet. 1989. Surface properties of microbial cells and their role in adhesion and flocculation. Colloids Surf. 42:313–329.

267. Hurst, H. M., and G. H. Wagner. 1969. Decomposition of ^{14}C labelled cell wall and cytoplasmic fractions from hyaline and melanic fungi. Soil Sci. Soc. Am. Proc. 33:707–711.

268. Webley, D. M., and D. Jones. 1969. Biological transformation of microbial residues in soil, p. 202–256. *In* A. D. McLaren and J. Skujins (eds.), Soil biochemistry, Vol. 2. Marcel Dekker, New York.

269. Cortez, J. 1989. Effect of drying and rewetting on the minealization and distribution of bacterial constituents in soil fractions. Biol. Fertil. Soils 7:142–151.

270. Martin, J. P., K. Haider, W. D. Farmer, and E. Fustec Mathon. 1974. Decomposition and distribution of residual activity of some ^{14}C-microbial polysaccharides and cells, glucose, cellulose and wheat straw in soil. Soil Biol. Biochem. 6:221–230.

271. Marumoto, T., J. P. E. Anderson, and K. H. Domsh. 1982. Decomposition of ^{14}C and ^{15}N labelled microbial cells in soil. Soil Biol. Biochem. 14:461–467.

272. Theng, B. K. G. 1979. Formation and properties of clay polymer complexes. Dev. Soil Sci. 9.

273. Greenland, D. J. 1965. Interaction between clays and organic compounds in soils. II. Adsorption of soil organic compounds and its effects on soil properties. Soils Fertil. 28:521–532.

274. Greenland, D. J. 1972. Interaction between organic polymers and inorganic soil particles, p. 897–914. *In* M. De Boot (ed.), Proc. Symp. on the Fundamentals of Soil Conditioning, Ghent, Belgium.

275. La Mer, V. K., and T. W. Healy. 1963. Adsorption—flocculation reactions of macromolecules at the solid liquid interface. Rev. Pure Appl. Chem. 13:287–297.

276. Slater, R. W., and J. A. Kitchener. 1966. Characteristics of flocculation of mineral suspensions by polymers. Discuss. Faraday Soc. 42:267–275.

277. Parfitt, R. L., and D. J. Greenland. 1970. Adsorption of polysaccharides by montmorillonite. Soil Sci. Soc. Am. Proc. 34:862–866.

278. Parfitt, R. L. 1972. Adsorption of charged sugars by montmorillonite. Soil Sci. 113:417–421.

279. Clapp, C. E., and W. W. Emerson. 1972. Reactions between Ca-montmorillonite and polysaccharides. Soil Sci. 114:210–216.

280. Olness, A, and C. E. Clapp. 1973. Occurrence of collapsed and expanded crystals in montmorillonite dextran complexes. Clays Clay Min. 21:289–293.

281. Guckert, A. 1973. Contribution à l'étude des polysaccharides dans les sols et de leur role dans les mécanismes d'agrégation. Thèse Doctorat d'Etat, Nancy, 124 p.

282. Cortez, J. 1977. Adsorption sur les argiles de deux lipopolysaccharides rhizosphériques. Soil Biol. Biochem. 9:25–32.

283. Guidi, G., G. Petruzzelli, and M. Giachetti. 1977. Molecular weight as influencing factor on the adsorption of dextrans on sodium and calcium montmorillonite. Z. Pflanzenernahr Boden. 141:367–377.

284. Chenu, C. 1985. Etude expérimentale des interactions argiles polysaccharides neutres. Contribution à la connaissance des phénomènes d'agrégation d'origine biologique dans les sols. Thèse de Doctorat de l'Université de Paris VII, 198 p.

285. Chenu, C., C.-H. Pons, and M. Robert. 1987. Interaction of kaolinite and montmorillonite with neutral polysaccharides, p. 375–381. *In* L. G. Schultz, H. Van Olphen, and F. A. Mumpton (eds.), Proceedings of the International Clay Conference, Denver, 1985. Clay Minerals Society, Bloomington.

286. Habib, L., C. Chenu, J.-L. Morel, and A. Guckert. 1990. Adsorption of maize root mucilages on homoionic clays. Consequences on the microstructure of the complexes. C. R. Acad. Sci. [Paris] 310:1541–1546.

287. Saini, G. R., and A. A. McLean. 1966. Adsorption flocculation reactions of soil polysaccharides with kaolinite. Soil Sci. Soc. Am. Proc. 30:697–698.

288. Aly, S. M., and J. Letey. 1988. Polymer and water quality effects on flocculation of montmorillonite. Soil Sci. Soc. Am. J. 52:1453–1458.

289. Clapp, C. E., and W. W. Emerson. 1965. The effect of periodate oxidation on the strength of soil crumb. Soil Sci. Soc. Am. Proc. 29:127—134.

290. Chenu, C., and J. Guérif. 1991. The mechanical strength of clay minerals as influenced by an adsorbed polysaccharide. Soil Sci. Soc. Am. J. 55 (in press).

291. Hayes, M. H. B. 1980. Role of natural and synthetic polymers in stabilizing soil aggregates, p. 262—296. *In* R. C. W. Berkeley, J. M. Lynch, J. Melling, P. R. Rutter, and B. Vincent (eds.), Microbial adhesion to surfaces. Ellis Horwood, Chichester.

292. Rees, D. A. 1977. Polysaccharides shapes. Outline studies in biology. Chapman and Hall, London.

293. Morris, E., and I. T. Norton. 1983. Polysaccharides aggregation in solution and gels. Aggregation processes in solution. Stud. Phys. Theor. Chem. 26:549—593.

294. Powell, D. A. 1979. Structure, solution properties and biological interactions of some microbial extracellular polysaccharides, p. 117—160. *In* R. C. W. Berkeley, C. W. Gooday, and D. C. Elwood (eds.), Microbial polysaccharides and polysaccharases. Academic Press, New York.

295. Sanford, P. A. 1979. Exocellular microbial polysaccharides. Adv. Carbohydr. Chem. Biochem. 36:265—313.

296. Sutherland, I. W. 1985. Biosynthesis and composition of gram-negative bacterial extracellular and wall polysaccharides. Annu. Rev. Microbiol. 39:243—270.

297. Barreto-Bertger, E., and P. A. Gorin. 1983. Structural chemistry of polysaccharides from fungi and lichens. Adv. Carbohydr. Chem. Biochem. 41:67-103.

298. Gaur, A. C., and R. V. Subba Rao. 1975. Note on the isolation of bacterial gums and their influence on soil aggregate stability. Indian J. Agric. Sci. 45:186—189.

299. Channey, K., and R. S. Swift. 1986. Studies on aggregate stability. I. Reformation of soil aggregates. J. Soil Sci. 37:329—335.

300. Muneer, M., and J. M. Oades. 1989. The role of Ca—organic interactions in soil aggregate stability. I. Laboratory studies with ^{14}C-glucose, $CaCO_3$ and $CaSO_4 \cdot 2H_2O$. Aust. J. Soil Res. 27:389—400.

301. Foster, R. C., and J. K. Martin. 1981. In situ analysis of soil components of biological origin, p. 75—111. *In* E. A. Paul and J. N. Ladd (eds.), Soil biochemistry, Vol. 5. Marcel Dekker, New York.

302. Cheschire, M. V. 1979. Nature and origin of carbohydrates in soil. Academic Press, London, 216 p.

303. Adu, J. K., and J. M. Oades. 1978. Physical factors influencing the decomposition of organic materials in soil aggregates. Soil Biol. Biochem. 10:109–115.

304. Williams, B. G., D. J. Greenland, and J. P. Quirk. 1966. The adsorption of polyvinyl alcohol by natural soil aggregates. Aust. J. Soil Res. 4:131–143.

305. Oades, J. M. 1984. Soil organic matter and structural stability mechanisms and implications for management. Plant Soil 76:319–337.

306. Dinel, H., M. Schnitzer, and G. R. Mehuys. 1991. Soil lipids: origin, nature, content, decomposition and effect on soil physical properties, p. 397–430. *In* J. M. Bollag and G. Stotzky (eds.), Soil biochemistry, Vol. 6, Marcel Dekker, New York.

307. Guidi, G., G. Petruzelli, and M. Giachetti. 1983. Effect of three fractions extracted from aerobic and anaerobic sewage sludge on the water stability and surface of soil aggregates. Soil Sci. 136:158–163.

308. Capriel, P., T. Beck, H. Borchert, and P. Harter. 1990. Relationship between soil aliphatic fraction extracted with supercritical hexane, soil microbial biomass, and soil aggregate stability. Soil Sci. Soc. Am. Proc. 54:415–420.

309. Shulten, H. R., and M. Schnitzer. 1990. Aliphatics in soil organic matter in fine-clay fractions. Soil Sci. Soc. Am. Proc. 54:98–105.

310. Haider, K., J. P. Martin, and Z. Filip. 1975. Humus biochemistry, p. 285–311. *In* E. A. Paul and A. D. McLaren (eds.), Soil Biochemistry, Vol. 4, Marcel Dekker, New York.

311. Schnitzer, M., and J. A. Neyroud. 1975. Further investigation on the chemistry of fungal "humic acids." Soil Biol. Biochem. 7:365–371.

312. Saiz-Jimenez, C. 1983. The chemical nature of the melanins from *Coprinus* spp. Soil Sci. 13:65–74.

313. Schnitzer, M., and Y. K. Chan. 1986. Structural characteristics of a fungal melanin and a soil humic acid. Soil Sci. Soc. Am. J. 50:67–71.

314. Senesi, N., T. M. Miano, and J. P. Martin. 1987. Elemental functional infrared and free radical characterization of humic acid-type fungal polymers (melanins). Biol. Fertil. Soils 5:120–125.

315. Bond, R. D. 1969. Factors responsible for water repellence in soils, p. 259–263. *In* Proceedings symposium on water repellent soils. Univ. Calif., Riverside.

316. Savage, S. M., J. P. Martin, and J. Letey. 1969. Contribution of soil fungi to natural and induced water repellency in sand. Soil Sci. Soc. Am. Proc. 33:405–499.

317. Bailey, A. I., and A. G. Price. 1970. Interfacial energies of mono molecular film of fatty acids deposited on mica in aqueous and non aqueous media. J. Chem. Phys. 53:3421–3427.

318. Jouany, C. 1991. Surface energy of clay-humic acid complexes. Clays Clay Miner. 39:43–49.

319. Jouany, C., and P. Chassin. 1987. Determination of the surface energy of clay–organic complexes by contact angles measurements. Colloid Surf. 27:289–303.

320. Ma'shum, M., and V. C. Farmer. 1985. Origin and assessment of water repellency of a sandy South Australian soil. Aust. J. Soil Res. 23:623–626.

321. Shipitalo, M. J., and R. Protz. 1988. Factors influencing the dispersability of clay in worm casts. Soil Sci. Soc. Am. J. 52:764–769.

322. Molope, M. B. 1987. Soil aggregate stability: the contribution of biological and physical processes. S. Afr. J. Plant Soil 4:121–126.

323. Haynes, R. J., and R. S. Swift. 1990. Stability of soil aggregates in relation to organic constituents and soil water content. J. Soil Sci. 41:73–83.

324. Emerson, W. W., R. C. Foster, and J. M. Oades. 1986. Organo-mineral complexes in relation to soil aggregation and structure, p. 521–548. *In* P. M. Huang and M. Schnitzer (eds.), Interactions of soil minerals with organics and microbes, SSSA Special Publication No. 17, Soil Science Society of America, Madison, Wisconsin.

325. Anderson, D. W., S. Saggard, J. R. Bettany, and J. V. B. Stewart. 1981. Particle size fractions and their use in studies of soil organic matter. I. The nature and distribution of forms of carbon, nitrogen and sulfur. Soil Sci. Soc. Am. J. 45:767–772.

326. Ladd, J. N., M. Amato, J. Jocteur Monrozier, and M. van Gestel. 1990. Soil microhabitats and carbon and nitrogen metabolism, Vol. 3, pp. 82–87. Trans. 14th Int. Cong. of Soil Science, Kyoto.

327. Gupta, V. V. S. R., and J. J. Germida. 1988. Distribution of microbial biomass and its activity in different soil aggregate size classes as affected by cultivation. Soil Biol. Biochem. 20:777–786.

328. Cheschire, M. V., G. P. Sparling, and C. M. Mundie. 1984. Influence of soil type crop and air drying on residual carbohydrate content and aggregate stability after treatment with periodate and tetraborate. Plant Soil 76:339–347.

329. Dormaar, J. F. 1983. Chemical properties of soil and water-stable aggregates after sixty-seven years of cropping to spring wheat. Plant Soil 75:51–61.

330. Lynch, J. M., and L. F. Elliott. 1983. Aggregate stabilization of volcanic ash and soil during microbial degradation of straw. Appl. Environ. Microbiol. 45:1398–1401.

331. Guckert, A., T. Chone, and F. Jacquin. 1975. Microflore et stabilité structurale des sols. Rev. Ecol. Biol. Sol 12: 211–224.

332. Martin, J. P., and K. Haider. 1963. Decomposition and binding action of a polysaccharide from *Chromobacterium violaceum* in soil. J. Bacteriol. 85:1288–1294.

333. Martin, J. P., J. O. Erwing, and R. A. Schepherd. 1965. Decomposition and binding action of polysaccharides from *Azotobacter indicus* and other bacteria in soil. Soil Sci. Soc. Am. Proc. 29:397–400.

334. Martin, J. P., J. O. Erwin, and S. J. Richards. 1972. Decomposition and binding action of some mannose containing microbial polysaccharides and their Fe, Al and Cu complexes. Soil Sci. 113:322–327.

335. Olness, A., and C. E. Clapp. 1972. Microbial degradation of a montmorillonite-dextran complex. Soil Sci. Soc. Am. Proc. 36:179–181.

336. Cortez, J. 1976. Rôle des argiles dans la biodégradation de deux lipopolysaccharides bactériens. Oecol. Plant. 11:243–256.

337. Guckert, A., H. H. Tok, and F. Jacquin. 1975. Biodégradation de polysaccharides bactériens adsorbés sur une montmorillonite. *In* Proceedings symposium on soil organic matter studies. Braunschweig 1:403–411.

338. Tiessen, H., and J. W. B. Stewart. 1988. Light and electron microscopy of stained microaggregates: the role of organic matter and microbes in soil aggregation. Biogeochemistry 5:312–322.

339. Paul, E. A. 1984. Dynamics of organic matter in soils. Plant Soil 76:275–285.

340. Tiessen, H., J. W. B. Stewart, and H. W. Hunt. 1984. Concept of soil organic matter transformations in relation to organomineral particle size fractions. Plant Soil 76:287–295.

341. Van Veen, J. A., J. N. Ladd, and M. Amato. 1985. Turnover of carbon and nitrogen through the microbial biomass in a sandy loam and a clay soil incubated with ^{14}C glucose and ^{15}N ($NH_4 2SO_4$) under different moisture regimes. Soil Biol. Biochem. 17:747–756.

342. Taylor, G. S., and P. E. Baldbridge. 1954. The effect of sodium carboxymethyl cellulose on some physical properties of Ohio soils. Soil Sci. Soc. Am. Proc. 382–385.

343. Wood, J. D., and J. D. Oster. 1985. The effect of cellulose xanthate and polyvinyl alcohol on infiltration, erosion, and crusting at different sodium levels. Soil Sci. 139:243–249.

344. Wallace, A. 1986. A polysaccharide (guar) as a soil conditioner. Soil Sci. 141:371–373.

345. Helalia, A. M., and J. Letey. 1988. Polymer type and water quality effects on soil dispersion. Soil Sci. Soc. Am. J. 52:243–246.

346. Ben-Hur, M., J. Faris, M. Malik, and J. Letey. 1989. Polymers as soil conditioners under consecutive irrigations and rainfall. Soil Sci. Soc. Am. J. 53:1173–1177.

347. Ben-Hur, M., and J. Letey. 1989. Effect of polysaccharides, clay dispersion, and impact energy on water infiltration. Soil Sci. Soc. Am. J. 53:233–238.

348. Ben-Hur, M., J. Letey, and L. Shainberg. 1990. Polymers effect on erosion under laboratory rainfall simulator conditions. Soil Sci. Soc. Am. J. 54:1092–1095.

349. Gordon, S. 1988. The use of marine products in land reclamation, p. 320–326. *In* S. Paoletti and G. Blunden (eds.), Proceedings workshop on phycocolloids and fine chemicals, Brussels, September 1988.

350. Metting, B., and W. R. Rayburn. 1983. Influence on microalgal conditioner on selected Washington soils: an empirical study. Soil Sci. Soc. Am. J. 47:682–685.

351. Metting, B. 1986. Population dynamics of *Chlamydomonas sajao* and its influence on soil aggregate stabilisation in the field. Appl. Environ. Microbiol. 51:1161–1164.

352. Metting, B. 1987. Dynamics of wet and dry aggregate stability from a three year microalgal soil conditioning experiment in the field. Soil Sci. 143:139–143.

353. Barclay, W. R., and R. A. Lewin. 1985. Microalgal polysaccharide production for the conditioning of agricultural soils. Plant Soil 88:159–169.

354. Metting, B., W. R. Rayburn, and P. A. Reynaud. 1988. Algae and agriculture, p. 335–370. *In* C. A. Lembi and J. R. Waaland (eds.), Algae and human affairs. Cambridge University Press, Cambridge.

Index

Printed and bound by CPI Group (UK) Ltd, Croydon, CR0 4YY

17/10/2024

01775700-0015